Books by Eric J. Chaisson:

Cosmic Dawn: The Origins of Matter and Life
La Relativita

The Invisible Universe:
Probing the Frontiers of Astrophysics (with George B. Field)

The Life Era: Cosmic Selection and Conscious Evolution

Relatively Speaking:
Black Holes, Relativity, and the Fate of the Universe

Universe: An Evolutionary Approach to Astronomy

Astronomy Today (with Steve McMillan)

THE HUBBLE WARS

ERIC J. CHAISSON

Astrophysics
Meets
Astropolitics in
the Two-Billion-
Dollar Struggle
over the
Hubble Space
Telescope

HarperCollinsPublishers

We gratefully acknowledge the following for providing illustrations:

Page 110: (Middle) From "Hubble's Legacy," by Tim Beardsley. Copyright © 1990 by *Scientific American, Inc.* All rights reserved. (Bottom): Copyright © 1990 by Scholastic Inc. Reprinted from *Scholastic News,* Explorer Edition, March 2, 1990, by permission of Scholastic Inc.

Page 195: Copyright © 1990 by *The Sydney Morning Herald Magazine.* Reprinted by permission.

Page 199: Copyright © 1990 by Jim Meddick, reprinted by permission of NEA, Inc.

Page 202: Copyright © 1990 by Dan Lynch, reprinted by permission of NEA, Inc.

Page 332: Copyright © 1991 by Time Warner. Reprinted by permission of *LIFE* magazine. Inset photo copyright © 1991 by Donna Ferrato. Reprinted by permission.

HarperCollins books may be purchased for educational, business, or sales promotional use. For information please write: Special Markets Department, HarperCollins Publishers, Inc., 10 East 53rd Street, New York, NY 10022.

FIRST EDITION

Designed by Alma Hochhauser Orenstein

Photograph insert by Barbara DuPree Knowles

Library of Congress Cataloging-in-Publication Data

Chaisson, Eric.
  The Hubble wars : astrophysics meets astropolitics in the two-billion-dollar struggle over the Hubble Space Telescope / Eric J. Chaisson.
      p. cm.
  Includes bibliographical references and index.
  ISBN 0-06-017114-6
  1. Hubble Space Telescope. I. Title
  QB500.268.C48 1994
  522'.2919—dc20   93-37468

94 95 96 97 98 ❖/RRD 10 9 8 7 6 5 4 3 2 1

To Bridget Aquila
who completes the Summer Triangle

Contents

Color photographs follow page 148

Preface

So this is where the great man is buried. Just look at the place. It's vast. Both the Duomo and Santa Croce have seating capacities of a major-league ballpark. The sheer power these buildings radiate must have been awesome at the time. No wonder he had so much trouble. He was torquing insecure colleagues amidst a powerful and conservative institution. I do admire Galileo for his tenacity.

But, Eric, don't you see the similarities? Nothing has really changed in four hundred years. You are now tasked with choosing the first science targets that *Hubble* will see and you are fielding heavy criticism from ego-minded colleagues, and especially from an institution so thoroughly bureaucratized as to give the Curial Inquisition a good name. The parallels are uncanny. *Hubble* is today's cathedral of science, the grand Duomo of our time. And NASA is as conservative and closed-minded as the Church under siege of the Reformation. Standing here before Galileo's tomb, I urge you to keep a clear record, however emotional, of the months and years ahead. You are positioned as few others to make a contribution to the culture of the modern scientific endeavor, to burst the myth that science is lacking in human values. Only a rocket scientist actively plying the trade can inform the next generation of the true conduct of space science today. Take the risk. Be as tenacious as the man you so admire.

—closely paraphrased from a conversation between the author and a philosopher-friend, Ervin Laszlo, in Florence, Italy, late 1980s

Scientists of the twentieth century have made great strides in extending our understanding of the size, scale, and structure of the Universe. These advances have been achieved largely through a succession of increasingly more sophisticated telescopes located at mountaintop observatories and operated by a staff of dedicated astronomers.

Today, modern astronomy is taking another quantum leap forward. As the flagship of NASA's Great Observatories program, the Hubble Space Telescope is designed to yield unprecedentedly detailed views of the Universe unobtainable with even the largest ground-based instruments. This space-borne observatory is named after Edwin Powell Hubble (1889–1953), whose research in the 1920s and 1930s greatly widened humanity's knowledge of the Universe and of our position in it. The intent for *Hubble,* the telescope, much like for Hubble, the man, was to further extend the frontiers of astronomical research, perhaps enough that by the end of the millennium we might need to rewrite significant portions of today's textbooks.

Unfortunately, early in the *Hubble* mission, a number of serious hardware problems surfaced, most notably a badly misshapen primary mirror at the heart of the orbiting telescope. The $2-billion "extraordinary looking glass" was deemed greatly out of focus—in effect, purblind. The resultant fuzzy images meant that light from some of the faintest objects in the Universe could not be captured. The engineers were embarrassed, the astronomers devastated, the politicians disgusted, the press highly critical; the public thought *Hubble* was broken. Scientists eventually learned to work around some of the difficulty and managed to acquire limited amounts of astronomical data, some displaying spectacular images of celestial wonders. But the originally intended and ambitious science agenda planned for *Hubble* cannot be addressed unless the telescope is successfully repaired and regularly maintained by astronauts, tasks that NASA expects to attempt throughout the 1990s.

Space Telescope now travels around the Earth roughly every hour and a half—some fifteen times a day—in a circular orbit of several hundred miles' altitude. Having been deployed from a U.S. space shuttle in the spring of 1990, this unmanned telescope was put through an exhaustive period of checkout and testing that lasted until nearly the end of 1991—a lengthy process known as commissioning. Much like sea trials preceding a vessel's oceanic operations, Space Telescope endured a lengthy array of space trials during which the communication links and myriad operating modes of the spacecraft were verified, its onboard scientific instruments calibrated, and its computer software programs revised. We on the ground also needed to learn psychologically how to observe effectively with *Hubble*. Humankind had never constructed such an ambitious piece of complex robotic equipment, and the commissioning activities were both frustrating and fascinating—frustrating because the cast of human characters on the ground sometimes seemed bent on worsening a space mission already troubled from the start, and fascinating because the early observations with *Hubble* enabled us to explore the Universe with greater clarity than had ever been achieved before.

Throughout the commissioning period, I was a scientist on the senior staff at the Space Telescope Science Institute, in Baltimore; it was one of the most exciting and turbulent times in the history of astronomy. Some ten years ago, at the urging of the National Academy of Sciences, NASA had established the Institute to serve as the scientific nerve center of the *Hubble* space mission—the place from which the astronomers, largely separated from the spacecraft engineers and government managers, would conceive, plan, and conduct the pioneering telescope's science agenda. There, I was also director of educational programs, a position that put me in charge of overseeing relations with the print and broadcast media. Thus, I lived this project for several years from many angles—as an astronomer, an educator, and an informant.

In the course of the narration, I aim neither to praise, criticize, name-drop, or embarrass any specific person or group, nor to whistleblow on any government agency or industrial contractor. Rather, it is the process and methodology of science—and especially the culture of "big science"—that I seek to disseminate, sharing the ups and downs of hard science and engineering along with nontechnical vignettes of what I felt and experienced while intimately involved in such a grand and difficult project. This is the real "scientific method," with all its human values and emotions not well described in sterile and specialized textbooks. To be sure, modern science has a decidedly subjective component and none more evident than in expensive, politicized, and publicly visible projects such as this one.

This book, then, is a personal view of *Hubble*'s deployment, commissioning, and initial experiments, told by a participant in the endeavor who kept a daily journal as history's most complex and prominent science project proceeded through its space trials—technical trials in orbit and public trials on the ground.

Eric J. Chaisson
Concord, Massachusetts
Autumn 1993

Note: No part of this book divulges sensitive military-intelligence material not previously having entered the public domain. I have been scrupulous about neither identifying reconnaissance assets unknown to the public nor disclosing the specific capabilities of any known yet classified project.

Launch of Space Telescope

> From our home on the Earth, we look out into the distances and strive to imagine the sort of world into which we were born. Today we have reached far out into space. Our immediate neighborhood we know rather intimately. But with increasing distance our knowledge fades, and fades rapidly, until at the last dim horizon we search among ghostly errors of observations for land-marks that are scarcely more substantial.
>
> The search will continue. The urge is older than history. It is not satisfied and it will not be suppressed.
>
> *Realm of the Nebulae*
> EDWIN P. HUBBLE, 1936

Reflecting floodlight so brightly that I had to squint, Space Shuttle *Discovery* seemed like another Washington monument, a gleaming piece of mostly metal anchored firmly to the ground yet pointed toward the stars. To imagine it rising into the sky the next morning was difficult even for a person intimately involved with space-science projects. The shuttle's white and shiny appearance on Pad 39B at the Kennedy Space Center on the eve of launch is an image indelibly burned into my brain. What made this shuttle special to me was its cargo, itself a monument to human innovation—the Hubble Space Telescope.

Shortly after sunset, less than twelve hours before launch, the late April air was balmy, the winds gusty. Overhead, the dusky sky was mostly clear with patchy clouds and somewhere up there Comet Austin, another dud of a comet whose brightness fell short of expectations and whose tail was virtually invisible. I recalled one of the arguments used to sell the Space Telescope to Congress and the American people well over a decade ago: Through *Hubble*'s sensitive cameras, we would never again lose sight of Halley's Comet. Not that Halley was so mysterious anymore—non-American spacecraft had reconnoitered the dirty snowball up close several years ago—but it was a commonly understood benchmark for illustrating Space Telescope's anticipated power and clarity of view.

A number of us had received clearance to visit the launching area one final time before liftoff—a few scientists and technicians, some NASA officials, the astronauts' families, and a pack of press photographers. I had gone on this evening vigil for no good reason. There was nothing more to be done regarding payload checkout and preparation for launch. The design, construction, and preflight testing of Space Telescope were complete after nearly twenty years of effort. Perhaps I just wasn't ready to leave the Kennedy Space Center on the eve of what seemed a whole new era in humankind's understanding of the cosmos.

Along a narrow coast road hugging the outer banks of the Canaveral National Seashore, I watched a Japanese film crew position their equipment on a grassy knoll aside a swamp. Their tripod-mounted cameras joined hundreds of other unattended still and video cameras aimed toward the shuttle, all to be triggered automatically the next morning by the fierce sound waves emanating from the ignition of *Discovery*. No humans would be present this close to the pad at the time of launch—about a quarter-mile—except for the astronaut crew atop the firecracker itself.

The four-person Japanese team went about their business with the same thoroughness and politeness they had demonstrated less than an hour earlier, when they had interviewed me during a live satellite broadcast for Japan's version of "Good Morning America." In contrast to the aggressive, investigative demeanor of their Western counterparts, the Japanese producers courteously bowed and requested my presence on their television show, their reporter briefed me on the specific questions that he would put to me, and once on the air to Tokyo the questions were served up like softballs to be knocked over the fence.

I had been interviewed forty-seven times that day at Kennedy's press site, and I was glad the press buildup prior to launch was coming to an end. During the previous few weeks at the Space Telescope Science Institute, we had been inundated with well over a thousand media inquiries from around the world. *Hubble* had become NASA's showcase scientific program, bringing the agency more favorable press than anything since the age of *Apollo*. Being a stickler for technical accuracy, I had worked closely with several colleagues at the institute, trying to help journalists decipher the intricacies of the onboard scientific equipment and the subtleties of black holes and remote galaxies—to say nothing of NASA's press materials about Space Telescope's capabilities.

I walked away from the roadway and onto a broad, unpaved path along which a giant, 4,000-ton mobile launcher platform had crawled, at about a half-mile per hour, transporting the shuttle orbiter, its huge fuel tank, and its mated booster rockets from the Vertical Assembly Building to the pad at Launch Complex 39. I wanted to get away from people, to relate mentally to the spaceship one-on-one. As I ambled along the dusty, eight-lane-wide crawlerway toward the pad, I found it peculiar that the shuttle's skin appeared bigger and brighter, yet dirtier.

At about two hundred yards, I stopped and saw clearly the magnificent flying machine and its high-tech environment. Some dozen powerful spotlights reached out as bright beams crisscrossed the guarded launch gantry rising hundreds of feet into the air, various red safety beacons flashed among scores of lit white bulbs throughout the tall scaffoldlike superstructure astride the shuttle, the lightning arrestor glowed weirdly orange atop the gantry, volatile gases vented from the vehicle's fuel tank to the ambient air, white-coated technicians scurried about checking this and that on numerous steel-grated catwalks, all the while loudspeakers yelled out numbers, test sequences, commands, and more numbers. The scene reminded me of Devil's Tower in the movie *Close Encounters of the Third Kind*. But there was one difference: The movie was fictional, the view before my eyes was real.

I took no pictures that evening. No camera could have captured the deep impression that came over me—deeper than I would experience the next day even at the moment of the launch. After some two decades of planning, building, testing, delays, and more testing in the wake of the Space Shuttle *Challenger* tragedy, the Hubble Space Telescope had finally been mated to Shuttle *Discovery* and was ready to be lifted from the surface of the Earth. It was about to be placed into its own element, away from humans and into a state of "zero-g," a free-flyer orbiting above Earth's atmosphere. The waiting was almost done.

Overhead, the celestial dome pulsed eerily with scores of bright sources, most notably Jupiter and Orion to the west, Vega and the Summer Triangle toward the northeast, and Arcturus and the Big Dipper almost directly above. Even the stars seemed to be waiting.

The Hubble Space Telescope is the largest, most complex, and most powerful observatory ever deployed in space. It is a project of the National Aeronautics and Space Administration (NASA), which built the telescope in cooperation with the European Space Agency (ESA) and numerous industrial contractors. Named after the late American astronomer Edwin P. Hubble, who perhaps changed our view of the Universe more than any other person since Galileo, Space Telescope is an unmanned orbiting vehicle designed to undertake a long-term international program of scientific exploration of the cosmos. Stated technically, Space Telescope should allow astronomers to probe the visible Universe with at least ten times finer resolution and with some fifty times greater sensitivity than any other machine built by humans. Stated nontechnically, Space Telescope provides for astronomers on Earth a decidedly new vista—by allowing us to look far back into the past with unprecedented clarity.

Launched in 1990 by the lightest of the U.S. space shuttles, *Discovery*, into the highest attainable Earth orbit (for a shuttle)—some 380 statute miles, 330 nautical miles, or 615 kilometers, and at an orbital inclination of 28.5° (which is fixed by a due-east launch from the Kennedy Space Center, whose latitude is 28.5° north of the equator)—Space Telescope is expected to operate within a few tens of miles of this "low" Earth orbit throughout its fifteen-year nominal lifetime. To achieve that full longevity, and perhaps

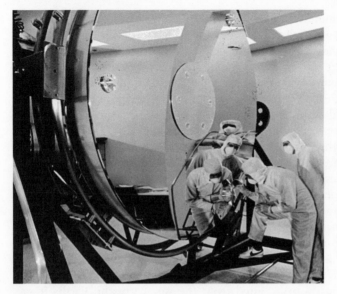

Technicians are shown here inspecting Space Telescope's 94.5-inch-diameter primary mirror prior to its installation into the telescope assembly. The metal plug at the center of the mirror was removed prior to launch, so that light from celestial objects can pass through the hole in the middle of the doughnut-shaped mirror on its way to numerous cameras and other science instruments aboard the telescope. [NASA]

more, space shuttles are tentatively scheduled to rendezvous with *Hubble* every few years; such servicing missions should allow astronauts to maintain, upgrade, repair, and redeploy the telescope to approximately its original orbit. Contrary to prenatal plans, *Hubble* will not likely ever be brought back to Earth for major overhaul, as was initially envisioned by overly optimistic shuttle designers.

The heart of Space Telescope is a 94.5-inch (2.4-meter) primary mirror designed and manufactured between 1977 and 1981 at the Hughes Danbury Optical Systems, Inc. (formerly part of the Perkin-Elmer Corp.) of Connecticut. While not the largest ever built, *Hubble*'s mirror is assuredly the cleanest and most finely polished mirror of its size. Gee-whiz statistics abound to describe its optical characteristics, one of which is this: If *Hubble*'s mirror were scaled up to equal the width of the North American continent, the highest hill or lowest valley would be a mere few inches from the average surface; by contrast, skyscraper-size imperfections would result if ordinary eyeglass lenses were so scaled from coast-to-coast. These statements are true despite a major optical flaw described in Chapter 4.

For several years during the mid- and late 1980s, Space Telescope underwent tests while mothballed at the other of its major industrial contractors, the Lockheed Missiles & Space Company of California, which integrated the myriad parts built by scores of subcontractors into a unified, flight-ready space vehicle. Those parts comprise the so-called Support Systems Module, consisting of the spacecraft's superstructure, its aperture door, and an extensive equipment section that houses many of *Hubble*'s vitals—storage batteries, gyroscope electronics, onboard computers, pointing control system, and thermal control devices. While at Lockheed, *Hubble* was kept in the world's largest clean room—a five-story vault reputed to be cleaner than the operating room in most hospitals. Called "the VATA," for Vertical Assembly and Test Area, its air was filtered and replaced every ninety seconds while maintaining constant humidity to ensure that the telescope remained as clean and as structurally stable as possible.

All the while the spacecraft did not just hibernate, as project scientists and engineers took advantage of the four-year delay caused by the 1986 *Challenger* tragedy to upgrade many of *Hubble's* capabilities. For example, its winglike panels were improved with new solar-cell technology to ensure that *Hubble* would have enough electrical power to operate its myriad onboard systems for several years; many of the spacecraft's component parts, such as capacitors, resistors, and integrated circuits were made more reliable; the outside of *Hubble's* hull was modified to make it easier for astronauts to change-out equipment during a servicing mission; its nickel-cadmium batteries were replaced with more powerful nickel-hydrogen types; a huge catalog of star positions needed to guide *Hubble* was completed; the ground computer system designed to command and control the telescope as well as the onboard computer system meant to keep *Hubble* safe were both greatly improved; and many useful operating modes were exercised for each of the telescope's onboard scientific instruments.

In addition, acoustic tests subjected the entire telescope to vibrations exceeding those expected during shuttle launch, and a two-month-long test in a huge thermal vacuum chamber, outfitted with a vast array of small lamps, simulated the vast temperature extremes such spacecraft encounter during each Earth orbit. Several end-to-end communications and commanding procedures ("ground systems tests") were practiced on the fully assembled vehicle, utilizing many of its more than 400,000 parts and its approximately 26,000 miles of electrical wiring. Ironically, given the faults to be later discovered regarding *Hubble's* hardware, the preflight software tests that simulated in detail the first several weeks of telescope commissioning were more exhaustive than for any other (civilian) space mission in history. As a result, the Hubble Space Telescope today is a more efficient spacecraft than it would have been had it been launched as originally scheduled in December 1983. It might not have worked then much at all.

Space Telescope's gross dimensions, which resemble that of a railroad tank car, are approximately these: length, 43.5 feet (13.3 meters) with its aperture door closed; diameter, 14 feet (4.3 meters) with solar arrays stowed, 40 feet (12 meters) with solar arrays deployed; weight, 12.3 tons (11,200 kilograms). Design and construction of Space Telescope (not including its science instruments), as well as much of the "Orbital Verification" period of vehicle testing and checkout during the first few months after deployment, were the responsibility of NASA's Marshall Space Flight Center, in Huntsville, Alabama.

The optical system and scientific instruments aboard *Hubble* are compact and pioneering. The telescope reflects acquired light from its large concave primary mirror back to a smaller, 12-inch (0.3-meter), convex secondary mirror the size of a dinner plate, which in turn sends the light through a hole in the doughnut-shaped main mirror and on into the aft bay at the spacecraft's stern. There, any of six major scientific instruments wait to analyze the incoming radiation; most of them resemble the size and shape of

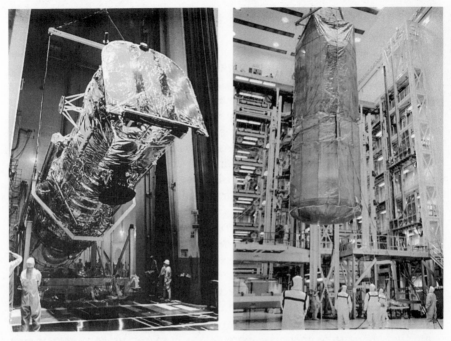

At left at the Lockheed Missiles & Space Co., the four-story-tall *Hubble* is being rotated from vertical to horizontal in preparation for shipping to Cape Canaveral. At right at the Kennedy Space Center, the telescope is encased in a cocoon—a large double plastic bag or sanitary liner to guard against contamination—while being prepared for transfer to Launchpad 39B. Despite these precautions, nine wasps did manage to penetrate the clean room at Kennedy, requiring NASA to hire an entomologist to aid in capturing or starving the insects; bug spray was out of the question, for fear that the propellant might penetrate *Hubble*'s protective covering and mess up its optics. *[NASA]*

a telephone booth. Those instruments include two cameras to image (or electronically "photograph") stars and other objects in space, two spectrographs to split (or spread) the captured radiation into its component colors, a photometer to measure the intensity (or magnitude) of light, and a group of fine guidance sensors to pinpoint the positions of stars on the sky. Together, these instruments comprise not merely a space telescope but a complete space observatory with a multitude of experimental options like those used by astronomers at ground-based observatories on Earth. Design and construction of *Hubble*'s onboard science instruments, as well as the year-long "Science Verification" period of instrument testing and calibrating following Orbital Verification, were the responsibility of NASA's Goddard Space Flight Center in Greenbelt, Maryland.

Although the scientific instruments aboard *Hubble* are explained in greater detail in later chapters, here are the design goals for each of the telescope's science tools. The following table lists each instrument's field of view, in arc minutes (') and arc seconds ("); wavelength coverage, in angstroms (A); angular, spectral, or temporal resolution, in arc seconds, angstroms, or microseconds of time, respectively; and design sensitivity, in limiting magnitudes.

In these computer-rendered views of the Hubble Space Telescope on orbit, one can clearly see its open aperture door (at top and left) and its dish-shaped high-gain antennas that are used for communications. The telescope's science instruments, computers, radios, motors, heaters, coolers, and numerous other onboard systems requiring power receive electricity from two movable solar arrays seen here to the port (left) and starboard; rechargeable batteries provide power during orbital night. Despite having the size of a city bus, *Hubble* is designed to move in space with the grace of a prima ballerina. *[D. Berry]*

DESIGN SPECIFICATIONS FOR *HUBBLE'S* SCIENTIFIC INSTRUMENTS

Instrument	Field of View	Wavelength Coverage (A)	Resolution	Limiting Magnitude
Wide Field/Planetary Camera	2.7'; 1.1'sq.	1150–11,000	0.1"; 0.04"	28
Faint Object Camera	11"; 22" sq.	1200–6000	0.02"; 0.04"	28
Faint Object Spectrograph	0.1–4.3"	1150–8000	3A; 20A	22, 26
High Resolution Spectrograph	0.25–2"	1100–3200	.03; .1; 1A	11, 14, 17
High Speed Photometer	0.4; 1; 10"	1200–8000	10 microsec.	24
Fine Guidance Sensors	0.69" sq.	4670–7000	0.002"	18

In principle, Space Telescope's optics are sensitive to wavelengths ranging from 1,100 angstroms (or 0.000004 inch), deep in the ultraviolet part of the electromagnetic spectrum, to about 1 millimeter (or 0.04 inch), in the far-infrared part of the spectrum. (Infrared radiation is nothing more than heat, whereas ultraviolet radiation is that which can cause severe sunburn. Both are largely blocked by Earth's atmosphere from reaching ground-based telescopes.) In practice, while most of *Hubble's* initial (first-generation) science instruments are sensitive to ultraviolet radiation, none of them penetrates much of the infrared spectrum, although the wide-field and planetary camera can acquire radiation in the near-infrared, out to a wavelength of 11,000 angstroms (or 1.1 microns). All but one of the onboard science instruments (the high-resolution spectrograph) are sensitive to visible light. One of *Hubble's* second-generation instruments, scheduled for installation by astronauts in the mid-1990s, will likely probe more deeply into the infrared spectrum.

A note on stellar magnitudes and the concept of sensitivity: The magnitude system for measuring the brightnesses of cosmic objects was devised during the second century B.C. by the Greek astronomer, Hipparchus, who arranged most naked-eye stars into six groups. He let the brightest and thus to him the most important stars be of the first magnitude, and the faintest the sixth magnitude. That is, the fainter the star, the *larger* the magnitude—a convenient little ploy to keep the beginners out. For example, Altair, the brightest star in the constellation Aquila, is of magnitude +1, whereas the majority of fainter stars comprising the nearby constellation Cygnus are of magnitude +3.

Of course, telescopes can collect light from stars much dimmer than even the faintest naked-eye objects; such "light buckets" have greater sensitivity. Hence astronomers have extended the magnitude scale beyond the six groups originally envisioned. For example, a good pair of binoculars can show stars as faint as the ninth magnitude and an amateur 6-inch telescope can reach about the thirteenth magnitude. The 200-inch telescope atop Mount Palomar, in California, can photograph objects as faint as the twenty-fifth magnitude, which is equivalent to seeing a candle at a distance of about 10,000 miles, a sensitivity roughly 10 million times greater than the human eye. (Earth's atmospheric emission, or nighttime airglow, contributes a dim background of

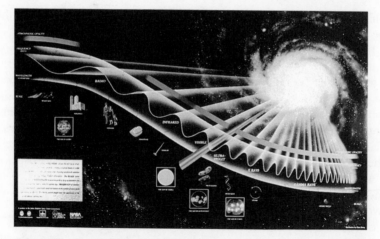

The electromagnetic spectrum of all known kinds of radiation, on Earth and elsewhere in the cosmos, stretches from radio waves at left to gamma rays at right. The icons display the huge range of wavelengths, from mountain-size radio waves (at the longest wavelengths or shortest frequencies) to subatomic-size gamma rays (at the shortest wavelengths or highest frequencies). The human eye is sensitive to only a small part of this vast spectrum, namely the visible or optical domain of red, orange, yellow, green, blue, and violet colors, shown spanning the middle of the chart, from 7,000 to 4,000 angstroms. (An angstrom is a very small unit of length measure; a billion such angstroms equal 10 centimeters or about four inches.) By analogy with pitch on the musical scale, the range of visible light covers not quite one octave, a frequency span of two, compared to some seventy octaves ranging across the entire electromagnetic spectrum. It is this visible radiation to which *Hubble*'s instruments are also sensitive, indeed much more so than the naked human eye. In addition, *Hubble* can sense most of the ultraviolet spectrum (longward of 1,000 angstroms) and a small part of the infrared spectrum (shortward of 11,000 angstroms), to both of which the human eye is blind. *[D. Berry]*

unwanted light to all ground-based astronomical observations equivalent to the brightness of a twenty-second-magnitude star; the virtual lack of such airglow in space is one of the main reasons why Space Telescope should be supremely sensitive, routinely reaching celestial objects as faint as the twenty-eighth and occasionally the thirtieth magnitude.) To accommodate the very brightest objects in the sky, the magnitude scale also extends into negative numbers; for example, the bright star Sirius has a magnitude of −1.5, Venus is brighter still at −4 (usually), the full Moon −13, and the blazing Sun has a magnitude of −27.

At the start of the twentieth century, even professional astronomers thought that our Sun held a special place in the Universe. Aristotle's ideas, and hence those underlying much church dogma, still held strong sway, even among intellectual circles. What we now regard as the Milky Way Galaxy was then reasoned by mostly everyone to be the totality of the cosmos, with the Sun at the center of that immensity. Enter Harlow Shapley, one of two dominant figures in the world of American astronomy during the first half of the century, who placed correctly our Sun and planetary system in the "suburbs" of the Milky Way. Shapley made this bold leap by leaning heavily on the photographic analyses of a small group of women who (eventually) worked for him at the Harvard Observatory.

If you know the absolute (intrinsic) brightness of an object and can measure its apparent (as-it-looks) brightness, then you can infer the distance to that object. The "computers," as those women were then called, led by Henrietta Leavitt in 1908, had found a correlation between the luminosity (or absolute magnitude) and the period of regular brightening and dimming among a certain class of variable (or pulsating) stars called Cepheids, so named for the best-known example, Delta Cephei in the constellation Cepheus. Such Cepheid variables became distant lightposts or "yardsticks."

Accordingly, astronomers could then determine the distances to tightly bound swarms of stars, called globular clusters, which contain most of the known Cepheids. And it was these globulars that enabled Shapley in 1918 to derive the full extent of our Galaxy—roughly 100,000 light-years across—and to show that our Sun is far removed from the center of it, roughly 30,000 light-years from the Galaxy's core. (A light-year is a distance, namely that traveled by light in a full year; it amounts to some 6 trillion miles, or 10 trillion kilometers.) Thus, hardly more than a half-century ago, Shapley endowed our Sun with the same mundaneness that Copernicus had done for our planet some five hundred years ago by dethroning Earth from the center of the Solar System.

Even Shapley, however, was unable to take the next step forward and to conceive that our Milky Way is only one such galaxy among innumerable other galaxies in the far-flung Universe. That fact was unambiguously established by Edwin Hubble, the other giant of American astronomy. He and his colleagues delved into this "realm of the nebulae" (as he would later title one of his books), and also went on to prove that the Universe is expanding as though ejected from a cosmic bomb, later dubbed the Big Bang, that must have occurred far back in the distant past.

Using in 1924 the recently completed and then world's largest telescope, the 100-inch reflector on Mount Wilson, outside Los Angeles, Hubble solved his first key problem in cosmology. He did so by using the new telescope's superb power and clarity of view to discover Cepheid variable stars in several puzzling celestial objects—called at the time nebulae—which were then the source of much scholarly debate. The problem was, were these nebulae local, that is, within the confines of the Milky Way, or were they distant galaxies in their own right, each with hundreds of billions of stars much like the Milky Way? By using the Cepheids as distance markers, Hubble decisively provided the answer: The objects, including what we now call the Great Galaxy in Andromeda, were full-fledged galaxies, much like our own and well beyond our own. The Milky Way Galaxy is not alone in space; it is surrounded by other galaxies, indeed many others, millions and millions of light-years away.

This breakthrough alone would have landed him in the scientific hall of fame, but Hubble did not stop there. Throughout the 1920s, he took, developed, and analyzed thousands of photographs and went on to classify the various shapes and sizes of galaxies, devising a morphological sequence—ellipticals, spirals, barred spirals, and irregulars—that is still used today. Continuing well into the 1930s, Hubble undertook an extensive observing campaign, attempting to map the spread of galaxies in progressively deeper realms of space; his work probed the far away and long ago to a then astonishing 500 million light-years. Accordingly, he showed that galaxies are more than a mere "next step" in the hierarchical organization of material systems; rather they are basic units or "test particles" in the distribution of all matter in the Universe.

Even these, though, are not Hubble's most important findings. Starting with his pioneering paper in 1929 and culminating with a series of publications in 1936, he and a close associate expounded what is now called Hubble's Law—that the red shift in the spectrum of a galaxy is observed to be directly proportional to its distance. Translated: The more distant the galaxy from Earth, the faster it moves away, since the observed red shift is interpreted as a Doppler effect, much like the lowered pitch of a receding train whistle. This is the basic observational work, well verified today, that bolstered the notion of an expanding Universe, a centerpiece of modern cosmology. Hubble did not invent the concept of the expanding Universe, nor did he discover the red shift of starlight, but he found the first direct evidence that most galaxies are racing away from us, and he worked out the basic relationship between red shift, distance, and velocity.

Indeed, these are the scholarly articles that made Albert Einstein realize, for all *his* greatness, that he (Einstein) had made a fundamental blunder in his General Theory of Relativity by arbitrarily inserting a "fudge factor" to keep his model Universes (solved theoretically on paper) from expanding or contracting. Apparently, Einstein, too, was an Aristotelian at heart, and it remained for Hubble's conclusive observational findings to convince him otherwise. (Contrasting in his work with Einstein, Hubble's closest collaborator

Edwin Hubble posing at the business end of a telescope on Palomar Mountain. "He always posed," said a friend of his ingenuously. Hubble was by nearly all accounts an aloof, self-centered, yet imposing personality—"not a very nice man," said a Caltech colleague who knew him well. Born of middle-class parents in small-town Missouri, he eventually went on to study languages and jurisprudence as a Rhodes Scholar at Oxford, whereupon he apparently shed his Midwest American twang for a thick English accent. *[J. Bedke and the Huntington Library]*

was the late Milton Humason, a grammar-school dropout who was, successively and successfully, promoted throughout the first two decades of this century from Mount Wilson mule driver, to janitor, to night assistant, to telescope operator, to full-time astronomer.)

The Hubble Space Telescope is designed to build upon Hubble's research, measuring distances in deep space with greater precision than ever before possible. Provided it can be repaired properly, the orbiting observatory is expected to specify to an accuracy of some 10 percent a fundamental constant of nature that we now know only to roughly 50 percent. This is fittingly known as the Hubble constant, equal to the ratio of the velocity of the receding galaxies and their distances from Earth, in other words a measure of the rate of universal expansion. This should further enable us to derive the age of the Universe also to unprecedented accuracy and, although it sounds presumptuous, to determine no less than the fate of the Universe—expansion forevermore or collapse to a point much like that from which it began. No small issues, to be sure, the time machine called *Hubble* has been built to attack, scientifically and logically, some of the most profound questions of our day or any other.

Asked what he expected to find with the newly commissioned 200-inch Mount Palomar telescope in 1948, Hubble answered, "We hope to find something we hadn't expected." And so it was also with the Hubble Space Telescope as it began its quest to solve some of the most elusive mysteries in the Universe. It may also answer questions we are now unable even to pose.

Returning from my evening visit to the launchpad, I headed back to my hotel room on Cocoa Beach. Driving out through the main gate, I noticed scores of cars already queued, awaiting entry to the Kennedy Space Center via the NASA Parkway at dawn the next morning. The risk of danger at their position (more than a half-dozen miles from the launchpad) was virtually nil—far less than inside a 3-mile radius where I would be the next day near Shuttle Launch Control, which in the event of a major rocket explosion, according to Air Force security, "would not ordinarily be hit with anything larger than a desk." In all, nearly a million people were expected to jam the parkway, more than perhaps for any space flight save the epochal Moon voyage of *Apollo XIII* in 1969.

Farther from Kennedy, I spied in the darkness scores of vans and mobile campers parked alongside the Bennett Causeway separating Port Canaveral from the Atlantic coast, many having accumulated for several days in order to guarantee an unobstructed view of the launch from a distance of some fifteen miles.

Although it was well before midnight, I began to worry that I might have trouble getting back to the space center in time for the launch shortly before 9 A.M. the next day. NASA public affairs officials had warned us that, by dawn, automobile gridlock would likely occur throughout the Cocoa Beach and Titusville areas on all main roads leading to the space center. I wondered several times on the way back to the hotel if I shouldn't turn around and make it an all-night affair at Kennedy.

Suddenly, I was shaken when an accident occurred on the opposite side of the divided roadway. A few travelers stopped near the accident site, partly out of curiosity and partly to offer aid. From a distance, a small boy who had been injured when thrown from his mother's car helped me to mentally put our space science efforts into perspective. While waiting for the arrival of a local ambulance, I came to realize the relative importance of the next morning's launch. *Hubble* is a billion-dollar robot, but one that pales in comparison to the value of a living human being. So what if the telescope was punted into the Atlantic, or if it didn't work in orbit? Nothing we humans build is likely to surpass the importance of humanity itself.

Sure, machines aid our human senses, enabling us to build cathedrals, telescopes, weapons, and the like. Indeed, *we* had built *Hubble*. But it was very clear, at least to me that evening, and even at a time of great emotional attachment to Space Telescope, that the bond among humans will not likely ever be exceeded by that between man and machine. Why then would our technological society risk astronauts' precious lives tomorrow to launch and deploy a robot that could otherwise have been lifted more cheaply and more effectively into orbit by means of an unmanned, expendable rocket?

Back at the hotel, I tried to get some sleep, but couldn't. I rose, walked along the beach for a while, encountering several colleagues experiencing the same insomnia, our adrenaline clearly peaked. Returning to the hotel, I arranged with the midnight receptionist to wake me at 4 A.M., I set two alarm clocks as a backup, and I called a friend to have him ring me as a triple-check against oversleeping. I did not intend to miss this launch. As things turned out, I hadn't needed to make such extravagant arrangements.

By 3 A.M., my telephone was ringing repeatedly. A senior scientist expressed worries about last-minute technical changes that had been executed in the vastly complex software needed to command and control *Hubble* on orbit. A junior scientist shared his second thoughts about the viability of one of the science instruments aboard Space Telescope, claiming that it hadn't been tested well enough. A young postdoctoral fellow wanted to discuss the political significance of the long-awaited event slated to occur later that morning, and again agonized over the extent to which the bird had been prepped. And yet another science colleague simply wanted to know—at 3 A.M.—my recommendation for the best viewing site.

Contrasting with the increasing worries expressed by scientists leading up to launch were the supremely confident attitudes of NASA management, often expressed publicly. Take this spinning statement uttered only hours earlier on ABC's *Nightline* television talk-show by Charles Pellerin, the astronomy head at NASA Headquarters in Washington, who was fond of calling *Hubble* "the crown jewel of space astronomy":

> I'm not really worried. This is a very, very fine piece of machinery. It's been thoroughly tested, it has no liens against the hardware, it performs according to our calculations better than its requirements in almost every area. It's been trouble-free, we've got lots of run time on it, and I'm very confident this telescope's going to work just fine.

What came to mind most, and repeatedly so, in the middle of this particular night, was a deeply held concern expressed to me many times by Riccardo Giacconi, the director of the Science Institute: Twenty years earlier, while designing and building a spectacularly successful spacecraft, known as the *Small Astronomy Satellite #1* and nicknamed *Uhuru* (meaning "freedom" in Swahili), he had become so obsessed with the vehicle and all its potential problems that, as launch neared, he often dreamt about being inside the spaceborne satellite, occasionally waking in a sweat upon realizing that some technical widget needed addressing; he had even slept on the mobile launchpad off the coast of Kenya on the eve of flight. Now, he was wondering who was worrying about *Hubble*. Who was obsessed with the vehicle to the point of awakening in cold sweats? Who, this night, was out on the pad?

All the while these issues danced through my head, I imagined the roadways surrounding the space center filling up with myriad onlookers and gridlock setting in prematurely. Impatient, I dressed, made a few calls to others whom I had promised to wake later as part of our early-riser network, told them I couldn't wait any longer, and began making my way back to Kennedy. But not before shoving into my pocket, for no good reason, a key-chain kaleidoscope that the systems analysis and design group at MIT's Lincoln Laboratory had rigged with a slide of an artist's conception of *Hubble* in orbit—"an emergency Space Telescope," they told me, smiling, when I left the group in the mid-1980s, "in case things don't go quite as well as expected."

Outside the hotel, the main highway was absolutely deserted, but as I neared the Bennett Causeway, I nearly panicked as the high-tech pilgrimage had now swollen to hundreds, perhaps thousands, making parts of the roadway nearly impassable. Vehicles of all sorts, including bikes and buses and pedestrians, too, were converging on the center, seemingly drawn toward the powerful beams of light shooting up into the nighttime sky. Early-morning fog had set in over the area, and the launchpad floodlights that I had found startling the evening before were now creating an absolutely eerie image.

A tremendous throng of vehicles clustered about the space center's brightly illuminated outer gate, including security vans with lights flashing and radios squawking, amid much commotion. Fortunately, I noticed that SWAT teams had kept a very narrow lane clear, so I took it. I had secured well ahead of time all sorts of passes and I knew upon reaching the main gate that one or another of them would gain me entry. Once inside, I aimed for the press site. Although the VIP site is slightly closer to Pad 39B, its view of the shuttle is obstructed by the gantry, which blocks the rocket until it has cleared the tower. The press site, a planetarium-like domed structure adjacent to Shuttle Launch Control and just behind the staging area for the armored personnel carriers poised to rescue the astronauts in the event of a pad emergency, affords an unobstructed view of the launch.

Even in the predawn darkness, the place was already busily engaged. Some people had indeed spent all night there, although most were only now beginning to straggle in. A few phones were ringing and several overhead television monitors were focused on the astronauts, recently awakened and now eating breakfast. Soon these five veterans of spaceflight would be suiting up for launch, and soon I would have to deal again with the media—the price to pay for viewing a launch from the press site—but right now I wanted to talk to no one. I grabbed a cup of strong coffee from the "roach coach" mobile snackbar behind the dome and slipped into the unreserved press booth #1, adjacent to the *New York Times'* desk #2, wondering all the while if the #1 position had been purposely left unassigned.

With the pace of activity in the dome beginning to rise, I was nonetheless able to revisit Ralph Waldo Emerson's 1836 essay on nature. I was determined to give the *Hubble* mission a larger, historical perspective that I thought many others associated with the project had lacked. Although I was convinced that I was alive and present at a truly historic moment in humankind's attempt to fathom nature, I wanted to force myself to appreciate the idea that we really were standing on the shoulders of others before us. Somewhat by contrast, a compatriot astronomer from the Science Institute, Colin Norman, had no illusions. He wiled away the time rereading Shakespeare's *Henry V,* arguing that what surrounded us that early morning more resembled Agincourt, with science's eventual triumph despite the odds against the government getting its act together. As things turned out, Colin was especially perceptive.

The rate of phone ringing gradually accelerated, the army of reporters assumed its desk-bound position, and the amount of incoming electronic mail ("e-mail")—the computerized communications method of choice among today's technically minded—rose dramatically. Noting the monitors overhead showing the astronauts entering a van for delivery to Pad 39B, Colin and I left the dome for the roadway across from the Vertical Assembly Building. As the caravan approached in the predawn darkness, the crew transport van was led and trailed by dimly flashing security vehicles on the ground; all the while a helicopter glided overhead. No appreciable light shone save for the aircraft's ground-looking search beam illuminating the roadway afront the lead vehicle, nor was there much sound save for the sluggish swish and beat of the chopper's blades.

The unofficial ceremony was short and unscripted: the caravan came to a stop in the middle of the road astride Shuttle Launch Control; the helicopter and its downward-searching light made a tight circle around the area; and we gave the "thumbs up" sign to convey our good wishes to the window-peering astronauts whom we had gotten to know in the years of preparation for Flight STS-31—the thirty-fifth mission of what NASA officially calls its Space Transportation System but which everyone else calls "the shuttle." (The preflight numbering system is out of sync with the order of actual launches both because the assembly of some payloads was delayed and because some military missions have preempted regularly scheduled shuttle launches.) Almost immediately, the helicopter rejoined the caravan, illuminating once more the lead vehicle's future roadway, and off sped the procession toward the pad.

By the time we made our way back to the press dome, dawn had broken. Some three hours prior to scheduled liftoff, media activity had now reached advertised proportions. It seemed that a fair fraction of the nearly two thousand journalists who had been accredited for this launch were setting up cameras, stringing cables, filing stories, or were out and about interviewing people—pursuing the "human-interest" angle, while preparing to feed their editorial chiefs who maintain that it is people who make the news. "How do you feel on the morning of launch?" "What's going through your mind now?" "Do you have any second thoughts?" "Were you really reading Emerson a few hours ago?" "Can I have a sound bite?" The dome's many phones were now ringing unceasingly, much like in those annoying but necessary fund-raising periods on PBS television. I swore that the whole domed structure itself was beginning to vibrate. I also swore that I was going to try to get away from this mass of media in order to experience this historic launch alone.

Time seemed to tick away more rapidly now, but of course a second is a second. The huge, 10-foot-high digital clock at the press site became the focus of much attention—a little like the trance often brought on by the flames in a wood-burning fire. Then, almost simultaneously, the clock stopped—a scheduled hold, everything still "nominal"—and a colleague yelled for me to join him. I had almost fogotten that there was one more important prerequisite to witnessing the launch. We needed to journey to the VIP site to press the flesh.

NASA takes the VIP business seriously, not suprisingly given its multi-billion-dollar federal budget. At the VIP site, I noticed a hierarchy of very important persons, each group in roped-off areas, apparently so as to avoid the embarrassment of very VIPs having to interact with lesser VIPs. In all, there must have been about a thousand people there, including leading politicians, government bureaucrats, international diplomats, astronauts' relatives, descendants of the Hubble family, university regents, industrial chief executives, prominent scientists, officials of the European Space Agency, and a few journalists and camera crews who had wormed their way into the VIP area to get the "human-interest" stories.

The highest honor goes to those who arrive at Kennedy on the morning of launch on the "Administrator's plane." This is a high-level group of a hundred or so special guests of the head of NASA, including key senators, cabinet secretaries and ambassadors, a few relevant scientists, among a handful of others who manage to finagle their way onto the invitation list by savvy means indeed. The plane leaves Andrews Air Force Base just outside Washington several hours prior to launch and lands directly at the Shuttle Emergency Landing Strip a few miles from the pad, after which its riders get a mission briefing that the scientists find boring and that the nonscientists don't seem to understand. At least the vice-president wasn't there to ask again, "By the way, which planet is *Hubble* going to visit?" These "vVIPs" then appear at their appointed viewing stations about an hour before launch, looking fully as tired as the rest of us who had been up most of the night worrying about what time bombs might be onboard the spaceship.

My "targets" that morning were an influential politician whom we had befriended during the year before launch, the retiring president of Johns Hopkins University where I taught, and my boss, the eminent astronomer Riccardo Giacconi. Scanning the crowd, I spotted Senator Barbara Mikulski and spent a few moments discussing with her and her aides nothing in particular—

it was the gesture that counted, although NASA in the ensuing weeks would worry excessively about what I had really said to her. My university president had apparently missed the plane, and so I thought had Giacconi, until I found out later that he had become virtually incognito while (unlikely for him) slumped under a newly purchased baseball cap with the (even more unlikely for him) acroynm "NASA" blazoned across the top.

One of the most peculiar sights—*glasnost* notwithstanding—was a group of a half-dozen leading Soviet astronomers, sitting rather sheepishly in the VIP stands while waving little American flags. The flags were too large to put in their pockets, yet the (mostly) Russians were too politically astute just to put them down somewhere.

Unlike at the press site, where most reporters were veterans of previous shuttle launches, few of the VIPs had ever before seen a rocket launch live. A very definite excitement yet tension pierced the dawning air; people were thrilled to be there yet subdued, a sort of unvoiced concern that astronomy's greatest gambit could go awry. Nearly everyone had aimed themselves toward the launchpad, thinking or talking in a concentrated manner about the shuttle, the telescope, the astronauts. Many of us assembled as witnesses seemed on edge, both physically and mentally. By contrast and ironically, a few weeks later, at dinner in the Hubble Room at a hotel restaurant near Johns Hopkins, one of the onboard astronauts, Kathy Sullivan, confessed to me that at just about that time she had fallen asleep.

The Sun was low and spectacular, "rising" in the east. The fog was beginning to lift. From the VIP site, the view of Pad 39B intersects a shallow lagoon laden with dead, defoliated trees. What an Ansel Adams–like contrast these people were about to see: a brilliant rocket springing to life with tremendous force juxtaposed against a depressing, black-and-white swamp partly polluted by the exhaust products from many of those same rocket launches. Even so, I knew that the VIPs would not see the shuttle until it cleared the tower, and, given the prevailing winds, that they might have to evacuate the area soon after launch lest they be covered with a highly acidic

The VIP site at Kennedy, shown here about an hour before liftoff, was jammed for the *Hubble* launch. *[NASA]*

misty residue (an "acid rain") from the solid-rocket wastes—mostly hydrogen chloride and aluminum oxide, not too unlike the annoying sulfurous chemical that irritates eyes and respiratory tracts while cutting onions. (Thousands of fish are estimated to die of suffocation after each shuttle launch since many of the surrounding shallow-water lagoons become for a time extremely acidic, with a pH of 1 or less; an official Air Force document states, "These affected animals should be considered dedicated in the interests of the mission.")

It was now less than an hour to launch, so I began making my way back to the press site where I had left several Science Institute colleagues. Approaching the huge infield beside the dome, I surveyed the throng of mostly reporters peppered with scientists, engineers, and sundry NASA officials. The Sun was now shining brightly, the fog gone, only fluffy clouds overhead. It was a spectacular day for a spectacular event. But strangely, I still didn't feel as though it was really going to happen. Even with hundreds of tripod-mounted still cameras and dozens of television minicams trained toward a single target, we all realized that the shuttle is an extraordinarily complex machine, not to mention its human cargo, and as such any small irregularity in the countdown telemetry could cancel the launch that day.

Two weeks prior, we had gone through this entire drill and had been very close to launch of STS-31, only to encounter a faulty APU—one of three auxiliary power units that drive hydraulic pumps in each of the shuttle's main engines. Merely four minutes before expected liftoff, the launch director had not hesitated to scrub the flight. And rightly so, for if *Discovery* had lifted off with an erratic backup engine valve, although not of immediate danger to the crew (since only two APUs are needed for critical spaceship systems such as steering flaps during takeoff), the shuttle would likely have been ordered to make an emergency landing—either back at the Kennedy Space Center, which never before had been attempted, or downrange at one or another of the transatlantic abort sites at the Banjul International Airport in the Gambia or Ben Guerir Air Base in Morocco. Or perhaps a single orbit would have been authorized—an "abort to orbit"—landing more sedately at Edwards Air Force Base in California. In any event, Space Telescope would not have been deployed, and would have returned to Earth inside the cargo bay of the crippled *Discovery*.

Given an oft-quoted gibe used to describe the aerodynamic performance of the space shuttle—"it comes down like a brick," owing to its steep descent back to Earth—Space Telescope project officials had often expressed doubts about *Hubble*'s ability to survive a shuttle landing. In the years leading up to launch, many shake and vibration tests had simulated the multiple stresses on a *Hubble* cargo during a shuttle's launch and landing, yet the results seemed ambiguous as to how structurally robust *Hubble* really was. To my mind, the concerns about landing were not well founded, as Space Telescope had been successfully delivered to Kennedy inside the belly of the world's largest aircraft—a specially modified U.S. Air Force C-5A Supergalaxy used to transport a class of huge spy satellites similar in size and shape to *Hubble*.

During the late stages of the Vietnam era while on assignment in Europe, my diplomatic passport had won me a passenger seat on the flight deck of a

C-5A cockpit when, with the bird having lost much of its hydraulic fluid, we came down hard at a Royal Air Base in Mildenhall, England (the European home of the stealthy SR-71 Blackbird), blowing out twenty-two of the aircraft's twenty-eight tires. The aircraft commander, a veteran pilot of a nearly identically shaped though smaller transport called a C-141, approached the flight line manually yet forgot in the stress of the moment that a C-5A's cockpit is fully three stories above the tarmac. I had always thought that if the several hundred Marines survived that day in the cargo bay of the C-5A—which incidentally is rated to carry the Army's largest piece of inventory, a 17-ton bridge fully assembled—then *Hubble* was not likely to be harmed appreciably in the event of a normal shuttle landing.

Admittedly, retrieving Space Telescope from a dusty landing strip in northwest Africa after a faulty launch might present a challenge. But the real fear of ever bringing *Hubble* back to Earth is more political than technological; given the paucity of shuttle flights and the backlog of grounded spacecraft sitting in the U.S. inventory, we might never get *Hubble* back on orbit. "If *Hubble* ever returns to the ground, it will most likely end up in the Smithsonian Museum," Giacconi used to say.

In early October 1989, *Hubble* was secretly flown from Lockheed's Sunnyvale plant, in Silicon Valley, to the Kennedy Space Center. Actually, the cloak of secrecy was pretty thin, up to a point. We knew when the telescope was to be crated at Lockheed and I was alerted by a friend at NASA's Ames Research Center when two pregnant-shaped C-5As arrived at neighboring Moffett Field outside San Francisco. Thereafter, I kept calling my Ames friend several times a day, saying, "Has it left yet?" One day, just after dusk, one of the C-5As departed, arriving the next morning before dawn at Kennedy—a sleek transfer in the dark of night. It all happened only a week or so before the Richter-6.5-magnitude earthquake hit the San Francisco area at the start of one of that fall's World Series baseball games. As Lockheed's Sunnyvale plant is a mere few miles from the main San Andreas fault—an odd place to have stored for several years humankind's most expensive scientific gadget—we sci-

Shrouded here inside its classified shipping container, in turn draped in canvas and other camouflage, *Hubble* had just been off-loaded from a huge C5-A Air Force cargo plane. In the background is one of the world's most voluminous structures, the Vertical Assembly Building, originally built to assemble *Saturn* Moon rockets and now used to mate space shuttles to their fuel tanks and booster rockets. The two-story structure with the large picture windows between the VAB and *Hubble* contains the "firing room" of Shuttle Launch Control. *[NASA]*

entists, who like to think that we eschew superstition, nonetheless considered *Hubble*'s safe and timely trip to be a good omen.

The secrecy surrounding the shipment of *Hubble* was unwarranted, in my view, especially when the Space Telescope has absolutely no military role whatsoever. I acknowledge, however, that a great deal of confusion surrounds this topic, especially among the general public. Even a Soviet friend once joked about *Hubble* being a spy satellite—at least I think he was joking—as we walked, ironically, through Lubyanka Square, headquarters of the Komitet Gosvdarstvennii Bezopasnosti. But the facts are these: Although most portions of the Earth save its sunlit limb are not too bright for *Hubble* to look at, without any onboard propulsion system for tracking or navigational means to lock onto a terrestrial target, photos of Earth taken by a swiftly orbiting *Hubble* are hopelessly blurred. In a single second, *Hubble* overflies some 5 miles (8 kilometers) of Earth's surface, reducing all terrain detail to visual mush.

At any rate, I am prohibited from divulging the exact date of Space Telescope's shipment, and I cannot show a direct photograph of the outside of its tank car–shaped shipping container, known to the cognoscente as "the casket." Presumably this secrecy arises because several other "space telescopes" do in fact exist, and we were using the high-tech cargo crate of these downward-looking vehicles whose targets are indeed on the Earth. In truth, a fair amount of *Hubble*'s technological design and fabrication derive from Project *Keyhole*, a "darkside" enterprise that has during the past two decades orbited a virtual fleet of telescopic intelligence gear. But fears that the outside of a *Keyhole* shipping container could reveal something that has not already been reported in *Aviation Week and Space Technology* seem silly at best and probably a little paranoid.

All these things, and more, fluxed through my head as launch preparations proceeded. The clock was stopped again at nine minutes prior to liftoff—"L-9," a

Hundreds of cameras and lots more eyes at the press site were aimed toward the launchpad amidst the prearranged hold at nine minutes to liftoff. The huge digital clock can be seen at center, and the Shuttle *Discovery* in the far distance to the left of the flagpole. *[NASA]*

scheduled hold before any shuttle launch in order to give technicians in the "firing room" at Shuttle Launch Control an opportunity to canvass once more the most crucial elements on the preflight checklist. Communications: Go. Hypergolics (fuel): Go. Safety officer: Go. Medical (astronaut) status: Go. All systems synchronized: Go. More preprogrammed checks: Flaps and wing elevons, speed brake, rudder, and ailerons all waved, via television monitors, to the controllers as computers put critical shuttle surfaces through their paces.

My mind began involuntarily flashing back, and then fast-forwarding, in no particular order, to the scores of *Hubble*-related events that had left deep grooves in my memory. Cruising through my head were many of the peaks and valleys along the path trod by the estimated 10,000 people who had worked over two decades to bring *Hubble* to the launchpad—its long and troubled history of multi-rescheduled launch dates ever just over the horizon, its heated meetings and endless discussions and personality clashes and management politics, and, despite it all, its ostensibly exquisite technology for space astronomy. Negative pressures and positive energies pushed and pulled on my psyche, rendering a rapid replay of our oscillating hopes and aspirations for *Hubble,* but in the end humanity had done it—had built a robot seemingly capable of, at will, unlocking secrets of the Universe. The button was soon to be "pushed." We were poised at a critical point in humankind's rite of passage. Everyone who had anything to do with *Hubble* must have been aware at that moment of his or her own contribution to this immensely large and complex project.

I recalled the huge crowds attending public talks across the country, eager to hear about the promise and power of *Hubble* for discovery, about this preeminent symbol of humankind's curiosity. In the months and weeks before launch, the Science Institute had been unable to keep up with the increasing demand for speakers, and when we did send out our scientists— typically once per day—they invariably returned with glowing accounts of how well and closely the public was prepared to follow this space mission. The intense anticipation was a real high for me. Lay citizens far and wide seemed uncommonly excited about science and supportive of the fun we scientists were expecting to have. Attendees everywhere had been friendly and noticeably keen to hear of the telescope's science goals and objectives, even though I often made a point of downplaying the hype associated with them. The impending mission of the Hubble Space Telescope was easily the most eagerly anticipated event in the history of astronomy. It seemed like the whole world was watching.

Of course, much of the buildup for *Hubble's* vaunty eye on the sky was orchestrated by the popular press, and I knew this would be tricky territory. Were we going to regret some of the headlines accompanying the media frenzy in the weeks and days leading up to launch? I had recently entered in my journal a handful of the many superlative statements and bold headlines printed across the nation's front and editorial pages. For example, the *Baltimore Sun* had proclaimed: "Nothing like *Hubble* has been done before and, if all goes well, its discoveries are certain to be unparalleled in the history of astronomy. . . . *Hubble* promises to send the human mind to the very limits of space and time." *Florida Today:* "Serious study [of the cosmos] began centuries ago, but 1990 will go down in history as the year the quest for answers started a long-term stay in space." *Albany Times Union:* "The Hubble Space Telescope is going to make every astronomy book obsolete. . . . And, oh, what celestial sights it is expected to see." *New York Times:* "Landing on the Moon was a giant technical step for humankind; the *Hubble* telescope promises giant intellectual strides toward understanding the structure of the Universe." *Washington Post:* "Once the week or so of routine checking is finished, and the *Hubble's* array of instruments start transmitting images free of the atmosphere's distortions, the Universe known by stargazers is going to start getting bigger and clearer and more interesting, fast."

I snapped out of my trance when I overheard my deputy, Ray Villard, mutter what many of us then felt: "I just hope the damn thing works."

Without warning, the words flashed over the intercom from Shuttle Launch Control: "All systems are Go. The countdown has resumed." Simultaneously, the monster digital clock at the edge of the press site came alive. We were now inside nine minutes, little more than counting to five hundred. Silicon-based computers had assumed control of the launch. My inner attitude changed while scanning the sky. I saw nothing problematic except for one rather small fluffy white cloud at a few thousand feet; the winds were light, the weather at the overseas abort sites clear. I felt for the first time that this launch might well occur.

The cable networks were now airing live from the Cape, their commercial counterparts about to interrupt regularly scheduled programming in order to carry this celebrated launch live, in "real time" as we say. Adjacent to the press site's dome and inside the permanent huts labeled NBC News and CBS News, workers were frantically washing the wall-size picture windows behind which familiar anchor persons would sit framed alongside the distant shuttle. ABC News had a cleaner idea—hours before, they had completely removed their huge picture window, thereby eliminating entirely any possibility of dirt or reflection on the glassy panes behind the anchorman.

My anticipation rose dramatically as I saw parts of the service tower retracting and the gantry catwalks swinging away from the orbiter, allowing the spaceship and its mated rockets to stand upright on their own, anchored only by eight 2-foot-long bolts near the tail of the shuttle. The spent gases from the super-cold liquid oxygen stopped venting atop the shuttle's external tank, meaning that its supply had been topped off and capped, allowing the pressure to build necessarily for maximum thrust. The words, "close and lock your visors . . . and ya'll have a good trip," came over the intercom from the launch director to the astronauts. The flight team in the firing room was all business now, intensely eyeballing the monitors displaying *Discovery's* vital functions. *Hubble's* quick spurt to orbit indeed seemed likely to happen.

I wanted to be in two places at the same time—there at Kennedy at the moment of launch and back at our Science Institute in Baltimore with so many of my colleagues who were watching a closed-circuit satellite transmission directly from the Cape. I knew that after launch they would actually have a better view, especially of the crucial solid-rocket-booster separation at high altitude, much as the home television viewer often has a better view of a key play of the ballgame than a fan at the ballpark. But I knew from my youth that the crack of the bat and the roar of the Fenway crowd had something a television signal simply cannot capture; I needed to be at Kennedy to witness this particular launch live, to see and hear for myself the star-bright flame and the deafening roar of the rocket engines, to feel the ground vibrate.

I couldn't stand still now, and I certainly couldn't sit calmly in a set of grandstands filled with journalists, scientists, and a mix of NASA and industry officials. I began walking out onto a huge grassy area fully the size of a soccer field, to visualize the launch in isolation and let it be burned into my memory forever. Alas, as soon as I started across the field toward the mesmerizing digital clock, a battalion of reporters followed. Pencils and paper ready, and including an Orlando TV crew with a shoulder-mounted mini-camera transmitting live, my impromptu entourage was probably disappointed as I didn't say a damn thing. I just kept one eye on the clock and another on the shuttle, all the while ambling around the infield. A colleague back in the bleachers later told me that I looked like a pied piper and not just a little ridiculous.

The moment of launch was near—some twenty years of design, construction, and testing had converged to less than a minute. All systems seemed fine, when at L-31 seconds, the clock stopped. I stopped, too, as I looked back to the bleachers, about a hundred yards away, where many colleagues were poised as though frozen. Something was wrong with the launch procedure and it was as though time itself—not just the countdown clock—had quit. Even the wind lulled, as the Cape air became uncommonly calm. I thought of the tense ground crews, the cloistered astronauts, the pent-up rockets. I thought of *Hubble*. Was this bird jinxed?

In less than a minute, engineers in the Shuttle Launch Control building next door informed us on the outdoor intercom of an apparent system failure regarding the transfer of vehicle management from the launch control computers in the firing room to the onboard shuttle computers. Was it another engine power problem like the one that had postponed the mission two weeks prior, or was it the fault of one of hundreds of other widgets critical to a successful shuttle launch? Still, everyone within view remained suspended, only eyes and lungs moving. The air now quieted completely, anticipating propagation of an important announcement toward our ears: "A faulty piece of software has been identified regarding a fuel valve and a work-around was successfully effected. We have a 'Go' to proceed. Resume the launch." The unscheduled hold had lasted but three minutes, yet it was the longest three minutes I can ever remember.

With sixteen seconds to go, hundreds of thousands of gallons of water began flooding the mobile launcher platform and its underlying "fire bucket," a huge cavity at the base of Pad 39B made of brick and concrete about the size of a three-story building, partly underground and beneath the rocket's engines. (A few months earlier, during a Science Working Group meeting at Kennedy, a few of us had ventured into the fire bucket below a shuttle being prepared for another mission. Standing beneath the unfueled shuttle's engines at a place that would be an absolute inferno filled with fire and water at the moment of launch was both breathtaking and terrifying.) The sudden gush of water is not used to quench the flames but rather to dampen the rockets' acoustical vibrations that might reflect from the concrete pad back up vertically, thus tampering with the structural integrity of the shuttle's wings and tail. After all, sound is energy, too, and in this case the unbridled waves of sheer noise could be powerful enough to harm delicate parts of the shuttle itself or of its fragile payload. Government documents state that the accumulated noise made by the five rocket engines is louder than several locomotives heard simultaneously.

Several seconds prior to liftoff, *Discovery*'s three main engines ignited, spewing forth more than a million pounds of thrust—enough power at that moment to supply the state of New York. The thick, billowing clouds of white smoke and steam that quickly engulfed the pad were clearly visible from my vantage point, although nothing was moving and nothing could yet be heard. Then, a couple of seconds after L-0, the two solid-fuel boosters fired, produc-

Virtually the only cloud in the Florida sky is penetrated by *Discovery*, on its way to a higher altitude than any space shuttle to date. This photograph was taken from the press site, some 3 miles away from Pad 39B. *(See also the color insert section.)* [NASA]

ing within half a second an additional 5 million pounds of thrust—enough to electrify the entire eastern U.S. seaboard.

At this point, there was no way to stop the launch; the boosters, the largest solid-fuel rockets ever constructed and fully as tall as a fifteen-story building, cannot be turned off or throttled back. Nearly simultaneously, explosive charges shattered the eight bolts holding the whole assembly onto the pad, umbilicals released, and the vehicle finally began rising—the powered orbiter and its fuel tank hardly connected to one another, soon to be virtually flying in formation. Said the Kennedy public address system animatedly, "Lift off of the Space Shuttle *Discovery* with the Hubble Space Telescope, our window on the Universe." By my watch, it was 8:33 A.M. Eastern Daylight Time, Tuesday, April 24, 1990.

The press coterie was kind enough to back off at the instant of launch. Either they wanted to let me experience this event as I had wished, or they themselves found it preferable to witness rather than to report at this historic moment. Or perhaps because I had not become emotional I was not newsworthy. Indeed, for me, the launch was simply awesome, but not emotional.

The shuttle seemed to leap from the seaside launchpad like a sprinter leaving the blocks, in seconds reaching a speed of nearly a hundred miles per hour before clearing the adjacent gantry tower. Within a minute, it would break the sound barrier while surpassing 30,000 feet, the typical altitude of a commercial airliner. "Once those solids light up, you really know you're going someplace," the shuttle pilot, Marine Colonel Charlie Bolden, later told me.

Since sound travels more slowly than light, the first fifteen seconds or so after ignition was only a visual event for those at the press and VIP sites. The white-hot, 600-foot-long flame from the main engines and booster rockets was brighter than any human-made light I have ever seen—comparable in luminosity to the Sun and much more ablaze than in any televised or still picture of a shuttle launch. But the onrushing sound wave, whose shock could actually be seen racing at us in the tall-standing grasses, left perhaps the most telling impression. The sound made by a space shuttle at launch is indeed tumultuous, easily the equivalent of several locomotives. Not an even, smooth sound, but a crackling, staccato, rapid rat-tat-tat type of roar, apparently caused by the acoustical interference of the two boosters' exhausts, resulting in a kind of beat. And the ground did vibrate.

After it had rolled onto its back and pitched appropriately to put the vehicle in the proper launch plane and to facilitate separation from its bolted-on rockets, the shuttle headed straight for the lone cloud in the sky. Effortlessly slicing through the low-lying, fair-weather cloud, the spaceship nearly blew it away. Owing to the excellent atmospheric ("VFR") conditions, some two minutes later when *Discovery* was 40 miles downrange on its oblique trajectory due east, I saw clearly and with my naked eyes the twin reusable rocket boosters separate from the orbiter and its main fuel tank, and fall away gracefully into the Atlantic.

As the shuttle rose into the sky, I hardly noticed it. All I could "see" was the Hubble Space Telescope inside its cargo bay. As far as I was concerned, the shuttle was invisible, and *Hubble* alone was thundering skyward atop a long and slender flame. Fluxing through my head were many of the grand themes, claims, plans, and benefits associated with this long-heralded space mission.

Would *Hubble* really make epochal contributions to the world of science? Would it discover new things unimagined? Might it have a cultural impact on our way of life, thus altering our perception of our position and role in the Universe? How about philosophically—could the findings of the Hubble Space Telescope change our cherished philosophies, our cosmological views of the Universe? These issues some of us had debated occasionally at lunch at the Science Institute, without answers of course. Generally, there are two schools of thought.

Some scientists maintain that Space Telescope will not likely discover anything dramatically new; thus, it is inconceivable that its findings will alter our basic philosophical outlook. This group argues correctly that our ground-based telescopes have already dimly perceived the virtual limits of the observable Universe. Indeed, Space Telescope cannot see appreciably farther than existing telescopes—a popular myth, NASA's repeatedly incorrect claims to the contrary notwithstanding—although it will enable us to observe many phenomena more clearly and therefore help to refine our existing theories of the cosmos. Still, I wonder how this school of thought can be so sure that *Hubble's* explorations will yield little basically new. Surely, Galileo himself, before turning his pioneering telescope skyward in Renaissance times, had little idea that what he would see would help to change humankind's philosophy of life. Observing mountains on the Moon, spots on the Sun, moons around Jupiter, phases of Venus, and a variety of other "wondrous things," he proved that the celestial sphere is not perfect and immutable, thus dealing a devastating blow to Aristotelianism.

The opposing view holds that Space Telescope has a chance to leave a genuine legacy, fundamentally new science results that our children's children will read about in the history books, knowledge that will stand the test of time in encyclopedias centuries hence. At the least, I suspect that *Hubble* will advance the science of astronomy enough to be remembered well after most other current scientific programs have long been forgotten. And it is the concept of increased resolution—angular, or spatial, resolving power—that will most likely generate most of the telescope's greatest findings, whatever they might be.

Resolution means clarity of seeing, sharpness of vision, the perception of fine details on an object; the physicist might state, more formally, the ability to discriminate between adjacent sources of light. For instance, the human eye can barely distinguish two objects about 1 arc minute apart—two car headlights, for example, at a distance of 3 miles. (One arc minute is $\frac{1}{60}$ of an arc degree, of which there are 360 in a full circle.) This explains why our naked eyes can discern some features—such as the mythical "Man in the Moon"—on the surface of the full Moon, which covers an angle of about a half-degree or 30 arc minutes.

When Galileo, who incidentally did not invent the spyglass, used a magnifying lens as a telescope for the first time in 1609, he overnight (quite literally) improved our human vision by about a factor of ten. (The term *spyglass* was Galileo's, as the word *telescope* was not coined until 1611.) I am uncertain precisely how well the best of his crude 2-inch (5-centimeter) oculars could resolve cosmic sources, but I would estimate from carefully reading his *Sidereus Nuncius* that he was probably able to see features as small as about 6 arc seconds. (One arc second is, in turn, $\frac{1}{60}$ of an arc minute, or equivalently the angle subtended by an American dime when viewed at a distance of 2 miles or the ability to read a license plate at 30 miles; one arc second also equals about $\frac{1}{1,800}$ of the apparent diameter of the full Moon.)

On at least two occasions, I have attempted to visit, testing gear in hand, the Galileo Museum on the banks of the Arno in Florence, where several of Galileo's pioneering lenses are kept. But each time I have found at the front door a posted guard who invariably produces a hand-lettered sign announcing that the museum hours have changed and that it is now closed. It's as though a conspiracy has kept me from determining the angular resolution Galileo was able to achieve. Frankly, given the subsequent troubles with *Hubble*'s mirror, I'd like to know how well he ground his lens!

At any rate, larger optical telescopes have been constructed in the nearly four centuries since Galileo made his revolutionary observations, but none of these telescopes has improved dramatically the clarity of our view of the Universe. For example, the "monster" 4-foot (1.2-meter) telescope built in the late eighteenth century by the Britisher Sir William Herschel and the 200-inch (5-meter) Hale telescope built in the mid-twentieth century on Palomar Mountain in California did not make great leaps forward in resolution. To be sure, these and other large pieces of glass collect more light than did Galileo's first such device, thus making them far more sensitive to faint objects in the Universe, yet they do not allow us to see the Universe much more clearly. It's not a question of size but of location; virtually all imaging telescopes are sited on the ground and therefore must observe the Universe through Earth's murky atmosphere.

This same atmosphere—the very air we breathe—is the bane of all astronomers for several reasons. Earth's atmospheric layers are constantly in motion owing to unpredictable weather patterns, the result being a chancy jiggling of air pockets that not only make stars *seem* to twinkle but also make images of celestial objects look blurry. Such air turbulence, much like the disturbed, shimmering air currents above a hot toaster or radiator or even the "clear-air turbulence" sometimes experienced during bumpy flying conditions, is mainly caused by myriad, small, randomly moving cells of air, mostly at altitudes between 3 and 6 miles. To our naked eyes, a star's image at any one time falls onto different parts of our retinas, thus causing the star's perceived intensity to vary or flicker up to hundreds of times each second, namely to twinkle—a phenomenon celebrated by poets but cursed by astronomers. Accordingly, all stars' images as seen by most of today's professional ground-based telescopes are spread out from a sharp point to a fuzzy patch about 1 arc second in diameter. Hence the reason why astronomers go to great lengths (or heights) to position telescopes at high mountain sites, above as much of our turbulent air as possible.

In addition, ground-based telescopes suffer from an illuminated murki-ness in the air, as noted earlier, so looking through it is much like trying to see things with a dirty pair of eyeglasses. The layers of tenuous air above us per-manently glow at night, both with scattered light from the Moon and the stars, and also by producing their own light as charged particles from outer space collide with the gases of Earth's upper atmosphere. The resulting sea of background light termed *airglow* is very faint (about the twenty-second mag-nitude), but becomes a crucial (yet not impenetrable) barrier for astronomers trying to observe the even fainter light of distant objects. Accordingly, even on the darkest mountaintops, airglow can drown out dim stars and galaxies, making astronomical observations of them as difficult as studying intricate rock formations at the bottom of a turbulent and murky stream of water.

A third drawback affecting astronomers on Earth is the atmosphere's tendency to reject certain types of radiation. While visible light well pene-trates to our planet's surface (assuming it is not cloudy), much of infrared and nearly all of ultraviolet radiation cannot pass through the atmosphere. Water vapor, carbon dioxide, smog, and a whole host of other trace aerosols limit cosmic infrared radiation from reaching our ground-based telescopes, whereas the ozone layer at high altitudes keeps virtually all cosmic ultraviolet radiation from reaching Earth's surface. The atmosphere may well be crucial to our human way of life, but to astronomers striving to decipher hints and clues about the mysteries of the cosmos, it is just another obstacle that gets in the way.

For *Hubble*, soaring above most of the distorting veil of Earth's atmo-sphere, none of these drawbacks pertains. There is no unsteady air causing stars to twinkle, no airglow to limit sensitivity, no barrier to sensing nonvisible

This artist's conception depicts the increased resolving power of the Hubble Space Telescope within a fanciful field of view outlined by the large circle. A color version of it was painted by Brian Sullivan of New York's Hayden Planetar-ium and is reproduced here by courtesy of the Association of Universities for Research in Astronomy, the parent organization of the Space Telescope Sci-ence Institute, which holds the copyright thereto. After printing thousands of copies of this painting as an educational poster, at no cost to NASA, the Sci-ence Institute received a formal, written reprimand from NASA to destroy or reprint all of them because the NASA logo was not prominently displayed.

radiation. While the best ground-based telescopes can routinely resolve down to about 1 arc second, and occasionally twice as good as that on nights when the "seeing" at high mountaintops is exceptionally clear, *Hubble* is able to achieve a resolution of 0.1 arc second, and often twice better than that when observing deep in the ultraviolet. This remains true despite the primary mirror's major optical flaw. And since acuity scales linearly with distance, 0.1 arc second can be equivalently described as the ability to resolve the period at the end of this sentence from a distance of about a mile, discern a dime at 20 miles, read a license plate at 300 miles, or discriminate between a car's right and left headlights at 2,000 miles. Thus, *Hubble* provides a significant advance over current ground-based telescopes—on average an advantage of about a factor of ten in acuity, or roughly the difference between the large and small letters at the top and bottom of an optometrist's eye chart.

In fact, the numbers are even more impressive in that *Hubble*'s tenfold advance in resolution applies to both the horizontal and vertical dimensions of a picture; compared to typical ground-based performance, *Hubble* data therefore yield ten-squared or a hundred times more detail. Technical caveats aside, we can legitimately make the bold claim that in the history of optical astronomy there have been only two great leaps forward in the clarity of humankind's view of the Universe: Galileo's telescope in the autumn of 1609 and the Hubble Space Telescope in the spring of 1990. I believed before launch, as I continue to do so now, that this dramatic advance in angular reso-

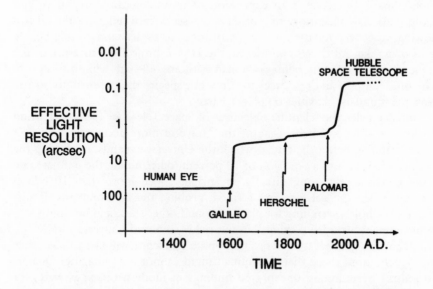

Only two great advances in angular resolution mark the history of optical astronomy. Galileo's first use of the telescope provided a tenfold improvement in resolution, but no appreciable advances in resolution from ground-based observatories have occurred in the past several centuries (owing to their location under Earth's atmosphere). While the use of special techniques—such as active and adaptive optics, to be discussed later in this book—has sometimes afforded superb resolution toward bright objects in recent years (hence the slight tilt of the drawn rising curve to the right), it is only the Hubble Space Telescope that routinely grants us another full order-of-magnitude leap in resolving power. (The author first made this diagram for a talk given in Italy by the director of the Science Institute to help celebrate Galileo's epoch-making discoveries in the early seventeenth century.)

lution is the single most important scientific justification for having built Space Telescope.

Throughout the year preceding launch, Senator Mikulski and her staff had put it to me: "What's the big deal," she would say bluntly. "Why are you astronomers so excited about the Hubble Space Telescope?" Normally, I responded along the lines that, "It is not just any civilization, but the great civilizations, that leave legacies." And, like the vanity we all stroked around this project, she liked that argument and she used it in some of her speeches. Yet as we got closer to launch, one day she asked me to articulate more formally the socially relevant benefits, as I saw them, to flow from the Space Telescope project. I responded in writing that the "payoff" was likely to be intellectual, educational, and cultural. No doubt, there is the crass economic benefit so often heard around observatory hallways: "Money allocated for space science is not spent in space; it's spent on the ground." But I stressed to this influential senator that, to my mind, direct technological spinoffs were unlikely, or at least intangible.

Although NASA repeatedly pitches the economic benefit angle—"for every dollar invested on the space program, seven dollars are returned to society"—I judge this to be a specious argument for which there is no evidence. *Hubble*'s vaunted technology will not likely produce a new can opener or better washing machine. Nor should it. This giant eye in the sky was built neither to improve our economy nor to challenge our Cold War enemies. Rather, the Hubble Space Telescope is an expression of technological poetry, pure and simple. This magnificent orbiting observatory seeks knowledge for the sake of knowing, discovery for the sake of discovering, to understand cosmic beauty for its own sake. All things considered, the *Hubble* project is a measure of the extent to which humane beings—human with an "e"—are willing to devote some time, money, and resources to allow our species the opportunity to rise above the mundane activities on planet Earth.

Intellectually, the scientific objectives of Space Telescope include charting the distance scale and hence size of the Universe more accurately than ever before, mapping better the rate at which the Universe expands, measuring the age of the Universe to a precision of 10 percent, determining the ultimate fate of the Universe, exploring directly an ancient epoch more than 10 billion years ago when the galaxies first formed, probing the environments of suspected black holes, searching for planets around other stars, and no doubt seeing many *more* "wondrous things" in our richly endowed Universe.

As for the greatest discoveries that Space Telescope will likely make, they are, by definition, those that the human mind cannot now imagine. Quasars and pulsars were among unexpected findings in modern times, as were, for example, Columbus's "discovery" of America or Galileo's pioneering observations centuries ago. Even so, astronomers have tried to plan for the unexpected, to probe directly for that which is not yet known. Here, in one of *Hubble*'s key projects, called the Medium Deep Survey, the telescope would be commanded to look hard in heretofore empty regions of the Universe— regions of utter darkness where no humans have ever seen anything at all. In an attempt to "dig out" the unknown, this series of long-exposure observations is explicitly designed to seek the unexpected. It is in this sense that *Hubble* had been hailed, at least before launch, a "discovery machine"—a device at the cutting edge of what humankind can achieve intellectually and technologically, not a trivially important ingredient in the life of a robust, inquiring civilization.

Educationally, the Hubble Space Telescope project can help enormously to enthuse children to consider careers in science and technology, as well as to inspire nontechnical persons of all walks of life—the intelligent taxpayer—to appreciate better the technological society that we all share. Clearly, we live at a time when the majority of the public does not understand what science is about or what scientists do. The general public is, by and large, scientifically illiterate, and precollege teachers are both confused about our fast-moving subject and lack the confidence to teach it. All manner of alarms have been sounded in recent studies, including, for example, those (in a government-sponsored survey) showing that nearly 30 percent of adult U.S. citizens think that the Sun revolves around the Earth, while an additional 28 percent do not know that it takes one year for our planet to orbit the Sun. Unfortunately, "gee-whiz statistics" like these do not help to ameliorate the fact that the youth of America are failing to learn adequate science and math skills. The shortfall undermines our economy, which depends on advanced technology and industrial motivation. The result, in my view, is an erosion of our technological work force, which might in turn become a threat to our democratic way of life.

What can be done about this problem? Clearly and foremost, we, the practitioners of science, must recognize that all of us have an obligation, a professional responsibility, to share our knowledge with the world's populace, to inform the public and to educate our students, in short to teach. Currently, the precollege educational community should perhaps be the greatest beneficiary of any newfound involvement in education among the traditionally reluctant research empires. Attitudes toward science are formed in the early school years and astronomy, in particular, can provide positive experiences and spark interest in further learning. Who among us has not looked at the nighttime sky and wondered? Above all, we want to carry a simple message to our kids: Science is fun. That message says that learning is an adventure, a pattern of discovery, a process of exploration, and that these activities can be rewarding and enjoyable. It encourages children to be imaginative and curious about the world around them. By capitalizing on astronomy's capacity to engage the mind, stimulate thought, and directly experience the grandest scale nature provides, we make the lives of children richer and improve our fortunes with the best competitive edge: investment in the future. The Hubble Space Telescope should also be an "education machine" that can help us to achieve these objectives.

Culturally and socially, the benefits of a Hubble-type project are potentially much greater than many of us realize. To be sure, in the years leading to launch, a number of us at the Science Institute regularly debated the cultural spinoffs from space science, and what we found most intriguing is the nature of a "frontier." For about a hundred years following the birth of our nation,

America's exuberant, seemingly limitless growth had a mighty and positive effect on the prospects of its young people. The essence of American education then prepared our youth with the skills and initiative to find prosperity in the wilderness, regardless of where he or she would actually dwell. There was something powerful and compelling about the Western frontiers that was uniquely American—frontiers to be conquered, frontiers that made us great as a nation.

Today, our terrestrial frontiers are gone. But the heavens loom overhead, providing for our imaginations what the Western horizon once was. As such, America's leadership in astronomy and space science reflects the continuing significance of exploration and discovery for our democratic institutions and for our citizens personally. America needs an astronomical frontier—not to make our country vibrant but to make us great into the indefinite future. In short, we need something akin to a "Strategic Education Initiative" to secure for individual citizens the cultural, educational—indeed perhaps spiritual— benefits of our having an astronomical frontier. A decidedly nationalistic argument, to be sure, for its intended recipient was a U.S. senator, yet one easily rationalized for all peoples of the globe.

All these issues raced through my mind as *Hubble* rose skyward—puncturing the midst of a lonely, white and fluffy cloud, climbing beyond solid-rocket booster separation and the "throttle-up" condition that had destroyed *Challenger*, and out of sight over the Atlantic. My trance broke when NASA's Ed Weiler, the *Hubble* program's lead scientist in Washington, came trotting across the field and yelled to me excitedly, "We've done our part. Now *your* work really begins!"

1

Deployment and Early Operations

It would be altogether a waste of time to enumerate the number and importance of the benefits which this instrument may be expected to confer, when
used by land or sea. But without paying attention to its use for terrestrial
objects, I betook myself to observations of the heavenly bodies; and first of
all, I viewed the Moon as near as if it was scarcely two semi-diameters of the
Earth distant. After the Moon, I frequently observed other heavenly bodies,
both fixed stars and planets, with incredible delight; and, when I saw their
very great number, I began to consider about a method by which I might be
able to measure their distances apart, and at length I found one.

Sidereus Nuncius
GALILEO GALILEI, Venice, 1610

"It's not the launching but the deployment that will have me biting my nails."
A glib remark to be sure, yet pithy enough for my purposes at the Kennedy
Space Center shortly after launch. Surprisingly, it raced around the globe on
the international news wires. I meant it as a note of caution that the most
demanding (and exciting) part of getting the Hubble Space Telescope on
orbit was likely to be the day after launch, when it would be gingerly removed
from the shuttle's cargo bay and begin extending its many robotic
appendages, much like the petals of a tulip responding to sunlight.

Part of my concern stemmed from the complexity of the telescope and
the intricate procedures needed to command it. Commissioning *Hubble*
would not be a turn-key operation like throwing a light switch; it was more
like fitting all the pieces into a jigsaw puzzle, during more than a year of
testing and calibrating, thereby transforming a collection of metal and glass
into a telescope masterpiece. But my worry also had to do with the multi-
centered and already strained management team on the ground—not unlike

the rival deMedician city-states of Galileo's time. Once launch had occurred, Kennedy had transferred control to the Johnson Space Center in Houston, which would thereafter maintain jurisdiction over the shuttle and also be responsible for the astronauts during any dangerous space walk. Johnson in turn passed control of *Hubble* to the Marshall engineers—employees of the Space Telescope project office in Huntsville—who would begin methodically to activate the observatory's myriad electromechanical systems and align its mirrors. Adding to the project's complexity, the Marshall contingent was operating "on foreign soil" at arch-rival Goddard's control center, whose own separate set of engineers—employees of the Space Telescope project office in Greenbelt—were eager to enliven the telescope's science instruments.

Hovering around all these operations teams, among several groups of Hubble contractors, especially from Lockheed and Hughes who were also based largely at Goddard, were officials from NASA Headquarters—employees of the Space Telescope program office in Washington, and not exactly welcome in the trenches. Not least, technical representatives from the European Space Agency were also on hand, mostly at Goddard, should problems arise with the equipment they had provided, especially *Hubble*'s vital solar arrays. And then there were the rest of us at the Science Institute, largely backseat drivers until the telescope completed its first few days of manual checkout, after which our schedulers would begin preparing meticulously coded tasks to be radioed up every day to the orbiting observatory. The result was many people with many personalities and many agendas, and not much coordination, or systems management—all of which spelled disorder if *Hubble* ran into trouble.

By contrast, the federal bureaucrats responsible for managing the *Hubble* project seemed unconcerned. NASA officials often stressed being part of the "*Hubble* team," not realizing that the agency's version of total quality management damps innovation while breeding mediocrity. Consider this typical exchange during a prelaunch press conference at Goddard between a knowledgeable science writer, Kathy Sawyer of the *Washington Post,* and the NASA manager with line authority for *Hubble,* Lennard Fisk—whose words would later return to haunt him:

QUESTION: There have been a number of criticisms of the *Hubble* project, the complexity of design, the complexity of the management structure, its dependence on the space shuttle, the complexity of its observing schedule, and its data management, and so on. Dr. Fisk, can you please tell us what you have learned that you are going to apply to future Great Observatories or other big science projects of this kind?

ANSWER: Well, you had quite a litany there of things that there's been some criticism about. There is certainly some lesson to learn from the complexity of the management structure that was put together to do *Hubble*. But we all should remember that in about four weeks from now, when we begin to see what *Hubble* can do, I bet there aren't a whole lot of questions about how we got here.

* * *

Upon achieving orbit nearly ten minutes after liftoff and circularizing the orbit (by firing small thrusters) about thirty minutes thereafter, the crew of shuttle *Discovery* first went about some standard mandatory tasks to prepare their spaceship to return to Earth in the event of an emergency. They then opened the payload bay doors, allowing pent-up heat and moist air to escape to prevent short-circuiting its precious cargo when power was supplied, and began a visual survey of Space Telescope via several onboard television cameras. They also put through its paces the 50-foot-long Remote Manipulator System—or "Canadarm" for short, so named since this electromechanical arm-and-claw was a contribution to the U.S. space program from the Canadian aerospace industry. The arm is articulated much like a human's, having "shoulder," "elbow," and "wrist" joints that are controlled manually by an onboard astronaut while looking out over the payload bay through small windows at the rear of the orbiter flight deck. At the end of this arm is another miniature television camera that allowed further inspection of Space Telescope berthed in *Discovery's* cargo bay.

One of the astronauts surveyed *Hubble's* hull, including its shiny thermal blankets and instrument covers, and radioed that all seemed well: "Near as we can tell, everything looks perfect." Nothing obvious had jostled loose during the rigors of launch. A few hours later, telescope controllers at Goddard began activating some of its onboard systems. Said Mike Harrington, the Marshall engineer charged with early checkout of the bird, "It gives you goose bumps when you send those first commands in the blind and the telescope talks back to you."

The mission was proceeding splendidly. The launch had been perfect and the early "uplink" and "downlink" transmissions to and from *Hubble* were going smoothly—in fact, better than for any preflight test conducted in the years prior to launch. "We should launch more often," quipped one of the astronomers at the Science Institute. But it was the calm before the storm.

After some routine housekeeping duties, the astronauts turned in. As they had been up since before dawn that morning, their sleep cycle required

Shortly after arriving on-orbit, three of the five *Hubble* astronauts took a moment from their duties to enjoy the view through the "Earth roof" atop *Discovery's* flight deck. *[NASA]*

them to get a full period of rest starting early that evening. This would enable them to be fresh and alert for the crucially important day of deployment. It also gave those of us who had duties at either the Kennedy or the Johnson space centers during the launch a chance to return to either Goddard or the Science Institute, from which we could closely follow the deployment activities. In all the rush, I heard a colleague complaining good-naturedly about the excess concern for the astronauts' rest: "What about the sleep cycle of the engineers and scientists on the ground?" many of whom had been feverishly preparing for launch nearly round-the-clock for weeks.

In brief, the telescope would be deployed like this: A radio command would disconnect the umbilical cord (an electrical cable) carrying power from the shuttle to Space Telescope, after which the Canadarm would grasp *Hubble* and lift it out of the cargo bay. An array of solar panels would unfurl on each side of the telescope and a pair of high-gain antennas would extend from their stowed (shipped) positions. Other than the manual manipulation of the Canadarm by one of the mission-specialist astronauts, virtually all deployment activities would occur robotically upon command from the ground. Two other astronauts would be ready to don their spacesuits and to aid in these vital actions, as well as a later opening of the aperture door, if necessary, but there would be no direct human involvement with *Hubble* if all went according to plan.

NASA officials and concerned scientists had debated at length, for more than a year prior to launch, the minimal acceptable altitude for deployment of the Space Telescope. Our atmosphere's thickness decreases smoothly with altitude: At 6 (statute) miles, or 31,000 feet—the height of a typical commercial flight—the density of air is about half that at sea level, at 75 miles it is a millionth, and at 250 miles a trillionth. Satellites in "low-Earth orbit"—which can be anywhere between the minimum altitude of roughly 125 miles to as much as several hundred miles—plow through the remnants of rarefied gas (mostly hot oxygen atoms) at great speed, orbit after orbit, thereby accumulating the effects of friction that can eventually bring them down.

Typically, space shuttles operate between 150 and 250 miles altitude and stay up for short periods of time lasting hardly more than a week. By contrast, satellites at a much higher altitude of 22,400 miles (or about 5.6 Earth radii) above Earth's surface are said to be in "geosynchronous orbit." This is the

At 380 miles altitude, about twice the height of usual shuttle flights, the view of Earth is sweeping even to veteran shuttle astronauts. This picture of the "big blue marble," much like those taken in the 1960s by *Gemini* astronauts, was made on one of *Discovery*'s first revolutions around our planet. The change in continental perspective is so dramatic at this altitude that *Discovery*'s astronauts, in another view, were momentarily puzzled by the presence of huge lakes to the northeast of Mexico—until they realized they were looking much farther along Earth's rim at the Great Lakes near the U.S-Canadian border. [NASA]

altitude at which any object orbiting Earth naturally revolves at the same rate at which Earth spins, thereby making the object "geosynchronous" or "geostationary," while hovering above a fixed point on the planet's surface. At this "high-Earth orbit," there is no atmospheric drag and they effectively stay up forever.

Atmospheric drag is what caused the pioneering *Skylab* space station to break up and fall back to Earth in 1979. Ironically, the United States is now struggling to build a permanently crewed space station, yet we had in orbit well more than a decade ago the backbone of a perfectly good manned outpost. *Skylab* in fact was a marvelous platform from which to make many types of scientific observations—of the Sun and stars, and of Earth itself. And it was a laboratory superbly suited to study the biological and psychological effects of extended spaceflight on human beings. But the U.S. space program laid fallow by the late 1970s, mostly for want of leadership during several previous presidential administrations, and it simply lacked the will—and the technological means—to rescue *Skylab* before it fell to the ground. Indeed, the *Saturn* rocket technology used to boost *Skylab* into orbit, as well as to win the race to the Moon, is essentially lost—if not the blueprints then surely the technological work base, the aerospace infrastructure having so withered that our country would, for all practical purposes, now need to start over in order to build a *Saturn* rocket again.

Atmospheric drag on lowly orbiting space vehicles is particularly troublesome when, every decade or so, the Sun experiences its most active phase. For reasons not well understood, the number of sunspots—dark blemishes on the photosphere or "surface" of the Sun—increase and decrease in a cyclical pattern. Throughout much of the last few hundred years during which systematic solar observations have been made, this so-called sunspot cycle has averaged about eleven years, although the cycle has been known to vary from eight to thirteen years.

As the sunspots rise toward their peak number each cycle, other forms of solar activity also emerge. Flares and prominences, often associated with sunspots and having the power of multiple atomic bombs, increase dramatically, spewing forth huge quantities of particulate matter. This is not the radiation ("sunlight") that leaves the Sun at the speed of light (186,000 miles per second) and reaches Earth about eight minutes later, but particles—mostly fast-moving protons and other elementary particles—that literally escape the pull of the Sun's gravity at speeds of merely hundreds of miles per second and therefore arrive here a day or two later.

Sunspots, flares, prominences, and the like have little effect on the Sun itself. Although we estimate that about 4 million tons of solar matter—the mass equivalent to a dozen (loaded) supertankers—escape the Sun *every second,* this mass loss has no appreciable bearing on the Sun as a star, no known influence on its evolutionary path from birth to death. Even at this astounding rate of mass reduction, and even over the course of the 10-billion-year lifetime of our star, the Sun will have lost far less than a tenth of one percent of its total mass.

By contrast, the multimegatons of matter leaving the Sun each second—

called the "solar wind"—have a decided effect on its attendant planets. As clouds of protons and other elementary particles arrive at Earth a few days after each solar outburst, many of them become caught up in the huge Van Allen belts that girdle our home in space. These belts are regions of organized magnetism, and since the solar particles are electrically charged they become trapped above our atmosphere—a first line of defense, if you will, from those energetic particles that could rip through and thus mutate the biological cells within life-forms on Earth.

At times of peak solar activity, when the Sun's spots and prominences are maximized, Earth's Van Allen belts become overloaded with subatomic particles arriving in "gusts," and some of them tend to leak from the belts, wreaking havoc on Earth. They do so mostly near the north and south poles of our planet, causing, in addition to minor disruptions in power grids and some communication blackouts, the aurora borealis or "northern lights," which are spectacular plumes of colorful and wispy light streaking across our atmosphere—our second line of defense.

Aurorae occur because the fast-moving particles collide with and excite air molecules of nitrogen, oxygen, water vapor, and others, causing them to glow during geomagnetic storms. But the incoming solar particles also have another effect on Earth's atmosphere: they heat it. (The amount of energy absorbed by our planet's atmosphere from just a single major solar flare typically equals the energy content of a megaton hydrogen bomb.) And when our atmosphere heats, it expands, swelling outward to greater altitudes above the surface of our planet. The rising of Earth's atmosphere to higher altitudes causes spacecraft to experience increased drag that can, in some cases, cause them to reenter, break apart, and burn up. This is what brought down *Skylab* in 1979, the year of the previous sunspot maximum—called "solar max" for short. And it was mounting solar activity that more recently (1989) brought down another satellite, called *Solar Max*, which had been sent up in the early 1980s, ironically, to study the details of the Sun's many active phases.

All these varied forms of solar activity—spots, flares, prominences—were surging as we prepared for launch of Space Telescope. The current solar cycle was expected to maximize somewhere around spring of 1990, and disruptions of transmission lines were becoming increasingly frequent by mid-1989. Major, solar-induced electrical outages had occurred at power grids in Quebec and Sweden, the American military sporadically lost control of its communications network for brief times, and just months before *Hubble's* launch, the northeastern United States from Washington through New England came very close to experiencing a total blackout. Most early indicators suggested that this particular solar cycle ("Cycle 22," since astronomers of Galileo's time began tracking them) was likely to be very active—perhaps the most active in centuries. (An alternative explanation was that the peak of solar activity had arrived earlier than expected, which, we now know, is more or less what did happen.) After all the technological, political, and management delays in the construction of Space Telescope, it seemed most unfortunate that the Sun would conspire to further delay the launch just as we had finally become ready to go.

Following ad hoc discussions among solar experts, *Hubble* scientists, and the astronaut office at Johnson, we reached a consensus that the effects of solar activity could be minimized if Space Telescope were deployed in the highest possible circular orbit. So the lightest of the shuttle fleet, *Discovery,* was programmed to achieve a maximum altitude, some 380 (statute) miles. No space shuttle had ever been that high before, and to do so its engines would have to go full throttle until they were running nearly on fumes. Actually, the flight plan called for having just enough reserve fuel, after reaching this high orbit, to drift a few tens of miles away from *Hubble* after its deployment, to be able to rendezvous once with *Hubble* in the event it needed assistance, to participate briefly in a defense-related experiment, and to return safely to Earth.

Regarding minimum acceptable altitude for *Hubble*'s deployment, 325 miles was considered the "line in the sand." It would be roughly a year before another shuttle could be prepared and launched, capable of rebooting *Hubble* to its higher, normal altitude, and with solar activity rising toward an expected maximum at just about the time of launch, NASA officials were unable to guarantee that a shuttle rescue mission could be mounted in time to prevent a low-altitude *Hubble* from reentering, perhaps as soon as six months after launch. If *Discovery* could not achieve this minimal height, then the astronauts were prepared to activate a preplanned contingency to return the shuttle with *Hubble* on board. Given not only the technological concerns— Would *Hubble* survive a shuttle landing?—but also the political ramifications—When might we get back into the troubled shuttle queue?—most of us cringed even at the thought of *Hubble* returning to Earth.

Thus, the options were threefold: If *Discovery* achieved its expected cruising altitude of around 380 miles, all would be well, for *Hubble* would be safe from solar-induced atmospheric heating for several years—enough time to elevate the orbiting observatory during a regularly scheduled shuttle servicing mission in the mid-1990s. If *Discovery* developed a problem during launch and had to abort-to-orbit at less than 325 miles altitude, then *Hubble* would not be deployed but would be brought back to Earth. And if *Discovery* reached significantly less than the nominal 380 miles, but more than 325 miles, its crew would nonetheless proceed to deploy *Hubble,* knowing that such a precarious orbit could be subject to the unpredictable whims of the Sun.

Just prior to launch, I took some heat from overly sensitive NASA officials for saying, during a press conference, that should *Hubble* be deployed well below its normal altitude—the worst-case scenario should, say, the shuttle orbiter run into difficulty and need to dump *Hubble* overboard—we at the Science Institute (and by implication the scientific community) would "exert pressure on NASA to arrange for a subsequent shuttle to give it a boost." NASA preferred to think that it need not be pressured to rescue a troubled Space Telescope, but the historical record clearly shows that the space agency, then fixated on space shuttle development, did in fact allow two superb and unique spacecraft to fall to Earth. And by knocking NASA on this point, I was also criticizing my own scientific colleagues, for the demise of the *Skylab* space station was a national tragedy that the U.S. science community never should have allowed. In fact, when *Hubble* ran into much difficulty within months after launch, the Science Institute, on behalf of the world's science community, would indeed need to pressure a demoralized NASA, and heavily at that, to correct *Hubble's* blurred vision.

In the end, scholarly presentations about solar heating, detailed scenarios for the shuttle crew in the event of an abort-to-orbit or suborbital abort, and prelaunch controversies about rescue missions were moot, for the launch was flawless. In a refreshing demonstration of technical expertise, Space Shuttle *Discovery* arrived on orbit very close to its intended altitude, indeed the highest ever attained by a shuttle. Its initial orbit was nearly an ideal circle, approximately 381 by 383 (perigee and apogee) miles above Earth's surface, a most favorable position from the standpoint of a long and stable orbital lifetime. Months later, while we shared lunch on a floating barge on Clear Lake outside the Johnson Space Center in Houston, Air Force Colonel Loren Shriver, the commander of *Discovery*, looked longingly skyward and perhaps only semi-facetiously intoned, "I could have taken it even a little higher, but my crew-mates wanted to come home."

How long will Space Telescope likely stay in orbit? Even at its high altitude, *Hubble* encounters slight but noticeable drag due to minute amounts of Earth's atmosphere brushing against the vehicle. Within a couple of months after launch, the telescope had lost about a mile of altitude; by the time science operations began in earnest, more than a year later, it had come down as expected nearly 10 miles. *Hubble's* rate of descent is hard to predict not only because of irregular solar flare activity but also because of the constantly changing orientation of the telescope's long hull and solar arrays relative to the direction of the spacecraft's orbital motion. As *Hubble* loses altitude, its speed increases and its orbital period shortens. Schedulers trying to predict the telescope's position weeks in advance could find the spacecraft miles ahead or behind. Consequently, Space Telescope must be tracked regularly and new "orbital elements"—technically termed its ephemeris—derived to update periodically its orbit and its position within that orbit.

When the telescope's altitude has fallen to about 325 miles (roughly every half-dozen years), a space shuttle mission is expected to rendezvous with *Hubble,* capture it with the Canadarm, bolt it vertically within the shuttle's cargo bay, jettison its old solar arrays and affix new ones, replace any malfunctioning equipment, install some "next-generation" science instruments, and boost the telescope back up to its normal altitude so that it can continue its pathbreaking work. In this way, *Hubble* should endure for more than its natural lifetime, thereby operating as a forefront astronomical observatory for at least fifteen years and perhaps a good deal longer well into the twenty-first century.

Speaking of longevity, what about the oft-asked query, Could *Hubble* suffer a terminal collision in space? The answer, of course, is yes, but the probability is small. At the time of *Hubble's* launch, the North American Aerospace Defense Command—NORAD, the Air Force unit inside Cheyenne Mountain that keeps track of such things—estimated that about 3 million artificial objects were orbiting Earth, including fragments of rockets, discarded equipment, and other nuts-and-bolts debris. About 24,500 of these objects are detectable by ground-based surveillance techniques to have sizes larger than a centimeter, some as small as an aspirin tablet, others as large as an abandoned *Hubble*-class spacecraft; more than 7,000 of these have diameters of at least 4 inches and are reliably tracked by a network of military radar and optical sensors. The majority of such human-made trash orbits at altitudes lower than *Hubble,* yet even a single collision involving the smallest of these could cause catastrophic damage; a half-inch (1-centimeter) aluminum sphere traveling at orbital velocity (~18,000 miles per hour, or 8 kilometers per second) strikes with an energy of a 300-pound (135-kilogram) weight hitting the ground at 60 miles per hour.

Overall, statistical estimates suggest a one in a hundred chance that *Hubble* could be destroyed by a large object, greater than a few inches across, at some time in its fifteen-year mission. As for natural debris, such as meteoroids or asteroid fragments, a head-on collision with *Hubble*'s main mirror requires a very small angle of incidence, given the telescope's robust arrangement of concentric baffles inside its forward tube, so the chances are slim; furthermore, the telescope normally leads in orbit with its aft end, thereby minimizing the amount of debris scooped up by its open forward end. Incessant impacts by minute micrometeoroids (and even subatomic particles) are another issue, however, and no one quite knows the long-term viability of *Hubble*'s optics over the full course of the mission.

On the morning of the second day, nearly twenty-four hours after launch, a mutual decision was made by the astronauts and their handlers at Johnson, as well as by the robot's ground controllers at Goddard, that all systems aboard both *Discovery* and *Hubble* were "go for deployment." The Canadarm had been flexed and tested some more, the spacecraft-to-ground telemetry looked "nominal" (space lingo for good), the bulky spacesuits needed for an emergency "space walk" were unstowed and readied, two of the astronauts were at middeck prebreathing pure oxygen in the event they had to venture outside if the telescope malfunctioned (to prevent dysbarism or "the bends" as they transitioned from an Earth-like, oxygen-nitrogen, 14.7-pounds-per-square-inch cabin atmosphere to their spacesuited 4.1-psi, lower-pressured pure-oxygen gas needed to sustain life), and the shuttle commander was ready at the controls to maneuver *Discovery* to help *Hubble* out the door.

Once the umbilical cord was cut, there were few options for backtracking. The umbilical cord is a commonplace electrical connection (much like a household extension cord) between the shuttle and *Hubble,* providing power to the telescope via *Discovery*'s electrical system. Although we on the ground were in direct contact with *Hubble,* neither its huge solar arrays nor its principal antennas had yet been unstowed or activated—and thus its power to communicate rapidly, to run its onboard computers, and to keep itself healthy and safe, derived mostly from the shuttle. With the umbilical disconnected, *Hubble* would then have to rely on its own internal batteries. And if there is any piece of low-tech equipment that humankind has not yet learned how to make well, in my opinion it is a battery.

The six heavy batteries onboard *Hubble*, which together total 960 pounds, are the best that engineers can manufacture. Just months prior to launch and after extensive discussions that had dragged on for nearly a year, a decision with unknown implications was made by officials at the Marshall Space Center to replace *Hubble*'s nickel-cadmium ("NiCad") batteries with nickel-hydrogen types. These newer batteries can store more energy and should have a longer lifetime in orbit, perhaps as long as six years—a useful power margin, given NASA's private acknowledgment that frequent shuttle maintenance visits to *Hubble* could not be guaranteed. But the newer, nickel-hydrogen batteries had been bench-tested not nearly as much as had the original NiCad batteries, and were thus not yet certified as "flight ready." By contrast, the military-intelligence community has used for years these more advanced batteries in some of their spacecraft, but they were unwilling to share much of their experience with the civilian sector.

In any event, and although the batteries aboard Space Telescope had been "trickle charged" on Pad 39B up until a couple of days before launch, their discharge time was short. In telescopes as in today's portable laptop computers (let alone the kitchen flashlight), batteries are the greatest limitation. Without additional energy entering *Hubble* via its solar arrays, the telescope's power would become much depleted within several days. By deployment day, its batteries had already drained to about 50 percent of their total energy capacity. Once the umbilical cord was cut and *Hubble* began utilizing its stored electricity, the telescope would remain "healthy" on batteries for only about six hours and fifteen minutes. Provided the solar arrays extended properly and began absorbing sunlight, then the batteries would charge up again and all would be well. *Hubble* could vegetate there essentially forever, perfectly safely. But before the solar arrays could be extended, Space Telescope had to be hoisted out of *Discovery*'s cargo bay.

What were the implications had power not been restored to *Hubble* within the allotted 6-plus hours? No one really knew for sure. Engineers had calculated that, beyond that time, its onboard batteries would have been unable to keep *Hubble*'s vital control systems running, and its flight computers would have gradually become useless. Our ability to communicate with this sophisticated robot also would have progressively diminished. *Hubble* might have started drifting or even tumbling uncontrollably, and after several days as an unpowered free-flyer, it presumably would have become brain-dead, though not necessarily physically damaged. Had its solar arrays eventually restored power after days or weeks—or even after a rescue mission years later—many of its innards might have been resurrectable, such as its many electro-mechanical devices and even, perhaps, most of its computer chips and relays.

Of all the many consequences of battery failure, there was a fundamental fear, expressed only in hushed tones by project officials, that *Hubble* could be crippled forever by sheer contamination. This was the knowledge that, despite the best clean-room techniques on the ground, *Hubble* had gone into space with terrestrial gases trapped inside—in small cavities within its scientific instruments in the aft bay behind its main mirror, around the baffles and within the tube of the forward bay in front of the main mirror, in the graphite-epoxy support structure of the mirror assembly itself, even within microscopic crevices in the metal comprising the hull of the spacecraft per se. Once on orbit in the vacuum of space, such contaminants would begin to escape, the technical term for which is outgas or desorption.

We most feared the prime terrestrial pollutant— ordinary water vapor. If outgassed water vapor got near *Hubble's* cold mirrors, unheated for lack of onboard electrical power, it could coat the mirrors with a thin film of ice— very much like the frozen dew or frost that forms on automobile windshields on winter mornings. In space, of course, there would be no way to scrape the frozen water vapor from *Hubble's* optics, and icy films might continue to collect on its various mirrors. If the heaters later regained power sufficient to defrost the ice, chances were good that the telescope's glassy surfaces would still be littered with dirty residues of opaque contaminants—the water having mixed with Earthly dust known to be on the main mirror.

At about 8 A.M. Eastern Daylight Time, April 25, the shuttle was on its fourteenth orbit—"rev 14"—precisely the time to commence deployment, as specified in the preflight planning documents. Everyone felt good though apprehensive about umbilical disconnect; once done, the combined team of astronauts and ground controllers had until about mid-afternoon to get power to the spacecraft.

The Canadarm was manipulated ever so carefully to grapple the telescope, the umbilical was disconnected, and the arm began lifting *Hubble*— gingerly. This exercise had been simulated hundreds of times on the ground by Steve Hawley, one of the onboard astronauts, who is also an astronomer. (We liked to joke at the Science Institute that we had arranged for an astronomer to be among *Discovery's* crew, for he would surely have an extra motivation to get the telescope out in one piece.) Still, at some points mere inches separated the inside of *Discovery's* 15-foot-wide cargo bay and various appendages on *Hubble's* hull. These first few minutes of the hoisting operation were the most nerve-racking, what with "first-inning jitters" and the fact that the emerging telescope's front end completely blocked Hawley's view from *Discovery's* rear window. The angle and glint of the Sun's light changed continuously, too, making it hard to guide *Hubble* visually.

During these earliest maneuvers, Hawley relied heavily on the television cameras mounted at each corner of the payload bay and worked much of the time "by the numbers," consulting the digital readouts indicating the positions in inches and degrees of the various "elbow" and "wrist" joints of the Canadarm. Ironically, he had an easier time watching his Canadarm manipulations during the nighttime portion of the orbit, when artificial floodlights in the shuttle's cargo bay illuminated *Hubble* uniformly and steadily. Accordingly, and as was his prerogative, Hawley took an extra orbit to hoist *Hubble* slowly and well, slaved attentively "to the wheel" for a couple of hours. He told me later that, given his responsibility for one billion-dollar vehicle hovering less than a foot away from another billion-dollar vehicle, orbiting in formation at about 17,600 miles per hour, he thought it best to be a little cautious.

The preflight script next called for the wing-like solar arrays to be

Space Telescope is seen here emerging from *Discovery*'s cargo bay—the "low hover" position some 15 feet "above" the shuttle—the first step in a tricky deployment sequence. (The word "above" is in quotes, since at the time, the shuttle was nearly upside-down relative to Earth.) Notice *Discovery*'s rear engine pods at the top of the picture. At this point, *Hubble*'s high-gain antennas (one of which is clearly visible at middle top) had not yet swung into position, nor had its solar arrays (black cylinders to right and left) yet unfurled. The thin bars are handles and rails for astronauts to use while servicing *Hubble*. *[NASA]*

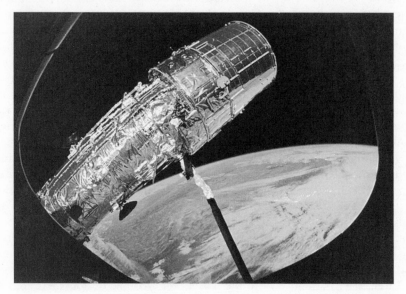

Here, *Hubble* is in the "high hover" position, held about 28 feet beyond the shuttle, which is not seen and to the bottom. At this point, the telescope is far enough away from *Discovery* to begin extending its antennas and solar arrays. In the weightless environment of outer space, the Canadarm, which is centrally visible in this picture, can routinely manipulate the 25,000-pound, city-bus-size *Hubble*. *[NASA]*

unfurled. But a decision was made—by whom I'm uncertain—instead to unstow first the principal, high-gain antennas. These are two 15-foot-long booms each topped with a 4-foot-wide metal dish, mounted a little like insect antennas on opposite sides of *Hubble's* hull. As the words *high-gain* imply, these antennas allow us to communicate with *Hubble* much more effectively than through any of its other (low-gain) antennas. Both high-gain antennas deployed correctly (albeit more slowly than expected) upon command from the ground—a reassuring, if not inspiring, sight as we witnessed it live via television cameras mounted on the shuttle. Even so, the batteries continued discharging, draining their remnant of precious power, and all the while *Hubble* dangled at the end of the robot tether.

Then came the moment of truth. Deploying the solar arrays would be a tricky task, if only because it needed to be done in the most timely manner. The 48,800 solar cells of both arrays absorb the light of the Sun, convert it into electricity, and thereby provide all the power needed by the spacecraft. That power totals 4,100 watts, of which 2,400 watts are used by the telescope, the remainder needed to charge the craft's batteries; each of *Hubble's* six scientific instruments utilizes a remarkably small 100–150 watts—just the power of a three-way light blub.

The two solar arrays, made by British Aerospace and contributed to the Space Telescope project by the European Space Agency, were stowed away like a pair of tightly rolled window shades. The portside array mast was the first to swing out perpendicular to *Hubble's* hull, its cassette latches unlocked, and the panels on each side of the mast began to unfurl. Upon command from the ground, motorized devices pushed out spring-loaded rods framing several panels from cassettes on each side of the central mast. The metal rods, called bistems, are not cylindrical in shape like a pencil-shaped antenna mounted on the hoods of some automobiles; rather they are slightly curved thin strips of steel much like in a compactly coiled metal carpenter's ruler. The bistems are slender enough to coil up flexibly along with the panels in their cocoons, yet become sufficiently rigid once deployed to hold each ultrathin array in place.

In fact, each 8-foot by 40-foot array of silicon wafers is so thin and fragile that back on Earth it had to be constructed partly while floating in a tank of water. Our "1-g" environment on Earth is too harsh for the arrays to be built in any other way, lest they collapse under their own weight. (Despite a thickness of only ½ of an inch, or less than a millimeter, each square foot of the flexible plastic-backed "blanket" contains some 150 postage-stamp-size solar cells that together weigh about a pound. Although we tend to think of each array as a flimsy fabric, its huge area accumulates a total weight of about 330 pounds, which is too heavy for such a thin structure to support itself on Earth. *Hubble's* arrays contrast greatly with those onboard the Soviet "space telescope," a pale equivalent of *Hubble,* named *Granat,* whose solar arrays are made of inch-thick cast iron.) And since they were meant to work in the "zero-g" environment of a free-falling spacecraft, *Hubble's* arrays were never fully tested due to technical and financial reasons—an issue that would return to haunt us.

Hubble's two solar arrays, one shown here, each measures 8 feet x 40 feet x ½ inch, and are so flimsy that they had to be assembled while floating in a tank of water. Some of the postage-stamp-size solar cells can be clearly seen; nearly fifty thousand of them are needed to power the spacecraft. *[ESA]*

Watching intently live television from the shuttle, we at the Science Institute counted as the individual panels of *Hubble's* first solar array extended slowly over the course of several minutes. When the fifth panel came out to each side of the central mast, the extension stopped, whereupon a NASA manager announced that we had a problem. Despite years of preparation, several ground controllers at Goddard were under the impression that the full extent of a solar array consisted of fourteen panels, not ten as we were seeing. But after hurried consultations with European Space Agency technicians who were huddled at Goddard and standing by in England, it became clear that only ten such panels comprise the total array; hence the first set of solar panels had deployed precisely as expected. At the institute we nervously watched the telemetry indicating that *Hubble's* onboard batteries had reversed their previous discharge and were now showing a sudden increase in power. Solar-generated current was beginning to course through the telescope's complex circuitry—not to mention the surge of adrenaline through many of us on the ground. The Hubble Space Telescope was coming alive!

About four hours had passed since the umbilical had been cut and only one more solar array remained to be unfurled that day. The second was crucial

Hubble, still connected to the shuttle's Canadarm, is shown here with its port solar array fully extended some 40 feet lengthwise. The extender rods, or bistems, used to guide the panels from their cocoon, are clearly seen to either side of the array. The view here is of the back of the solar cells, whereas the previous figure shows the front, or sunlight-absorbing, part of the arrays. One of the high-gain antennas is also visible midway between the solar array and the Canadarm. *[NASA]*

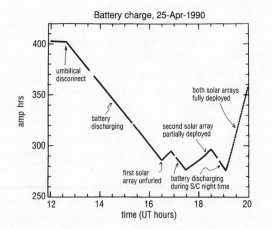

Battery charge, 25-Apr-1990

As downlinked telemetry was received at the Science Institute indicating the state of *Hubble*'s onboard batteries during the telescope's deployment, a computer monitor traced out this chart, showing battery charge, in ampere-hours, plotted vertically against Universal Time (UT) or Greenwich Mean Time horizontally. During the course of a hair-raising six hours, the steady decline of battery charge (shown to the left) was reversed three times—toward the middle of the plot as the port solar array unfurled, about an hour later as the starboard array was partially deployed before becoming stuck, and to the right as the starboard array finally began taking in its full complement of energy from the Sun. From this point on, power to the spacecraft remained more than adequate throughout the commissioning period. *[Space Telescope Science Institute]*

to maintaining a sufficient power margin, lest the bird potentially freeze up. We were well more than halfway across the six-hour, fifteen-minute "desert" needed to get *Hubble* fully energized, and there seemed no reason to expect any difficulty now. The Canadarm was still holding *Hubble* at a steady attitude in the arm's-length high-hover position, the shuttle commander was adroitly repositioning *Discovery* so as to avoid too much solar heating on key parts of *Hubble*'s hull, and two mission-specialist astronauts were fully prebreathed and partially suited in the event they had to aid in the deployment.

A signal was then radioed to *Hubble* to begin unfurling the individual panels of its starboardside array, and the television monitors clearly showed it activating as designed. One panel on each side of the mast came out, again propelled open by the motorized uncoiling of a tightly wound metal tape. But then, suddenly, all motion stopped. The bistems had stuck. No more panels of the array would unfurl, and there was silence on orbit and on the ground.

European Space Agency officials were now on the spot, puzzled. Only about 20 percent of the second array had opened. And this was not enough to power *Hubble* properly, as the downward telemetry trace of battery charge clearly indicated to us on the ground. Fortunately, with the high-gain antennas already up and running, the same series of commands was again quickly sent up to *Hubble*, but the starboard array would not budge. More silence, more thinking, more concern for *Hubble*'s safety. I began flashing back to the oft-quoted statement of Bob O'Dell, a former *Hubble* project scientist: "If we make it through the first few days successfully, we'll be okay."

We were about to lose contact with the communication satellite loop through which *Hubble* and humans talk back and forth—the Tracking and

Data Relay Satellite System, or TDRSS for short (pronounced "tea dress"), scarcely a geosynchronous network of two satellites. Because TDRS-East (hovering over the equatorial Atlantic Ocean at 41° west longitude, just off the northeast coast of Brazil) and TDRS-West (over the equatorial Pacific at 171° west longitude, south of Hawaii) were not orbiting exactly on opposite sides of Earth, during *Hubble*'s early operations the telescope routinely lost contact with both relay satellites for about 10 percent of each orbit. This pause gave ground controllers time to debug the solar-array dilemma while *Hubble* was out of range for ten minutes or so. Was it a hardware problem, namely a difficulty with the mechanical equipment on this particular array? Or was it a software problem arising from an incorrect computer program that commands the extender rods? Both the telemetry and up-close television pictures revealed nothing abnormal about the state of the hardware, which if broken could become, in NASA lingo, a "show stopper," seriously threatening the mission. And if it were a mere software problem, why hadn't it affected the port solar array?

Some five hours had now passed since the umbilical had been severed. *Hubble* project officials at Goddard and Marshall formally requested that the Johnson Space Center order the two mission-specialist astronauts, Navy Captain Bruce McCandless and geologist Kathy Sullivan, to don the remainder of their spacesuits and to enter the airlock for possible extravehicular activity— EVA—or "space walk." We were still not overly concerned at this point, but the dire implications for *Hubble* began to play on our minds in studied silence. I remember whispering to a colleague that we were "entering the woods."

When contact was reestablished with *Hubble* over the far Pacific, engineers were ready with a revised set of commands to pry loose the recalcitrant array. They had in the meantime interrogated a pressure sensor, which implied that the problem might be excessive tension on the delicate solar array, and had surmised that a piece of software was automatically (and correctly) interrupting the deployment sequence. So the ground controllers lowered the sensitivity of the computer's readout of the troubled array's tension, and we held our collective breath as *Hubble* was newly instructed. But the stubborn steel rods did not budge. The starboard array, of which we had a remarkably clear

All dressed up and no place to go. This helmet-less *Hubble* astronaut, Kathy Sullivan, nearly suited up for a possible space walk, was about to enter *Discovery*'s airlock. Her "Snoopy headgear" contains communications earpieces and a microphone. However, extravehicular activity was not needed during *Hubble*'s deployment, as the robot managed to unravel itself, as planned, without human intervention. [NASA]

televised picture on the ground, showed little if any movement or any sign of extreme stress in the array.

Minutes later, the order was uplinked for the spacesuited astronauts to begin depressurizing the shuttle airlock. The simple piece of audible English—"Prepare for EVA"—had a dramatic effect on the scientists and technicians with whom I was clustered at the Science Institute. The air suddenly became hot and sticky, people now more obviously serious and tense. A few colleagues got up from their chairs and began pacing.

Less than a half-hour remained before the secure six hours and fifteen minutes would be up, and we could only guess how much additional time the successful earlier deployment of the port solar array had bought us. *Discovery*'s airlock was now virtually depressurized, and the astronauts were about ten minutes from opening the exit hatch when the solar array experts decided to make one more try, gambling that the tension-reading sensor was faulty and that they could effect a software work-around with an alternative command. This time they essentially overrode the onboard computer altogether and drove the array's electric motor by a kind of high-tech brute force. And this time it worked. Our monitors at the Science Institute clearly and carefully followed the steady extension of the remaining eight panels of the starboard array, after which the telemetry showed *Hubble*'s batteries to be fully charging. The spacecraft had become "power positive." A spontaneous applause arose around me in the operations area, as this particular drama on the high frontier came to an end.

The only remaining task on the agenda for the second day of *Hubble*'s life was to release the telescope from the Canadarm. This was accomplished without a hitch on the twentieth orbit at 3:38 P.M. Eastern Daylight Time, while the shuttle was approaching the west coast of Peru—one orbit later than planned but no one cared. Through the eyes of *Discovery*'s television cameras, we watched transfixed as the Hubble Space Telescope became a free-flyer—a liberated spacecraft in its own right. No still photographs of the actual moment of release were taken, for the crew was busy to say the least. Astronaut Bolden was manning an IMAX movie camera, Hawley was working the controls of the Canadarm, McCandless and Sullivan were still confined to the shuttle's airlock, and Shriver was readying to fire about thirty seconds after release a set of vernier jets (small rocket thrusters) to permit the shuttle to drift very slowly away from *Hubble*. (Contrary to popular belief, even among some scientists, Space Telescope was not ejected from *Discovery* or "pushed overboard"; rather *Hubble* was lifted from a "tail-down" *Discovery*—its cargo bay facing Earth—and let go "stationary," after which *Discovery* slowly backed off.) The shuttle thrusters used no large bursts of fuel since project officials wanted to avoid any gaseous contaminants from polluting *Hubble*'s finely polished optics. Mesmerized, we witnessed before our eyes the image of the fully deployed and slowly receding *real Hubble,* engineering drawings and artists' conceptions of which we had seen innumerable times before. The Hubble Space Telescope was now truly in its own element, on-station some 380 miles above the surface of Earth.

*   *   *

This photo of the Hubble Space Telescope orbiting serenely over the west coast of South America was taken from *Discovery* while drifting several hundred feet away from the telescope. The landform at the top left is Peru, at the bottom left Chile, and to the right Bolivia. North is at the top, part of the Pacific Ocean seen at bottom left. Lake Titicaca in Peru is noticeable to the west of the shuttle's Canadarm, not yet retracted at top. The white patches are dry lake beds, the one immediately below *Hubble* being Uyani in Bolivia. *(See also the color insert section.) [NASA]*

When the astronauts were awakened on Thursday morning, April 26, they found that their shuttle had, just as planned, drifted into a parallel orbit some 48 (statute) miles above and behind *Hubble*. There, the shuttle would "station keep," essentially flying escort and standing by while the Marshall ground controllers at Goddard commanded *Hubble* to execute and test many of its gross motions in space. All these early robotic tasks—the initial stages of commissioning known as orbital verification—would climax with the opening of *Hubble*'s aperture door, a huge metal lid at the front end of the telescope tube, much like a camera's lens cap. In the event of difficulty, *Discovery* had preserved enough fuel to return to *Hubble* once to repair any malfunction.

The aperture door, of course, was paramount on our minds. Unless the door opened, no light would be captured and hence no science accomplished. The mission would be over before it began. The astronauts had spent hundreds of hours during the previous five years in a huge water tank—to approximate a "zero-g" environment—practicing removing the aperture door in the event it refused to budge robotically. A tool bag–equipped astronaut would suit up and "space walk" from *Discovery* to *Hubble,* and either pry open the door hinge manually with a large Allen wrench, much like cranking an old model-T car, or use a special set of low-torque ratchet tools to unbolt the door at its hinge and physically remove it. Either rescue scenario would cause future heartburn, since the door is intended to act as a sunshade to minimize sunlight entering the telescope; during every orbit, the spacecraft is rolled so that the door is between *Hubble*'s bow and the Sun. What's more, the door is designed as a failsafe device. If *Hubble* mistakenly points too close to the Sun, a sensor is tripped and the door closes, thereby protecting the telescope's optics. Yet if it were wrenched open, the scope would remain highly vulnerable.

Once *Hubble*'s orbit was determined accurately—its "ephemeris tables" calculated—the Science Institute could predict the craft's precise position in space and begin building observing calendars and instrument schedules to be uplinked to *Hubble,* which would then respond to the human-made tasks. The earliest commands from the ground simply asked *Hubble* to reorient itself—to pitch up and down, to yaw side to side, and to roll around an axis paralleling its long hull. These were large-scale maneuvers (generally 90-arc-degree "slews") that the telescope would need to perform throughout its multiyear mission, and ground controllers wanted to verify the "polarity" of these movements. Whether a spacecraft rolls clockwise or counterclockwise, or moves up or down, and so on, depends on a plus or minus sign in a mathematical equation buried deeply within some computer program. Often when a new spacecraft is sent aloft, controllers find that an incorrect sign—a gremlin or machine "bug"—has crept into the vast computer codes used for command and control; such malfunctions are easily fixed by uplinking revised software to the onboard computers.

These initial checks with *Hubble* went well. Our engineering telemetry on the ground showed no anomalies, and the astronauts were occasionally able to double-check *Hubble's* dynamical behavior by simply watching it. This was particularly satisfying for me; a year before, when the astronaut crew had visited the Science Institute, I had asked if they would be able to aid the early tests by observing *Hubble* visually. To my surprise, they said that they did not normally have the right instrument on the shuttle, but that my idea was a good one. Thus, I was pleased when I heard over the intercom from *Discovery* that some of *Hubble's* acrobatics could be seen by means of an ordinary set of binoculars.

All sorts of initial procedures were now under way. The telescope's main power bus was up and running, its heaters activated in order to keep *Hubble* at an environmentally controlled temperature, and each of its six internal science instruments energized just enough to keep them thermally safe. *Hubble's* onboard computers were already formatted and initialized, its gyroscopes and its science and engineering tape recorders powered up. Now, among many other tasks, the first software load was uplinked to *Hubble's* flight computer, the onboard fixed-head star trackers used for gross pointing were checked out and aligned, the fine-guidance sensors for precise pointing drew current and warmed up, and more polarity maneuvers were verified, this time checking motion of the solar arrays.

In all, *Hubble* had to act favorably upon some sixty-five major prerequisite commands before its aperture door opened and any light was allowed to strike its main mirror. And although shuttle mission control in Houston referred to them as "routine maneuvers," the roll, pitch, and yaw polarity checks that required coordinated motions of the telescope's hull and its huge solar arrays were truly wonderful to watch. After the solar array deployment drama of the previous day, all was going well again and the intended schedules were being met. Mike Harrington, the director of orbital verification, known in the trenches as "the DOV," summarized his first two daily reports with the reassuring words, "All the expected activities were accomplished in the planned time line. . . . There are no open spacecraft problems."

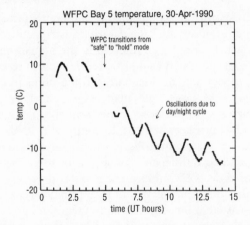

WFPC Bay 5 temperature, 30-Apr-1990

WFPC transitions from
"safe" to "hold" mode

Oscillations due to
day/night cycle

temp (C)

time (UT hours)

One small example of myriad information streaming to the ground regarding *Hubble's* status is this graph of temperature inside a compartment of the wide-field and planetary camera. Concerned that various sensors aboard Space Telescope could be damaged as the vehicle transitioned from a "comfortable" Earth environment to the harsh reality of outer space, ground controllers carefully monitored the cooling and outgassing of all of *Hubble's* scientific instruments. Plotted here is temperature in degrees Celsius (0°C = freezing point of water) plotted against fifteen hours of time nearly a week after launch. Each temperature cycle (wavy change up and down) is caused by a single orbit of the spacecraft. Coolers within this camera would later be used to lower its temperature to nearly –100° Celsius. *[NASA]*

Even so, all of us knew that the Hubble Space Telescope is the most complex piece of scientific equipment ever built by humans. "It's a little like being chief engineer for the pyramids," deadpanned Dave Skillman, the lead Hubble engineer at Goddard, during an earlier press briefing. Not just the most powerful telescope or the most expensive scientific gadget, *Hubble* is also hellishly complicated to schedule and operate effectively. The magnitude of the effort needed to build just the Science Operations Ground System, or SOGS for short, will suffice as an example. Comprising the computer instructions to command and control *Hubble* in orbit as well as to acquire and calibrate its captured data on the ground, all of SOGS amounted at launch to roughly 3 million lines of (mostly Fortran and C*) computer code. Given the additions and refinements made to this immense array of computer software during *Hubble's* commissioning, SOGS grew to nearly 4 million lines of code. As the "industry standard" maintains that a man-year of software development is needed to write 2,000 lines of (bug-free) code, SOGS contains the equivalent of almost 2,000 man-years of computer programming written by a small army of a couple hundred individuals at the TRW Corporation and the Science Institute, none of whom knows precisely how the entire system works.

In fact, *Hubble's* ground system exceeds in magnitude that of any computer program needed to operate any other known machine—at least in the civilian world, where a comparably coded device would be the worldwide AT&T network "switchers" essentially connecting every telephone of all the nations on Earth. Compare this to other examples of growing computer complexity: some 20,000 lines of coded instructions are needed to operate a typical supermarket bar-code scanner, 30,000 lines for a mobile cellular telephone, 130,000 lines for an average air-traffic-control computer, and 600,000 lines for a network of automated bank tellers. On the military side, reliable sources suggest that the *Aegis*-class cruisers operating in support of the naval fleet of U.S. *Trident* submarines are as comparably complex as *Hubble's* several million lines of code, as are (likely to be) the combat computers destined for the proposed *Seawolf* super-submarine, each of these sea ships surely compounded by their onboard humans and nuclear power plants, neither of which the *Hubble* spaceship has. In all, we estimate that *Hubble's* ground operational effort represents roughly 10 percent of the software computer code needed for a minimal Strategic Defense Initiative (SDI) System, which many computer experts claim would now be impossible to build, test, and operate effectively.

Given *Hubble's* inherent complexity and our inexperience operating such a system, many of us felt that the telescope would soon be balking. Said Europe's project manager for *Hubble,* Peter Jacobsen, at a press briefing, "HST is exploiting a twilight zone of parameter space"—a sentiment moments later translated by *Hubble* principal investigator Jim Westphal as, "No one has ever done this before." (Baffled by its fickleness during prelaunch tests, I had tried to bet some colleagues that *Hubble* would enter a temporary comatose state before completing its first week on orbit, but I had no takers.) Our lack of confidence was an expression neither of criticism nor cynicism as much as an honest acknowledgment of the likelihood of a software-induced meltdown in a spaceborne observatory latent with computer bugs lying dormant until some unforeseen event triggers them.

After all, no human or group of humans could reasonably analyze the logic of all aspects of such a megaprogram as the above-mentioned SOGS, which when first introduced into a realistic operating environment had a voracious appetite for mistakes. At least one NASA Headquarters official, openly worried before launch about the project's inability to master the telescope's myriad operating modes, let alone to identify the potential for untested conflict among literally millions of interacting machine instructions, put it bluntly: "Frankly, my problems would be best solved if we plop the damn thing directly into the Atlantic."

Equally pessimistic sentiments were often heard at Goddard: "We don't own *Hubble.* The high-strung bird owns us." Throughout the several years since the *Challenger* tragedy, an array of ground-system tests simulating many of *Hubble's* operating modes, among innumerable commands not well exercised, had clearly attested to *Hubble's* devilish complexity. Often during these "ground sims," there seemed no alternative to just hitting "Enter" on the computer keyboard and crossing our fingers. Almost always, software bugs and glitches emerged, often sufficient to send the telescope into what its builders call "safemode," lest it go berserk.

Just as the name implies, safemode is a minimal operating condition or "safety net" that keeps the vehicle safe and healthy—a sort of hibernation. Hedgehogs do much the same thing, rolling up into a ball if anything threatens them. But safemode is a state of being that *Hubble* assumes itself, using "smart sensors" and sometimes artificial intelligence techniques, without being so commanded from the ground. In other words, humans had granted *Hubble* the ability to protect itself, not only from devious bugs lurking in its myriad networks, but also in the event that those same humans operated the finicky robot improperly.

Thus, even in its design stages, the *Hubble* project had made a philosophical statement. Given that the computer code mimicking its central nervous system had become so large and unwieldy, possibly so close to the edge of what humans could handle intelligently, project officials also realized that *Hubble* needed to have some inherent means to protect itself from dangerous operating conditions. Events could potentially transpire so rapidly, and the Science Operations Ground System evidence its complexity so overwhelmingly, that humans might be powerless to save *Hubble* from a life-threatening execution. So the telescope was given powers—limited powers—to make some decisions on its own, to keep itself healthy and safe without human intervention. This caused us on a few occasions at the Science Institute to wonder if the project had built a machine that was, or could become, collectively smarter than humans. *Hubble*'s code is certainly faster than any human can think, in principle clearly able to "juggle many more balls in the air." Was it possible that we had built a Ferrari for the first time without having built a Ford? More to the point, had we built a HAL even before 2001?

Such self-sufficiency might not be unique to *Hubble*. The Soviets, for example, a few years ago propelled two unmanned spacecraft toward Mars to explore the larger of its two moons, Phobos. Both "bus" robots were designed to drop landers on the moon's surface whereupon the smaller probes would literally hop, like a 1950s pogo stick, from one place to another, making measurements of the soil, weather, and so on. Communications with one craft were severed not far from Earth when a ground controller made a fatal mistake, and contact with the second craft was lost while entering its approach pattern above the surface of Phobos—rumor has it, again, because of human error. Thus, both Soviet spacecraft are probably healthy, and having reached Phobos by a preprogrammed set of instructions, are possibly now hopping around the surface, earnestly learning all sorts of interesting things but unable to share them with us—the ultimate existential experience! Hence, we debated, tongue partly in cheek, whether *Hubble* might ever become smart enough to go about its orbital acrobatics while unlocking secrets of the Universe, yet unable—or even, like HAL in the movie *2001*, unwilling—to let us know what it had found.

Actually, *Hubble* can enter several levels of safemode, the most benign called "inertial hold" wherein the vehicle immediately terminates any further motion. In this case, *Hubble* essentially says to ground controllers, in telemetry bits of course, "I sense something is wrong, so I'm going to make steady my attitude." Any computer loads then commanding *Hubble*'s science instruments also cease, as do all its science observations. A deeper safemode condition called "software sun point" prompts *Hubble* to use a network of five photocells ("sun sensors") on its hull to orient its long cylinder and solar arrays perpendicular to the Sun, thus ensuring that no sunlight shines directly on any of the telescope's internal optics and instrumentation, while providing enough power to sustain thermal control and communications with the ground.

The main 8-foot mirror would not likely be damaged by such sunlight, since it is so highly reflective. But that very same reflectivity would concentrate a great deal of light and especially heat onto the secondary 1-foot mirror, possibly damaging it, much like a pocket mirror can focus sunlight to ignite a pile of leaves. Also, the onboard science instruments are so sensitive to radiation that direct sunlight could blind them, much as Galileo purblinded himself by looking directly at the Sun with his newly ground ocular centuries ago.

The least benign (or deepest) form of safemode—"hardware sunpoint"— is a rather dramatic computer statement that orders the telescope's aperture door to close immediately, and the vehicle to assume the technical equivalent of the fetal position. In this case, *Hubble*'s onboard computer (undesirably but necessarily) powers down much of the spacecraft and also stops sending out signals that it is alive and well. The near heart failure among humans on the ground is usually overcome when *Hubble* is aroused by sending short, sluggish streams of digital bits toward a small, low-gain spiral antenna near the telescope's rear end. Generally, although safemodes are good for *Hubble* in that they are designed to keep the spacecraft safe, such modes are worrisome to humans since a safing event could lead to other, irreversible difficulties while trying to recover from it. How many times can the aperture door slam shut before becoming stuck and refusing to reopen? How many times can we expect *Hubble* to effect a graceful exit from safemode? How many times did we want to risk the robotic vehicle essentially taking its state of being into its own hands?

The first substantive command loads began executing on *Hubble*'s computers at about midnight, Eastern Daylight Time, little more than two and a half days after launch. These were so-called SMS files—science mission specifications—generated at the Science Institute and uplinked to *Hubble*'s flight computer earlier in the evening via the Goddard Space Center's radio aerials. Less than a couple of hours later, we were back in the woods, with *Hubble* having entered safemode—fortunately its most benign inertial-hold state. The culprit was one of the high-gain antennas that had so flawlessly deployed on the previous day. Engineering telemetry implied that a gimbal device on the topside (No. 2) antenna was experiencing some extra pressure ("exceeding a torque threshold") as the antenna attempted to swivel in order to point automatically toward a geosynchronous relay satellite by which data are transferred to Earth. Normally, the flight software continuously adjusts the position of *Hubble*'s antennas to compensate for the changing orbital positions of the two spacecraft. Apparently, the rotating gimbal was sticking or binding, causing a break in the communications link with the ground, in which case *Hubble* sensed trouble and entered safemode until the problem was addressed.

Entry into safemode somewhat complicated life for us, if only by lowering its telemetry rate, thus making communications more sluggish. (*Hubble*'s forward and aft low-gain antennas normally send and receive information at merely 500 bits per second, compared to its high-gain antennas that can handle up to a million bits per second.) Soon thereafter, problems escalated when the safemode state deepened and contact with *Hubble* was lost altogether— the dreaded LOS state, meaning loss of signal to engineers yet implying potential loss of spacecraft to scientists. To our dismay, human error was the cause of the problem, as a ground controller at Goddard had switched incorrectly from one communications satellite to another. Fortunately, low-gain telemetry was reacquired about forty-five minutes later and high-gain telemetry some six hours later, in the process verifying that the vehicle had properly safed and that all subsystems (power, thermal, science instruments, etc.) were healthy and standing by. Nonetheless, tensions definitely do heighten when all contact ceases with a three-day-old, billion-dollar spacecraft.

Little did we know at the time, this relatively quiet third day on orbit would be our last straightforward period for many months. Anxiety in mission control ratcheted up another notch with a set of human decisions, driven I suspect by forces beyond the Space Telescope project. Rather than first systematically recovering from *Hubble*'s safemode state, puzzling through its erratic communications, and then resuming the commanding at the point at which the spacecraft had entered safemode, a decision was made to uplink a special real-time instruction to open *Hubble*'s aperture door earlier than anticipated and in a manner not planned.

Apparently, pressure was mounting (presumably from the military-intelligence community) to release shuttle *Discovery* from its idle station-keeping role, a prerequisite for which was that the spaceborne observatory's Cyclops-like eye be uncovered. What's more, while still in its initial safemode condition, *Hubble* was unexpectedly balking; it would talk to flight controllers but wouldn't move any of its parts. Consequently impatient, humans overrode *Hubble*'s safemode condition—a procedure that we now know resembles trying to extinguish a fire with gasoline—and sent to the bird the technical equivalent of the command, "open sesame."

Admittedly, the shuttle was depleting its slim fuel reserves while oscillating about a point roughly 48 miles down orbit from *Hubble,* something it would not be able to do indefinitely if it was to have enough fuel to fire its retrorockets to return to Earth. But a story was passed around the ground operations areas at Goddard and the Science Institute that one of the astronauts had become very sick and they were preparing to return to Earth prematurely. I doubted this at the time (and I now know it to be false) for we had been in regular contact with the crew, not only by radio but also via television. The astronauts of mission STS-31 were all veterans of previous spaceflights, and there was no evidence that any of them was feeling the ill effects of weightlessness or had experienced any medical emergency.

Instead, I suspect that the crew was ordered to address a largely unpublicized aspect of the STS-31 mission. After being released from Space Telescope duty but before landing, the astronauts and their shuttle were to participate in a test above the Air Force Maui Optical Site (AMOS). Few people in the *Hubble* project knew what this additional task was all about—not surprisingly—because AMOS is part of the ongoing experiments conducted by the Strategic Defense Initiative Organization—the "Star Wars" people.

AMOS is an ultrasophisticated optical and infrared observatory perched on the moonscaped rim of the world's largest dormant volcano atop Mount Haleakala on the Hawaiian island of Maui. Nineteen square miles in total area, the cratered terrain last erupted in 1790. Meaning "House of the Sun," Haleakala is where Maui's day begins; it is also the place from which I have witnessed the most spectacular sunsets. During previous work for MIT's Lincoln Laboratory, I had been involved in some classified experiments at the 10,000-foot-high military observatory, mostly having to do with measurements of dummy nuclear warheads and their accompanying decoys launched from Vandenberg Air Base in California. Before landing about 6,000 miles downrange within a small blue lagoon at the Kwajalein Atoll in the South Pacific's Marshall Islands, these multiple reentry vehicles ("RVs") pass high overhead above Hawaii and can be reconnoitered by a variety of impressive means, including the huge binocular telescopes at the AMOS site. (Ironically, as *Hubble* began running into major-league difficulties not long thereafter, it, too, was examined up close by spying AMOS eyes.) In the course of this work several years ago, I had also studied high-resolution, stereoscopic, infrared video images of the space shuttle showing a peculiar and troublesome semi-halo along the orbiter's leading edge—peculiar because researchers did not at the time understand the glow, and troublesome because Star Wars types were worried that enemy RVs might have similar asymmetric plumes—and that if you targeted the glow you might miss the vehicle.

Such an aura was also seen to accompany many large satellites and this was initially a source of concern to *Hubble* project personnel, because an atmosphere-like halo of glowing chemicals was precisely what the *Hubble* mission was striving to avoid by putting the telescope into space. Subsequent Air Force experiments, conducted after *Hubble* was already in orbit, have proved that the shuttle's orange glow results from trace molecules of nitric oxide combining with atmospheric oxygen atoms to form nitrogen dioxide, causing electrons in the NO_2 to become temporarily excited and the gas to discharge. We also now know that *Hubble* does not suffer appreciably from such chemical collisions, partly because it treks around Earth in a relatively high orbit where oxygen atoms are rare and partly because flight controllers normally orient the telescope aft-end forward in the direction of travel. (These two factors also serve to minimize any "sandblasting" or corrosion of *Hubble*'s precious mirrors by fast-moving debris, especially reactive atomic oxygen.) In any event, some of us became concerned that the shuttle's secondary military objective was beginning to dictate events in the *Hubble* mission, just as we seemed to be getting, as I was often wont to say, into a "deeper pile of yogurt."

Indeed, as the aperture door opened, problems multiplied. I remember this incident well, since at that moment, a few minutes after 9:40 A.M., Eastern Daylight Time, on Friday, astronomer George Field called me from Harvard. He had heard on the morning news that *Hubble* had entered safemode and wanted to know the reason for it. I took the call even though events were transpiring rapidly, partly because he was on a short list of mentors whom I had instructed my secretary to put through to me regardless of the circumstances, and partly because I was eager to give him a positive report that *Hubble*'s main mirror had just been exposed to starlight for the first time.

But even as I assured him that safemode was a positive state of affairs—

perfectly reasonable for the spacecraft to make itself safe in an emergency, in fact precisely as designed—the telemetry began to show another setback. The real-time command to open the huge aluminum eyelid had caused it to hit a resting brace while swinging around on its hinges, sending the entire Space Telescope into robust oscillation. Although we scientists had little to do with the engineering details of *Hubble*'s construction, it would seem at least from hindsight to be poor engineering practice to design a 10-foot-wide massive metal door first to flip open rapidly and second to stop quickly at the end of its travel, thus naturally prompting vehicle vibrations. It is not unlike kids shaking the whole house by slamming the front door. Low-tech devices known as "magnetic stoppers" that act on a torquing principle to damp movements in space could have been designed into the vehicle's aperture door assembly, thus slowing it while traveling outward and approaching its mechanical limit.

At this point, *Hubble*, already in safemode caused by its unsolved high-gain antenna problem, had its aperture door open but was oscillating. But by how much? We knew it must be quite a bit when, moments later, the issue further compounded. One gyroscope used to orient *Hubble* in space quit working (which, because of built-in redundancy, is permissible) and then another (which is not), after which I quickly apologized to Field that I had to hang up! With the bird heading toward a deeper state of safemode, sending intermittent distress signals toward Earth, and possibly gyrating wildly, it had managed to win our full and undivided attention.

The "gyros" had sensed the vehicle's unexpected motion tripped by the door opening, and since the flight computer (on *Hubble*) was in an inertial-hold state, it was not expecting to detect large changes in the gyros. The flight software constantly verifies that the gyroscope motions it reads agree with the way it is dictating the pointing of the spacecraft. If one gyro senses the spacecraft to be outside of some predetermined specified range, the computer presumes the gyro has erred and removes it from the mechanism—called the "control law"—that maintains the telescope's aim. But since *two* gyros sensed movement beyond their usual values, the onboard computer caused *Hubble*, on its own and without human intervention, to enter a more profound safemode state—"software sunpoint"—which does not rely on gyros to control spacecraft attitude. Fortunately, neither gyroscope had broken; rather the spacecraft's computer had merely taken them "off line." When *Hubble*'s unwanted vibrations died down, the gyros were later brought back "on line" and (except for two that permanently malfunctioned about a year after launch) continued to perform adequately throughout the telescope's commissioning period.

A note on gyroscopes: Vehicles in space must be stabilized, lest they wobble, tumble, or roll aimlessly. Recall those randomly "floating" objects—pencils, clipboards, foodstuffs—seen in films of astronauts frolicking in space. Just as we steer an automobile while successfully navigating a maze of city streets, spacecraft need to have a means of "steering," or aiming, toward specific locations on the celestial sky. Complicating the task, free-falling spacecraft operate in three dimensions—much like aircraft—compared to the two-dimensional

As *Hubble*'s aperture door opened, it sent the vehicle into jarring vibration, causing the gyroscopes to spin rapidly while trying to stabilize the spacecraft. An example of what we saw at the Science Institute's mission-control area, this plot graphs the change in gyro spin rate ("delta rpm") plotted vertically against two hours of time horizontally. As shown, the gyros reacted most rapidly when the aperture door swung open and hit a mechanical stop at 13.75 Universal Time, or 9:45 A.M., EDT, three days after launch. [*Space Telescope Science Institute*]

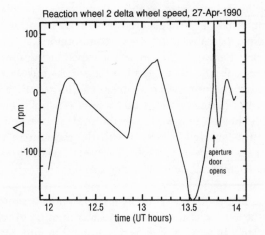

Reaction wheel 2 delta wheel speed, 27-Apr-1990

world of a car or truck. Both such craft use a rapidly spinning gyroscope, or "gyro" for short, which is just a universally mounted rotating wheel whose name derives from the Greek, meaning "to view the turning."

The basic property of a gyroscope, as we all know from playing as children with toy tops, is that if we mount it on a board and turn the board sideways the top does not change its orientation. In other words, a spinning gyroscope "remembers" which direction is "up" even though the surface supporting it is no longer level. This is merely an expression of Newton's first law of (inertial) motion: "A body in motion will remain in motion unless or until it is acted upon by some external force." Such a gyroscope, then, senses changes in the orientation of a spacecraft. That is all it does—detects and measures change. When gyros are also electrically or mechanically connected to other devices (such as rockets, motors, or in *Hubble*'s case, massive flywheels), that perceived change can be compensated for, or accelerated, depending on the desired command sent to the vehicle. These "torques" are expressions of Newton's second (force-acceleration) and third (action-reaction) laws of motion.

Hubble has six main gyros onboard, grouped in pairs within three replaceable shoe-box-size modules in the spacecraft's aft equipment bay. Each gyro weighs only 2 pounds and floats almost frictionlessly in a liquid mixture that is 90 percent hydrogen and 10 percent helium. Usually four of those gyros are activated at any one time; three are required to maintain normal attitude control, and two might be adequate for some science operations. Since, as with any vehicle in space, *Hubble* has three degrees of freedom—up and down, side to side, and back and forth, known in the trade as pitch, yaw, and roll—its fourth online gyro provides active redundancy in the event that the flight computer finds one gyro to be in error, in which case the computer ignores the faulty gyro, thus continuing the mission time line without interruption.

In retrospect, the gyros on that event-filled day were probably overly sensitive to vehicle oscillation because *Hubble* was still in an inertial-hold safemode, which calls for the vehicle to remain absolutely steady. Had the door been opened, as planned, while not in any safemode, the gyros would not likely have been so fussy about vehicle oscillations and no cause for alarm would have been registered by *Hubble*'s central nervous system. All the more reason, perhaps, not to have compounded events by attempting to open the aperture door in the midst of experiencing another problem. Maybe Space Telescope *is* smarter than humans! At any rate, troubles were clearly and rapidly escalating, one problem causing another. At the Science Institute, we felt like passengers in a bus rolling downhill, momentum increasing—all the while hoping that some capable engineer was in the driver's seat. And we constantly consoled ourselves with the thought that, despite this quick series of setbacks, *Hubble* was behaving exactly as designed. It was human beings who were acting a little irrational.

The first time the White House telephoned caused quite a stir at the Science Institute. I was at the time cruising among numerous technical meetings then under way when my secretary homed in and announced with her marvelous British accent that a man at the White House wished to speak with me. It so happened that President Bush was scheduled to give a commencement address the next day in Texas and he and his aides wanted to know when *Hubble* would be visible to the naked eye. I did some quick calculations, literally on the back of an envelope, and offered my best estimate of *Hubble*'s brightness and elevation over Texas during the following nights. I respectfully suggested that I could do a more thorough analysis and phone back later in the day.

What I found, upon further research, is that *Hubble* is brightest when transiting the equatorial zones of our planet. Depending on the tilt of its solar arrays and the orientation of its hull to our line of sight, the brightness of the telescope's star-like pinpoint of light directly above (at the zenith) varies from about 0 to +2 magnitudes. For comparison, Vega, the brightest star in the constellation Lyra, has a magnitude of 0, whereas Polaris, the North Star, is of the second magnitude. Given the rotation of Earth beneath *Hubble*, the spacecraft's orbital inclination angle of 28.5°—the angle its path makes with Earth's equator—means that *Hubble* will eventually pass over all territory between 28.5° north latitude and 28.5° south latitude.

Consequently, although the spacecraft will never be seen overhead for those of us living beyond about 30° latitude to either side of the equator, *Hubble* can occasionally be sighted at low elevations. For example, from the Washington, D.C., area (latitude 38.5°N), I have seen it traveling from southwest to southeast, reaching an elevation of 15° above the horizon midway through the event. (The width of a fist, held at arm's length, equals roughly 10°.) For several days each month, the ascending node of *Hubble*'s orbit crosses directly above the Florida peninsula from which it was launched, making it visible for several minutes. Owing to atmospheric attenuation at low elevations, the telescope's brightness is diminished to about +3 magnitudes, which is just a little dimmer than the three prominent stars in the belt of the constellation Orion.

Interestingly enough, one of our Science Institute astronomers missed the launch and deployment activities but was treated to another unique sight.

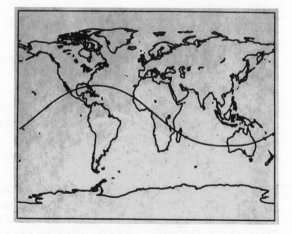

Owing to an easterly launch and the tilt of Earth's axis, *Hubble* travels along a sideways S-shaped trajectory when plotted on a flat map. Our planet's spin in turn causes the S-curve to project differently on each pass and thus the spacecraft overflies many different parts of Earth's surface, but never outside the zone ±28.5° north and south of the equator. *[Courtesy Space Telescope Science Institute]*

Long before, he had applied for telescope time at a leading ground-based observatory high in the Chilean Andes. There, he was able to watch the shuttle periodically orbiting overhead about every hour and a half. The topside of the shuttle reflects much sunlight and, depending on its orientation, can appear very bright in the nighttime sky—often as bright as Vega or even the planet Venus (magnitude −4). As deployment progressed and *Discovery* released *Hubble,* he was able to see a second bright object emerging from the first, much like the birthing act itself. For many orbits thereafter, he and perhaps countless others close to Earth's equator were treated to the fascinating sight of two bright artificial lights moving in formation across the sky.

I called back the White House, as promised, and offered more accurate information regarding *Hubble's* near-term brightness above Texas skies. As the news was beginning to leak that *Hubble* had experienced a series of setbacks, the man on the other end of the line took the opportunity to quiz me on the status of the mission. I paused momentarily, realizing how politically charged such answers might be, and then I replied carefully and accurately, apprising him of the deepening of *Hubble's* safemode condition. I put a positive spin on the concept of "safemode," all the while suggesting that it was evidently questionable engineering and human error that had provoked the current crisis.

We talked for some time about *Hubble's* profound science objectives, about the extensive commissioning period needed before we could do mean-

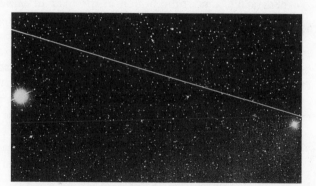

This ground-based photograph captures the Hubble Space Telescope traversing the nighttime sky above Chile. In this eight-minute exposure, *Hubble's* trail passes just north of the bright stars Delta Leonis (left) and 60 Leonis (right). *[Courtesy Cerro Tololo Interamerican Observatory]*

ingful science, and about the societal benefits likely to accrue from such a highly visible space mission—some of which President Bush touched upon in his speech the next day: "You are coming of age during a Golden Age of space. And there's no better example of this than the miracle now orbiting 380 miles above Kingsville—the Hubble Space Telescope. It will see to the farthest reaches of the Universe, to the edge of time . . . and help rekindle public interest in science and mathematics."

Before hanging up, I politely inquired why the White House had called me at the Science Institute. Why hadn't this influential man telephoned NASA?—at which point he entered into derisive laughter. I didn't know whether to join him in chuckling, for I sensed that his hilarity was caused by his previous inability to extract anything clearly and accurately from the space agency, or to be silent, not knowing who else might be privy to our call. I literally held my breath for an awkward moment, after which I heard the dial tone. My report of the two telephone conversations to the Goddard Space Center—the institute is contractually obliged to do so for each and every call from any professional group seeking information about the *Hubble* project—caused NASA to become apoplectic. The space agency was outraged that the White House had called our institute.

Some time later, I would be nearly crucified by NASA for saying in a background comment for *Aviation Week & Space Technology* that, regardless of *Hubble's* increasing ailments, its operation is a virtual miracle. For I did then, and still do, regard *Hubble's* performance as miraculous, given the long and sordid history of so many of humankind's grand ventures that did in fact end in utter failure. Lots of Babels have been attempted over the ages, but this telescoping tower really works—at least partly. I squirmed out of the controversy with NASA by noting that the president himself had earlier claimed that *Hubble* was just such a miracle—even if I had been the source of his words!

"*Discovery,* this is Houston. *Hubble* is open for business." Since the aperture door had swung aside successfully, albeit roughly, those words from Johnson Space Center's "capcom"—or capsule communicator, a term left over from Project *Mercury* in the 1960s—released shuttle *Discovery* from further duty on behalf of *Hubble*—in my opinion, prematurely—at the start of its fourth day in orbit. Given the multiplying problems that the *Hubble* project was then experiencing, amidst the notion in the back of our minds that a yet deeper form of safemode would likely close the aperture door, my expressed view was that *Discovery* should have been ordered to station-keep for the remaining two days of its planned five-day mission, SDI experiments notwithstanding.

In fact, as subsequent events clearly demonstrated, it might have been wise to have had *Discovery* nearly rendezvous with *Hubble* during the fourth or fifth day. A very careful reconnoitering from a few miles distance, including telephoto filming of the operational *Hubble,* might have shed light on some of the other difficulties the project was soon to experience—especially those plaguing one of its high-gain antennas and solar arrays. Nonetheless, *Discovery* descended to an orbit 2 miles lower and glided past the uptight *Hubble,* its crew the last humans for several years to see personally the now unveiled Cyclops in the sky.

The first Science Mission Specification load that the Science Institute had laboriously prepared for *Hubble* was now shot. Substantial and deep disruptions from the planned time line had occurred. When Marshall's Mike Harrington, then in charge of orbital verification, looked bewildered during a Goddard press conference, shook his head downward, and blurted out before the television cameras, "We have had a lot more problems than I anticipated . . . really, really stressful," the media had a field day. That was the sound bite of the day that flashed around the world, said by the same man who had earlier reassured the press that, "I have specialized in OV for the last ten years of my life." This was the point at which the press began to grow skeptical.

By contrast, the institute's operational shift summary for this period seemed more level-headed: "While no one in operations wants to experience safemode often, it is comforting to see the telescope properly protect itself against unexpected situations." To be sure, those of us at the institute resolved to stress the rightfully positive features of *Hubble*'s various safemodes, but not being NASA employees it didn't do any good since we were expressly forbidden by NASA to talk to the press about the vehicle.

Operations crews at Goddard spent much of the fourth day and into the weekend assessing the unforeseen and potentially threatening predicament confronting *Hubble*. The telescope had not been harmed; indeed its health and safety were "A-Okay." So it was time to stop commanding the spacecraft to do anything at all, to retrench as human beings, and to think through *Hubble*'s problems—or, perhaps better stated, the problems that we humans had partly caused *Hubble*. The operations and hardware teams from Marshall and Lockheed and their many subcontractors caucused in order to scrutinize the engineering data and to plan specifically and carefully how they would coax Space Telescope step by step back up through the normal sequence of safemode states to inertial hold, and thereafter to recover entirely to full operations.

As things turned out, the basic problem that had triggered many of the others during the early commissioning was an inch-thick electrical cable wrapped incorrectly around the malfunctioning high-gain antenna. Sophisticated technical analysis entailing repeated replaying of spacecraft telemetry captured at the time of safemode entry—as well as more convincing arguments put forth by an innovative Goddard engineer using a table-top model of the antenna made from a child's set of "tinker toys" and a lamp cord—suggested that the errant cable was the source of *Hubble*'s initial checkmate. Photographs taken by the astronauts and developed after the shuttle had landed clearly confirmed it.

The cable in question wraps around one of the 15-foot antenna masts, from *Hubble*'s hull to the 4-foot dish atop the mast, yet is a few inches too long beneath the dish, causing the rotating dish occasionally to rub against the obstructing cable. The result was impeded movement of the antenna and frequent interruption of communications with ground stations. Apparently, someone had botched the job either years ago while wrapping the cable at the manufacturer or more recently while packing *Hubble* into *Discovery* on the launchpad at the Kennedy Space Center. The other high-gain antenna showed no such cabling handicap. At issue was quality control, a management skill we would come to recognize as sorely lacking throughout the *Hubble* project.

Nonetheless, and although the problem was with hardware, a revision of the ground-operations software enabled controllers at Goddard and the Science Institute to effect a "work-around" by restricting the angle through which the antenna gimbal can swivel; instead of 90° side-to-side motion as intended, the antenna is limited to about 75° rotation in that axis in order to avoid touching the protruding cable. Although this caused some operating inefficiencies during the early months of *Hubble*'s commissioning period when the telescope and its engineering handlers required extensive real-time contact, negligible loss of efficiency has subsequently occurred during normal science operations, since real-time contacts are needed much less frequently.

Speculation had circulated at the time, especially given the speed with which the engineers had correctly diagnosed the antenna problem, that *Hubble* had been imaged in orbit. It is public knowledge that there exists a top-secret, "black" CIA–Air Force intelligence program—as noted earlier, colloquially called Project *Keyhole*, but whose joint directorate name and acronym are officially classified—that commands other "space telescopes" orbiting the Earth, in the process looking down at the ground for surveillance purposes. Conceivably, especially given their onboard propulsion systems, such clandestined vehicles could have been brought into play to aid the engineering analysis of *Hubble*'s recalcitrant antenna. This is especially feasible since the same Lockheed Company involved in building and commissioning Space Telescope has assembled several such *Hubble*-class spyglasses. I doubted then, and still doubt now for unsaid reasons, that *Hubble* was imaged *at that time* by some secret reconnaissance device. However, later in the mission, *Hubble* would in

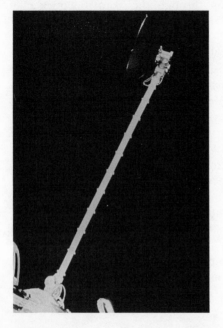

The astronauts on board *Hubble* took this close-up photograph of its topside high-gain antenna. The electrical cable connecting the hull of the spacecraft (at bottom) to the dish-shaped receiver/transmitter (at top) appears to be too long. An extra few inches of cabling bulges near the tip of the boom and occasionally caused the swiveling dish to "trip a limit sensor" early in the mission, thus impeding the antenna's movement and severing communications with Earth. *[NASA]*

fact be imaged on numerous occasions by what the U.S. intelligence community euphemistically calls "national technical means"—technological assets so black that they officially do not exist.

Throughout the course of the fourth and fifth days following launch, Space Telescope was gradually brought out of its safemode and its onboard flight computer fed routine "health and safety loads" that did little more than maintain communications contact. These commands were generated in real time solely to keep the vehicle in a stable condition that it would not regard as threatening. "We're petting the bird to keep it happy," announced a level-headed technician. No further transmissions would be uplinked to *Hubble* until the next week when the ground controllers felt they understood the reasons for the high-gain antenna problem and the resulting entry into safemode. Clearly, one did not want to enter safemode regularly, for each time *Hubble* safed during the commissioning period, so did its onboard science instruments. And no one wanted to risk the possibility that one or more instruments would not recover from a kind of hibernation.

On Sunday morning, April 29, Space Shuttle *Discovery* used virtually all its remaining propellant to decelerate from 17,600 miles per hour, reentered Earth's atmosphere in a fireball resembling a shooting star, and plunged to a safe landing at 220 miles per hour on a windswept lake bed runway at Edwards Air Force Base in California's Mojave Desert. Its crew of five astronauts had performed their mission superbly—the only aspect of the mission that had gone almost exactly according to plan.

Today, command and control of Space Telescope are the responsibility of two organizations close to Washington, D.C. One is NASA's Goddard Space Flight Center, in Greenbelt, Maryland, from which instructions are usually uplinked to the telescope thrice daily, and from which the vital signs of the

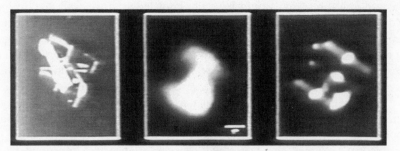

Space Telescope can be clandestinely monitored by a formidable array of "national imagery systems." The central picture is a raw image of *Hubble* as seen from a reconnaissance asset on the summit of Hawaii's Mt. Haleakala; in the right-hand picture, atmospheric blurring has been compensated for, thus showing increased detail. Both images were frame-grabbed from an infrared video of *Hubble* in orbit and the right one has been intentionally doctored, namely "fuzzed," so as not to reveal the full capabilities of the Air Force Maui Optical Site. The picture at left is an artist's conception of Space Telescope, drawn to the same size, scale, and orientation as that shown in the other two frames. Absolute scale is intentionally unspecified. *[Lincoln Laboratory and the Strategic Defense Initiative Organization]*

vehicle are constantly monitored by teams of engineers and flight controllers. The other organization, staffed largely by astronomers and computer scientists, is the Space Telescope Science Institute, an international research center located on the Homewood (Baltimore) campus of The Johns Hopkins University. Although funded by NASA, the institute is run by the Association of Universities for Research in Astronomy (AURA), a consortium of twenty-one major American universities having strong research programs in space science. These are the University of Arizona, California Institute of Technology, University of California, University of Chicago, University of Colorado, Harvard University, University of Hawaii, University of Illinois, Indiana University, Johns Hopkins University, University of Maryland, Massachusetts Institute of Technology, University of Michigan, State University of New York at Stony Brook, Ohio State University, Pennsylvania State University, Princeton University, University of Texas at Austin, University of Washington, University of Wisconsin, and Yale University. The intent of such a "big science" collaboration centered at the institute, 15 percent of whose professional staff hails from the European Space Agency, is for scientists to band together to conduct astronomical research on a scale impossible alone or even in small groups, yet to strive to maintain the traditional mode of investigation as individuals.

The Science Institute was founded in 1981 by NASA, although only after the National Academy of Sciences had made a strong recommendation to do so. Frankly, the astronomical community was fearful that NASA would not enhance the science return from the spaceborne telescope; rather that it would stress the vehicle's engineering feats, much as the space agency had done with other grand space missions, including the *Apollo* project. The global fellowship of astronomers wanted to be sure that they, and not just NASA personnel, had a chance both to use this powerful orbiting observatory, and to have a say as to how it would be used by others. The astronomers also felt that only a staff with strong links to the science community and an active involvement in astronomical research would have the knowledge, drive, and dedication to succeed. Simply put, the customer is always right. Thus, a separate *Science* Institute was established, amid much politics and a nationwide search regarding its location. And if all went well, according to British astrophysicist Martin Rees, who gave the annual *Hubble* Lecture in 1987, "Baltimore would become the center of the Universe for astronomers on Earth."

During the years leading to launch, as well as throughout the telescope's commissioning period described in this book, we at the institute aimed to emphasize the scientific aspects of the *Hubble* mission by striving in every way possible to maximize the telescope's science efficiency. Unfortunately, the institute was established too late, in fact several years after much of Space Telescope's hardware, including its crucial optical system, had been built, its metallic and silicon design frozen. Nor was science efficiency uppermost in the minds of those who did create *Hubble*. We also resolved to campaign for accuracy, whether we were speaking to our scientific colleagues, the general public, educators, or politicians. Again, unfortunately, each of these two central factors—maximizing the product and professing technical truth—got us into repeated trouble with NASA, for they slammed us right into the space agency's bureaucratic agenda of self-preservation.

In many ways, the Science Institute is an experiment—some say an exercise in behavioral psychology—to see how well a small army of technically trained individuals can conduct their affairs. At the least, it is an attempt by scientists to manage their own programs, their own careers, indeed to govern themselves—not a bad idea, in that by nature and training, scientists are notoriously difficult to manage. The institute's philosophy of approach has a doctoral astronomer heading nearly every division, a common-sense idea in principle; the high-energy physicists pretty much handle their own affairs at the world's accelerators and some hospitals are run by medical doctors after all.

The institute then and now, rightly or wrongly, maintains that only the full and intimate involvement of virtually the entire astronomical research enterprise can guarantee the success of a major scientific project like the Hubble Space Telescope—the end users have the greatest motivation to get things right. Our director, Riccardo Giacconi, put it succinctly several years back: "I'd like to see this place as a sieve, with people coming and going, with ideas flowing. HST is a terribly important resource. It goes beyond national boundaries. The limit is not observing time, but brains." Alas, much of this science-based modus operandi is foreign to today's NASA, which is basically a conservative engineering organization. The space agency has always viewed the Science Institute as too bold, too aggressive, too "out of control." "We have one too many institutes," was the way a former NASA administrator, James Fletcher, put it flatfootedly at an embassy party some years ago.

NASA funds the Science Institute, on a contractor basis, to design and conduct *Hubble*'s science mission during the expected fifteen-year lifetime of the telescope, and to engage the broad and responsible participation of the worldwide astronomical community. Like most major ground-based observatories, the institute is a research center where the telescope's scientific program is overseen by a highly trained resident staff of about 400 people, nearly a third of whom are doctoral space scientists, and another third of whom are skilled aerospace engineers and computer technicians. Here, the institute staff manages the selection of observing proposals submitted by astronomers from around the world and provides technical guidance to assist guest astronomers in preparing, executing, and analyzing their observations. For example, the institute created a huge star catalog of many millions of guide stars to facilitate telescope pointing and target acquisition. And it developed an extensive data analysis package to help interpret observations made with *Hubble*'s science instruments.

In all, for the first year of "general observing" following the commissioning of *Hubble*, the institute and its network of peer reviewers accepted 162 scientific proposals written by more than a thousand astronomers from twenty-one countries. (There are roughly 10,000 astronomers in the world.) This is only a fraction of the 556 proposals submitted, which together requested a total of about nine times more telescope time than is available in a single year. By comparison, most large ground-based telescopes are oversubscribed by about a factor of three.

At NASA's Goddard Space Flight Center, the Space Telescope Operations Control Center (known as "the stock," after its acronym STOCC) is staffed by about 150 engineers and technicians—mostly "telemetry readers," who track *Hubble's* vital functions every moment of every day. In all, the computers they watch over monitor some 6,200 specific items of information on the telescope's status—called telemetry points—each having a safe range of operation, outside of which an alarm automatically triggers. This is a staged photograph taken for a Lockheed public-relations campaign; ground controllers do not normally, if ever, wear suit jackets at their computer stations. *[NASA]*

Operating—around the clock, every day of the year—a machine hundreds of miles away and traveling at thousands of miles per hour is a vast undertaking, requiring a magnitude of effort considerably greater than using a conventional laboratory-based telescope firmly anchored to the ground. With *Hubble,* all science observations must be planned in advance, and precisely so to a fraction of a second, thereafter to be executed by a preprogrammed sequence of commands loaded into a pair of computers aboard the telescope. Target siting, instrument operation, and downlink communications are all driven by these onboard computers, though truth be known, both are far less powerful than a simple personal computer in most homes today.

From the early design days in the 1970s, the space agency insisted that *Hubble's* science instruments process data through its NSSC-I computer—NASA's standard spacecraft computer model no. 1, pronounced "nisskee"—which, some say only partly facetiously, derives from a "bug-hardened, cosmic-ray-proof antique" that went to the Moon as part of the *Apollo* project. The problem is that the memory of NSSC-I is a puny 64K bytes, in fact obsolete compared to more powerful personal computers that currently retail for a few hundred dollars.

The other, even smaller (48K-byte) computer onboard *Hubble,* called DF-224, a sort of mythical forebear of R2D2, largely handles pointing and control of Space Telescope as well as managing telemetry traffic with the spacecraft; once a series of commands is uplinked to this flight computer, *Hubble* operates autonomously, free of minute-by-minute control from the ground. (Small memory is the reason why *Hubble's* computers need to be spoon-fed instructions thrice daily; typically 12,000 commands are uplinked each day.)

By contrast, the ground-based computers at the institute that effectively act as *Hubble's* brain and that "talk" with *Hubble* daily are vastly more powerful and state-of-the-art, comprising some eighteen top-of-the-line VAX-class computers networked together in a cluster. Altogether, at time of launch, a network of some eighty-seven powerful computers at the institute and Goddard was needed for telescope operations, systems development and testing, as well as data analysis. Since the telescope orbits Earth once every ninety-six minutes, celestial targets come and go far too quickly to make real-time control practical. In order to maximize spacecraft efficiency, these computers act in consort to interweave parts of many different observing programs to such an extent that few astronomers complete their entire observation in a single block of time.

Planning and scheduling a given set of science observations is a daunting task at best, and sometimes nightmarish. Schedulers at the Science Institute must arrange and rearrange hundreds of unequal segments of observing programs, like oddly shaped field stones used in building a farmer's wall, to maximize spacecraft efficiency and hence science return. An astronomer's accepted proposal must pass through a demanding "pipeline" of computer codes that ultimately translate it into a set of commands that *Hubble*'s onboard computers and science instruments can understand and execute.

Before launch, institute programmers were prepared to partition several thousand observations within the approximately 3,000 hours available in each fifty-two-week observing cycle. Since there are nearly 9,000 hours in a complete year, and since Earth obscures cosmic targets about half the time, other spacecraft necessities such as instrument calibration, telescope repointing, and downlinking data were expected to lower *Hubble*'s science efficiency about 30 percent. During commissioning, however, *Hubble*'s efficiency rose to hardly a third of that.

As an aid to such scheduling of unprecedented complexity, especially as regards the preparation of long-range planning calendars, institute personnel developed a computer program known as the Science Planning Intelligent Knowledge Environment (SPIKE)—a "neural-net, LISP-language-based" algorithm that includes the latest and sundry kinds of smart software, expert systems, and artificial intelligence techniques. Actually, this was one program whose English words were later matched to an already existing acronym— "Hi, I'm Spike," was what the calendar-building computer first displayed on its monitor after arrival from the manufacturer.

SPIKE graphically displays for any time of the year a number of scheduling restrictions imposed both by nature and by *Hubble*'s science instruments, the orbital parameters of the spacecraft, the allocation of observing time granted to an astronomer by a strict peer review system, and any special requirements of the astronomer's proposal. Once, when trying to explain this novel computer code to the press, a NASA manager took to new heights the old saying that all professions are conspiracies against the laity: "With SPIKE, the institute has devised and analyzed novel probabalistic algorithms for scheduling under complex temporal constraints based on the concept of using local heuristics to iteratively eliminate constraint conflicts from an initial strawman schedule." In other words, English ones, SPIKE matches the intended science observation with the best available time slot that offers the highest chance of successful telescope execution.

The long-range observing schedule is fed, week by week, to the Science Planning and Scheduling System (SPSS). Here, a second-by-second time line is computer generated to describe every detail of *Hubble*'s science operation— navigational (guide-star) acquisitions, telescope positioning, target visibility, science instrument operation, orbital characteristics, etc. The main task of the

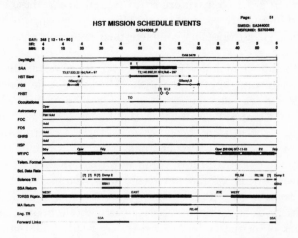

Detailed minute-by-minute planning for the *Hubble* robot is exemplified here horizontally for a mere two-hour interval in late evening of December 13, 1990. One can see among other factors noted vertically on the left that midway through this calendar, at five hours, zero minutes, the spacecraft exited orbital night and began orbit #3478 while transiting the South Atlantic Anomaly and talking to the ground via the TDRS-East relay satellite. *Hubble* had just finished slewing and rolling (away from Mars) and all of its science instruments were on hold, except for the wide-field and planetary camera that was standing by, pending a guide-star acquisition ending at five hours, twenty minutes, after which the camera shutter would open at five hours, twenty-eight minutes. Numerous other constraints on the orbiting telescope and its science instruments, not so listed, must be taken into account to ensure the viability of any observation at any one time. *[Space Telescope Science Institute]*

SPSS is to assemble these and other commanding requests that must be executed sequentially by computers aboard the spacecraft. The final output product of SPSS is the Science Mission Specification (SMS) file—the observing calendar itself—a piece of computer software comprising a specific and intricate calendar of telescope alignment times, exposure commands, filter settings, and a host of other computer instructions regarding science observations. It is this script that the science instruments and related spacecraft systems follow precisely and automatically. For those who toil around the clock in the stark rooms lit only by dozens of glowing computer terminals in the institute's science planning branch, the general sentiment often openly expressed is, "Don't expose them to bright light, don't ever give them water, and never, never feed them after midnight."

Normally, about three weeks before a scheduled observation, the observing calendar is transmitted from the Science Institute to the Space Telescope Operations Control Center, at Goddard, where the file is converted into the binary machine code that *Hubble*'s computers can directly execute. The SMS script is checked by another piece of software, called PASS, to be sure that spacecraft health and safety are not violated; adds yet more commands for operating the spacecraft's engineering systems, such as steering of the high-gain antennas toward the relevant communications satellites and alignment of the solar arrays toward the Sun; and tacks on many more spacecraft "housekeeping" chores such as tape recorder operation and ephemeris (orbit) updating. (The onboard tape recorders are used to store digital data when *Hubble* is unable to establish real-time contact with the ground, or when the telescope is passing through its communications "shadow" period for about ten minutes of each orbit when either relay satellite is out of sight.)

Commands are uplinked and data downlinked to and from *Hubble* through the Tracking and Data Relay Satellite System (TDRSS). From their high geosynchronous orbit (22,400 miles, or 35,800 kilometers altitude), such satellites can relay radio transmissions directly between *Hubble* and the primary ground terminal at White Sands, New Mexico. One more relay occurs, up to a Domestic Communications Satellite (DOMSAT), which is a commercial relay spacecraft also in geosynchronous orbit above the eastern United States, and then down to Goddard, after which the radio-wave signals reach by underground cable their final destination at the Science Institute. This circuitous, zigzag route totals some 90,000 miles, despite *Hubble* being a mere 380 miles above Earth's surface. "It's the way the government does things," joked an engineer when I first joined the *Hubble* project several years ago. Even so, when the link is up and running, back and forth communications transpire quickly, with a given bit of information normally taking only a few seconds to relay from *Hubble* to the institute. In the case of a bulky transmission such as a detailed image of some cosmic object, little more than a minute is needed for the image to be completely rendered on monitors at the institute.

Much as for the space shuttle fleet, priority use of the multisatellite TDRSS constellation is governed by the military-intelligence community. This is proved by at least two facts: First, TDRSS transmissions, which are encrypted, are not received directly by any NASA or other civilian facility. They go initially to Department of Defense (DoD) tracking stations at White Sands Missile Range, which then retransmit nonencrypted versions to the DOMSAT geosynchronous satellite parked above the eastern United States and to which civilians in the *Hubble* project have access. Even here, though, scientists are denied much of the raw engineering data streaming forth from *Hubble;* that information is stripped out of the data stream before reaching the Science Institute and, claiming national security concerns of a dubious nature, not readily made available to *Hubble* astronomers.

This level of secrecy became an issue of some controversy on several occasions as astronomers attempted to verify, after completion of an observation, precisely where *Hubble*'s nonimaging instruments had been pointing on the sky. Much of this sensitivity is perhaps brought into better focus by realizing that the TDRSS network is not owned and operated by NASA but by the shadowy Contel Corporation, a DoD contractor from whom NASA leases use of TDRSS (and whose CEO was reportedly killed in a little-publicized car bombing in northern Virginia some years ago).

Second, when the above-mentioned PASS system requests a communications link between *Hubble* and TDRSS, the government's TDRSS schedulers will reply to such a request only in binary form—"yes" or "no"; if it is affirmative, then *Hubble* gets the requested contact with TDRSS; otherwise we must ask again for a different time, and sometimes frequently so. TDRSS schedules are never published, nor can civilians determine if TDRSS is available at any given time other than by repeatedly making requests—a cumbersome guessing procedure that further complicates *Hubble*'s calendar building at the Science Institute.

Even with these restrictions, the TDRSS network was exceptionally cooperative with the Space Telescope project while *Hubble* experienced a parade of commissioning problems, and thus needed an abnormally large number of real-time TDRSS contacts during the first few months after launch. In fact, for a good part of the summer of 1990, *Hubble* monopolized use of this high-flying relay system, annoying quite considerably several other "customers" of America's most sophisticated communications network. All in all, radio traffic to and from *Hubble* was superb throughout the telescope's commissioning period, with error rates and lost data about a thousand times less than expected.

Several serious orbital constraints on Space Telescope serve both to confound mission planning and to lower efficiency of scientific observations. Given the vehicle's low orbit, the Earth blocks nearly half the sky; thus, most cosmic targets are occulted for up to half of each ninety-six-minute orbit. Continuous viewing zones, comprising about 20 percent of the celestial sphere having no appreciable Earth occultation, occur at the north and south poles of *Hubble*'s orbit; celestial targets within such zones are known as "circumpolar objects." These facts limit spacecraft efficiency to a maximum of slightly more than 50 percent. This is the greatest detriment to such a low orbit, compared to a hypothetical Space Telescope in a much higher geosynchronous orbit for which an efficiency of nearly 100 percent is theoretically possible.

Besides being an obstacle, the Earth is extremely bright and scatters sunlight that can ruin deep-sky exposures. This is why Space Telescope has such a lengthy shield forward of its main mirror; the 26-foot-long tube, as well as razor-sharp baffles and fins inside the tube, are designed to diminish scattered light entering *Hubble*'s optical systems. Most of this unwanted light is stray sunlight, although a good deal of it is also scattered sunlight from the Earth and Moon. Even with these hardware precautions, no science data of any kind are collected when the bright limb of the Earth is within 15° of *Hubble*'s field of view or its dark limb is within 5°.

Telescope schedulers are also careful to keep *Hubble* 50° and 15° clear of the Sun and the full Moon. As noted, in the event of human or computer error, the telescope can independently sense these bright objects and automatically "safe" itself by closing the aperture door and turning off its science instruments. With caution, lunar occultations—which are useful in studying, for example, a planet's atmosphere as the reflected light from the planet grazes the edge of the Moon—can be recorded.

Spacecraft motion is also restricted because *Hubble* repositions or "slews" very slowly about its own axes—some 6° per minute or about one hour to go full circle in pitch, yaw, or roll. Consequently, when Earth is occulting a given celestial target, *Hubble* cannot swing rapidly in the direction of another object. It is actually more efficient to remain targeted toward the original object and to wait half an orbit for Earth to get out of the way, at which point observations can resume. Again, seeking to maximize spacecraft efficiency, schedulers tend to group together targets in adjacent regions of the sky; thus, a given astronomer's observing program calling for images of several widely separated cosmic objects might have those objects observed months apart.

The science instruments aboard *Hubble* present several more operating constraints that must be considered when planning observations. One such restriction is the fact that the spacecraft does not provide electricity to all the instruments simultaneously. At best only three instruments can be used to "parallel observe" the cosmos at any one time. Switching instruments on and off is time consuming because it can take as long as twelve hours for an instrument to reach thermal equilibrium. To minimize on-off cycles, a given instrument is often used continuously for several days before being powered down. Trying to schedule *Hubble's* science operations around all these vehicle, instrument, and orbital constraints is analogous to negotiating a cosmic obstacle course—you get more efficient with practice, but you can still get tripped up badly from time to time.

Instrument performance is also adversely affected by cosmic-ray particles escaping from galactic sources (such as massive stars that have exploded as supernovae) and especially from the Sun during solar flares. Earth's magnetic field traps much of this radiation (really charged particles) in the doughnut-shaped Van Allen belts, and for about nine consecutive orbits each day *Hubble* sails through a region of space called the South Atlantic Anomaly where the

This artist's conception of *Hubble's* orbital trek around Earth illustrates how the telescope can observe a celestial target at upper left when moving along part of its orbit. During the "rear" part of *Hubble's* orbit at right, however, Earth blocks the target from the telescope's view, while during the foreground part of its orbit the telescope is acquiring guide stars, calibrating its onboard instruments, or otherwise preparing to make astronomical observations. It does not help, while in the rear of its orbit, to reposition *Hubble* toward another celestial target, such as the one at right, because by the time the telescope slews to the new target, it, too, would likely be blocked by Earth. [D. Berry]

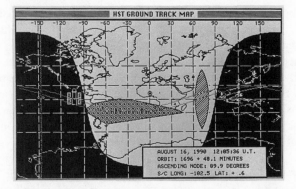

This is a map of charge particle concentrations well above Earth's surface, roughly at *Hubble's* orbital altitude of 380 statute miles. Depicted here on a computer monitor in the Science Institute's planning and scheduling branch, the South Atlantic Anomaly stands out boldly as a stippled, horizontally oriented, teardrop-shaped zone off the coast of Brazil and the southwestern Atlantic Ocean. A pair of *Hubble's* orbital, sideways S-trajectories are superposed as a guide to flight controllers to avoid scheduling telescope observations during passage through this anomalous region, which is laden with harsh radiation. The vertically oriented, stripped oval south of Asia is the zone of exclusion where *Hubble* is unable to make contact with any of the TDRSS relay satellites. The dark shading shows those parts of Earth experiencing nighttime when this map was made, namely August 16, 1990, at the start of the inaugural imaging campaign as described in Chapter 6. *[Space Telescope Science Institute]*

innermost belt dips unusually low, above eastern Brazil and out over the South Atlantic Ocean. Encounters with this decidedly odd area of high-energy particles can last as long as twenty minutes, during which time *Hubble* cannot make astronomical observations owing to "cosmic-ray strikes" that generate false signals, and thus corrupt data when fast-moving electrons and protons bombard the photodetectors of its science instruments.

All told, *Hubble* spends about 15 percent of its time in this so-called SAA, a significant "hit" to the science efficiency of the mission. *Hubble's* instruments can at least remain turned on while passing through the South Atlantic Anomaly, even though no meaningful science data can be acquired. On the down side, however, transit through this harsh region of space can disrupt *Hubble's* pointing control system, playing havoc with the telescope's aim. Eventually, we came to dread repeated passage through the particle barrage in the SAA, for there the high-energy protons especially began causing the spacecraft's electronics to generate digital garbage.

What is the origin of the little-known SAA? Geophysicists reason that Earth's magnetic axis must be offset from the planet's center by about 350 miles. In other words, the fixed, imaginary bar magnet inside the Earth apparently passes not through the very center of our planet, theoretically because of an irregularity in its spinning, liquid, metal core. Consequently, one side of the inner Van Allen belt extends closer to Earth's surface than the other, causing fast-moving charged particles, confined to the belt for hours to years, to penetrate the upper reaches of our atmosphere—occasionally as low as 150 miles altitude off the Brazilian coast. It is not that the magnetic field is greater there; rather it is weaker, allowing particles to congregate in a zone best avoided if possible. Most of these particles are electrons and protons, mainly those escaping the Sun in the form of a solar "wind," and the most energetic of these particles (especially protons) can give a spacecraft (and its ground controllers) fits, especially while impacting its vital electrical subsystems. The SAA, in fact, is one of the reasons why the space shuttle normally orbits lower than a couple hundred miles, seeking to avoid the thick curtain of high-energy particles that are unhealthy for humans as well as electronics.

Interestingly enough, some of the particles and some of the distortion in the Van Allen belts might also be man-made. In the late 1950s and early 1960s, before the Nuclear Test Ban Treaty prohibited such activity, the United States detonated in the upper atmosphere some of the most powerful multimegaton nuclear warheads ever made. Not widely known even among technical people yet no longer classified, Project *Argus* in particular created vast quantities of energetic particles in a series of nuclear bursts at altitudes as high as 300 miles above the South Atlantic Ocean—not far in latitude, longitude, and altitude from where the SAA now peaks. These weapons tests were not intended to experiment with destructive effects as much as to provide information on the trapping of electrically charged particles in Earth's magnetic field. Today's space scientists might well pay a price for the early successes of these outer-space nuclear explosions. "Sometimes we just sit there and cuss a little bit," said Gene Oliver, a NASA Marshall engineer, while commiserating about early pointing errors as the *Hubble* telescope cruised through this enigmatic zone, to some the orbital analogy of the Bermuda Triangle.

The aiming and repositioning of Space Telescope requires no fuel. The spacecraft's stabilization on a target, as well as its reaiming toward a new tar-

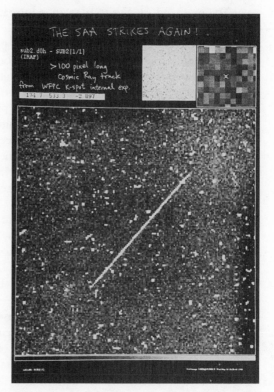

This image, made with one of *Hubble*'s cameras while passing through the South Atlantic Anomaly (SAA), shows a fast-moving cosmic-ray particle streaking across its detectors. The camera's shutter was closed at the time, clearly demonstrating the penetrating power of cosmic rays, in this case a highly energetic proton. Since this image represents only a small part of the camera's field of view, any image of an astronomical object taken during SAA passage would be heavily corrupted with many such cosmic ray strikes, much as rain streaking across an automobile windshield can deter seeing. Technically, each square centimeter of the camera's CCD chips is hit with as many as a hundred (high-energy) protons per second while *Hubble* passes through the SAA (compared to about one per second outside the SAA). The result of all this elementary-particle bombardment is a background noise count equivalent to a thirteenth-magnitude star, making the fine-guidance sensors virtually inoperable during SAA passage. [NASA]

get, are controlled by electronically activated reaction wheels (also called momentum wheels). Each of four, 2-foot-diameter, 100-pound flywheels essentially transfers momentum to the spacecraft, allowing it to maneuver in three dimensions without the need for rocket thrusters. In fact, Space Telescope has no onboard fuel or propulsion system of any kind. Chemical fumes from even small rockets could contaminate *Hubble*'s optics and its science instruments (especially as regards their sensitivity to short-wavelength ultraviolet radiation), and besides, jet fuel residue would likely put a miniature atmosphere around the orbiting observatory which, it must be repeated, is precisely what this mission is designed to overcome.

As implied by the name, reaction wheels employ the classic principle of action-reaction—Newton's third law of motion. When the wheels are spinning at a constant rate (as fast as 3,000 rotations per minute) and no other problems are evident, Space Telescope is steady to within 0.007 arc second. That is more stable than any Earth-based telescope that is subject to seismic activity, automobile traffic, and weather variations. It is also equivalent to keeping a laser beam launched from Baltimore steadily trained on a penny perched atop Boston's Hancock Tower several hundred miles away. In short, if a wheel's rate of spin changes, it introduces a force on the telescope in direct (but opposite) proportion to the amount of change in the wheel's spin rate. As such, there is a key trade-off in favoring principles of physics rather than of chemistry to reorient *Hubble* in space: the telescope repositions, or "slews," slowly—as noted above, at about the rate of the minute hand on the face of an analog clock.

To command *Hubble* to maneuver in space, ground controllers have neither a steering wheel nor joystick. Instead, they send human-designed yet computer-generated radio messages to the spacecraft requesting its pointing control system to send, in turn, a series of short pulses to one or a combination of three reaction wheel assemblies to allow the telescope to accelerate, coast, or decelerate to a new position. Onboard gyroscopes sense when the telescope reaches the desired coordinates in pitch, yaw, and roll, and instruct the telescope to hold steady at the new position. At the end of a long slew, *Hubble* checks and refines its position via three fixed-head star trackers used to confirm the spacecraft's position relative to the location of a set of well-known bright stars, such as Vega, Canopus, and North Star. This rough pointing routine aims the telescope at the appropriate target area with an accuracy of about an arc minute.

The final stage of target acquisition and telescope stabilization requires the spacecraft to use its fine-guidance sensors. When two of three such sensors find preselected guide stars along the perimeter of the telescope focal plane, light from the principal celestial target to be observed and studied is directed into one of several science instruments. But the use of the fine-guidance sensors is unprecedented, their operation tricky.

To achieve the big advance in angular resolution planned by its designers, *Hubble* must point and hold steady to an accuracy greater than for any previous space mission. It does so by using pairs of guide stars viewed around the perimeter of the celestial target; the guide stars essentially act as navigational

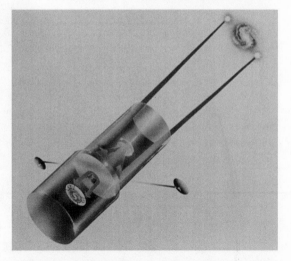

This illustration depicts *Hubble* acquiring light from two guide stars astride a galaxy that has been targeted for science observations. As shown, the light from the guide stars follows much the same path as would radiation from the galaxy itself. That light is first captured by the main mirror, reflected to a secondary mirror, and reflected again into the fine-guidance sensors in the aft bay of the spacecraft. *[D. Berry]*

beacons. To ensure that enough guide stars are available throughout the full-sky celestial sphere, personnel at the Science Institute built an immense new catalog of stars during the mid-1980s. A kind of genome mapping project in the sky, this *Guide Star Catalog* represents the largest stellar inventory ever assembled, nearly a hundred times the size of the previous largest (Smithsonian) star catalog.

To collect the required data, wide-field Schmidt telescopes were used to acquire survey photographs of the northern celestial hemisphere from the Palomar Observatory in California, and of the southern hemisphere from the Siding Spring Observatory in New South Wales, Australia. In all, nearly 1,500 photographic glass plates were scanned nearly twenty-four hours per day for about four years by an extremely precise microdensitometer (essentially a table-top robot whose eye is a very fine laser beam), mounted on a concrete pad sunk into bedrock and thus isolated from the rest of the Science Institute building. The whole catalog comprises some 19 million objects between the ninth and sixteenth magnitude, about 15 million of which are judged to be suitable guide stars. (Some entries are galaxies, extended nebulae, or multiple stars that cannot be used for precise guiding.) This new stellar census is so large that even if it is 99.9 percent accurate (as expected), the 0.1-percent positional inaccuracies, especially for stars photographed near the edges of the plates, mean that the catalog must contain nearly 20,000 errors!

Clearly, we are into the realm of astronomical numbers. In fact, a conventional 8-inch-wide hard-copy (paper) printout of the measured positions and brightnesses of the full content of the *Guide Star Catalog* would stretch for about 56 miles and weigh nearly 2 tons; remarkably, all of it can be condensed onto two compact discs (each with a memory of several gigabits), much like those now flooding the popular music market. However, a thousand such CD-ROMs (Compact Disc–Read Only Memory) are needed to store all the digitized imagery contained in the voluminous *Guide Star Catalog*. Since the laser-eyed robots examined the plates so finely, each plate is digitized with 14,000 picture elements ("pixels") on a side, meaning that the

One of 1,477 large, 14-inch square photographic (negative) plates, which provided data for *Hubble*'s *Guide Star Catalog*. Each glass plate covers an area on the sky roughly the size of the bowl of the Big Dipper. Taking a good part of a day to be scanned, each plate full of stellar blotches was rendered into a two-dimensional array of numbers itemizing the intensities and positions of virtually every speck of light—in all, a total of 18,819,291 celestial objects covering the whole sky. *[Courtesy Space Telescope Science Institute]*

entire catalog houses a whopping near-trillion bytes of data. In terms of a more conventional storage device—the traditional floppy disk used with most personal computers—such a large amount of data would require well more than a million floppies. Astronomical databases indeed!

Pairs of stars are needed to stablize *Hubble* in space: one to fix the spacecraft in two dimensions, the other to prohibit it from rolling about the first guide star. The fine-guidance sensors generally require several minutes of searching to identify and "lock" onto the desired first pair. If, however, *Hubble* recognizes one of the primary guide stars to be a close binary system (a fact that the poorer-resolution ground-based survey could not have known), the sensors can become confused and unable to acquire lock, so the second (and, if needed, sometimes a third) pair are then tried.

Once *Hubble* is pointed at a target, its fine-guidance sensors allow the telescope to remain precisely aimed for many hours. Even if Earth occults a cosmic object, allowing Space Telescope to observe for only half of each orbit, the telescope pointing system instructs *Hubble* to reacquire its target when the occultation ends, so that image or spectrograph exposures of many hours can be built up despite interruption.

How does a typical guest astronomer from around the world actually use the Hubble Space Telescope and its intricate ground system? With a lot of help from personnel at NASA and especially at the Space Telescope Science Institute. Many first-time visitors to the institute have no idea how complex *Hubble* is, or how tricky it is to operate. "They foolishly think they can just show up and start flying around the Universe by working a joystick," was typical of complaints heard inside the institute's user support branch.

While the engineering data stream is analyzed at Goddard, science data

An example of how Space Telescope, to conduct science observations toward the Horsehead Nebula (seen at center), would accurately aim itself by acquiring and "locking" onto a guide star in each of two fields of view (or "pickles") of its three onboard fine-guidance sensors. Not just one but a pair of guide stars is needed to hold the telescope steady in space, much like two nails are often used to secure and level a picture on a wall. For the orientation shown, the two bright stars at top and to the left might suffice, their acquisition amounting to a sophisticated navigational maneuver. *[Courtesy Space Telescope Science Institute]*

are sent by underground cable to *Hubble*'s ground-based nerve center, the Observation Support System (OSS), located at the Science Institute. In the OSS area, which is effectively the science mission control or "engine room" for the *Hubble* program, telescope operators on duty for twenty-four hours every day of the year continually monitor the status of the observing program and verify that the scheduled steps are carried out properly. The operational astronomers and technicians staffing the OSS have at their disposal an array of more than a dozen computer monitors displaying temperature, voltage, focus, pointing, and many other vital signs of the various science instruments—indicators that are used to judge the quality of the science data flowing from Space Telescope. Aided by a cluster of powerful mini-supercomputers, the operators watch for any fluctuation or irregularity that might be symptomatic of a potential instrument problem or malfunction. They can quickly display imaging, spectral, or photometry data and use a number of software tools to evaluate the status of *Hubble*'s instruments. They also examine the quality of the data, which can be lost or corrupted by observational gaps, transmission errors, and telemetry dropouts.

For some observations, a target's position might not be well specified, perhaps because it is an extended region like a nearby galaxy or a celestial object embedded within a rich star cluster along the galactic plane. In these cases, real-time human intervention is often needed to take direct control of spacecraft pointing, a possibility only when contacts with TDRSS—the relay satellites—are available. For such real-time observations, a "quick-look" image from one of the science instruments is transmitted to the OSS area, where telescope operators identify the target, measure its offset from the science instrument where it should be centered, and position a cursor on the computer screen to identify the new pointing location to which *Hubble* then slews.

In addition to normally scheduled observations, there occasionally occur "targets of opportunity," a Strangelovian term denoting unpredictable transient astronomical events of importance such as a supernova in the Milky Way or a new comet in the solar system. In these special cases, humans can intervene in real time, thus altering *Hubble*'s preprogrammed observing plan. Mission managers at the institute need at least twenty-four hours and sometimes a few days to revise and transmit commands to *Hubble,* and for the telescope to acquire its new target. Observing time for targets of opportunity has been set aside as part of the director's discretionary time, a small annual allocation that allows the institute's director some flexibility in commanding the spacecraft for scientific purposes. Such discretion can also be used to provide extra observing time following major discoveries and for unusual or risky projects that promise interesting results if they succeed.

Exiting the OSS, the flood of *Hubble*'s science data is automatically processed while passing along another computer "pipeline" called the Post Observation Data Processing System. First, the raw data need to be transformed into scientifically meaningful units: wavelength, magnitude, photon flux, etc. Second, calibration routines remove known instrumental effects and patterns, enabling us to distinguish real astronomical features from artifacts produced by instrument flaws and transmission errors. This system—colloquially known as "podops," from its acronym PODPS—also saves the data before any additional processing is done to them, in the event an observer needs to compare raw versus more heavily massaged data.

Guest observers can study their data in greater detail in the Science Institute's "Nemesis Room" (named after the Greek goddess who relentlessly persecuted the excessively rich, proud, and powerful) by using numerous computer workstations and electronic networks with names like *Sol, Shapley,* and *Scivax* that provide a panoply of image and graphics displays to help interpret

The Observation Support System (OSS) at the Science Institute is staffed by console operators and operational astronomers of the institute and of its subcontractor, the Computer Sciences Corporation. This is "mission control" for science operations with the Hubble Space Telescope, where guest astronomers can monitor their celestial observations as they are made. *[Courtesy Space Telescope Science Institute]*

the results. (My workstation, trademarked a Sun Sparcstation, I had earlier dubbed *Nuncius,* a sort of twentieth-century electronic "messenger" in some small way commemorating Galileo, the great seventeenth-century father of modern science.) They can then send the data to a film maker in the institute's photo lab for glossy prints and slides, or ship the data electronically to the Astronomy Visualization Laboratory, where scientists work with a computer illustrator to customize the presentation of results and sometimes embed the data within animation sequences, thus, perhaps, gaining further insight into nature's secrets. There, even higher-powered Silicon Graphics computer workstations, named *Paris* and *Cassandra,* are used to manipulate data in the most sophisticated ways (the latter a pun on Priam's daughter supposedly endowed with the gift of prophecy, but fated by Apollo never to be believed). Or the guest astronomers can remain at their home universities or institutions anywhere in the world and receive magnetic tape in the mail or electronic data shipped over an international computer network, such as the Bitnet or Internet worldwide telecommunications systems.

Receiving, archiving, and analyzing data from *Hubble* are not trivial undertakings. Each day, Space Telescope acquires nearly as much digital data as most major Earth-based optical observatories combined. Its telemetry stream can at times reach a megabit per second—incoming electronic transmission strings each second of a million ones and zeros, the pervasive "bits" of today's information age. This is roughly equivalent to filling up a traditional (360K-byte) floppy disk on a home personal computer every few seconds—as fast as one could interchange a disk in a soft drive. Needless to say, such "floppies" are not used to store *Hubble* data; rather the data are archived on optical disks and high-tech "jukeboxes" having vastly larger storage capacities.

On a typical day, the ground system routinely handles several hundred megabytes or a few billion bits of data, which is equivalent to roughly 100,000 pages of information in a large multivolume encyclopedia. (Each byte normally contains eight digits or bits, the smallest piece of data used by any digital computer, thereby affording a possible 256 different combinations, each one expressing, for example, a different amount of brightness or color.) "Drinking from a fire hose" is the phrase often used to describe such a deluge of information.

In truth, by state-of-the-art standards, these are not inordinately large data rates. The most powerful high-energy accelerators designed to probe the subatomic world of elementary particles can produce pulses of some 100 million bits of data each second. Even more impressive, for they are not hard-wired as with ground-based nuclear colliders, some of today's military-intelligence satellites can send to Earth stations bursts of imagery data of far greater volume than *Hubble*'s and in shorter periods of time. A single ("wide-band-width") satellite in fact can transmit a billion bits of information per second, fully a thousand times *Hubble*'s data rate—which might be more like contending with a tidal wave than a fire hose.

The bottleneck with *Hubble* is its high-gain antennas, which transmit at a frequency of about 2.5 gigahertz ("S band") and can trickle information to TDRSS hundreds of times slower than the best (civilian) state-of-the-art, higher-frequency methods (e.g., "Ku band" at 15 gigahertz); its onboard tape recorders are another limitation, as only a couple of dozen images can be typically stored on orbit at any given time. Still, *Hubble*'s maximum telemetry rate is greater than that familiar to most astronomers, and within a few months after launch we were drowning in the ubiquitous bits raining down upon us. All together, over the course of a decade, the Space Telescope mission will likely accumulate well more than 10 terrabytes (10^{13}), or some hundred trillion bits of data, thereby comprising a national treasure roughly equivalent to all the information now stored in the 20 million books of the Library of Congress, to be studied by generations of astronomers charting the Universe.

Scientists using *Hubble* have exclusive rights to their data for one year. No one else (not even at NASA, ESA, or the Science Institute) knows the computer "passwords" needed to gain access to science data without permission from the principal investigator. After this "proprietary period," the observational data (including images) enter the public domain, and are placed in the Space Telescope archive at the Science Institute where they are available for study by interested researchers. Any astronomer can write a proposal for archival research much as is done to gain observing time on Space Telescope. In a few years, when archival research has become more common and the proper technological devices are in place, any person throughout the world having a personal computer and telephone modem should be able to access the Institute's database and examine whatever of *Hubble*'s observations are publicly available.

With all these data being downlinked and commands uplinked amidst vast computing arrays on the ground, the *Hubble* project must guard against computer hackers and software viruses trying to penetrate the system—and thus potentially harming the multi-billion-dollar spacecraft. The Science Institute, to take just one part of the intricate command and control network needed to operate *Hubble,* years ago installed an invisible, National Security Agency–type privacy structure that effectively enables its extensive computer array to "disconnect from the rest of the world" whenever it senses the onset of illegal intruders trying to hack their way into *Hubble*'s central nervous system. This security mechanism, entailing more than a simple set of codes and passwords, also guards against electronic viruses approaching, for example, on the Internet or Bitnet international gateways, as was demonstrated in 1988 when the institute isolated its computer innards, allowing a well-known and widespread virus unleashed by a Cornell University graduate student to bypass us, yet to widely infect and thus jam thousands of other computers throughout the academic, industrial, and government sectors of society (including NASA).

The Hubble Space Telescope is not an extravagant toy just for erudite academicians and astute astrophysicists. Anyone can apply to use Space Telescope. Several years before launch, Riccardo Giacconi declared that he would grant small amounts of his director's discretionary time to individuals not formally trained in (or paid to do) science. He had in mind amateur astronomers who had, over the years, contributed greatly to our understanding of the heavens.

Amateurs, in fact, had made many basic discoveries in the centuries following the Renaissance, and even in modern times provide aid to professional astronomers by scanning the skies nightly for transient phenomena—comets, supernovae, planetary storms, stellar outbursts.

The intent was to give an opportunity to backyard sky gazers who might have a fresh idea or novel approach that might not dawn on a more conservative Ph.D.-level astronomer. We also wanted to tap into the vast audience of nontechnical citizens who, because of astronomy's universal appeal, at one time or another in their lives have looked to the skies and wondered about nature, to make a few of them an "astronomer for a day." A small part of an educational program that Riccardo had asked me to direct, this initiative's heart and soul strove simply to disseminate knowledge widely, and to share the joy of discovery with the general, nonscientific populace. Together we felt that science remains sterile unless it is integrated into contemporary culture, that Space Telescope should not be a "trifle for the rich." Although the allocated time was really a token amount (a few hours per year), opposition arose immediately in several quarters.

Professional astronomers complained, some quite bitterly. A telescope as valuable as *Hubble* should not have a mere amateur at the helm, they said, even for minute amounts of time. Claimed one scientist, "There is too much to be done by the professionals to allow any such amateurs to receive even less than one percent of *Hubble*'s total telescope time." Professional astronomers who must normally compete, sometimes under intense pressure, for access to already oversubscribed telescopes became naturally worried that they would now be up against the unwashed masses clamoring to get aboard *Hubble*. Sadly, many professionals did not seem to realize who had paid for Space Telescope, or that an amateur astronomer might indeed have a better idea for *Hubble*'s admittedly tight telescope agenda. NASA objected, too, largely for bureaucratic reasons. The space agency had not blessed this program of public outreach, nor apparently had even known much about it before Giacconi's announcement at an amateur astronomer's convention outside Boston in the mid-1980s. Reactive rather than proactive, NASA looked rather silly constantly objecting to the amateur use of one of its civilian-funded spacecraft.

Some of the hundreds of amateur proposals indeed reflected much original thinking: A New York high-school teacher proposed to capture a momentary brightening due to melting frost or snow on a moon of Jupiter by precisely timing the opening of *Hubble*'s camera shutter with the moon's appearance from behind its parent planet; a West Coast homemaker and mother of two children asked to use *Hubble*'s infrared capabilities to search for giant newborn planets around some nearby stars in such a way that professionals had apparently not thought possible; a museum volunteer in San Francisco wanted to observe a potential large cloud of comets swarming around a relatively nearby star by waiting for the star to experience a (nova) outburst, thus irradiating the outlying comets; a free-lance oil prospector from New England suggested studying the peculiar chemical and magnetic properties of a prominent Big Dipper star so close to Earth that deep-space astronomers had bypassed it in their thinking.

These objectives were honored with less than an hour of *Hubble* time each, brief and clean observations that would have made Galileo proud.

Many proposals were outright rejected, including a few ludicrous ideas that made us wonder more about the proposers than the proposals: to photograph dark asteroids in our solar system by using a ground-based laser as a flash attachment, thus requiring *Hubble* to be programmed to take a snapshot exactly when the laser pulse reached the suddenly illuminated asteroid; to write a poem while observing a specific celestial object, in much the same way as had been done earlier while commissioning the Brooklyn Bridge; to use *Hubble* to observe the most distant object in the Universe and thus (somehow) to confirm humankind's belief in paganism. (The latter proposer, when asked on the application form which of *Hubble's* scientific instruments would be needed for the intended observation, replied, "It doesn't matter; any one will do.")

As things turned out, a special Time Allocation Committee comprising the presidents of the nation's seven major amateur astronomy societies conducted a (very professional) peer review of the many proposals and recommended to the institute director that a mere five hours of exposure time be given to a handful of ordinary citizens during the first two years following launch. The Science Institute saw to it that these amateurs were accorded every right normally granted to professional astronomers; in fact, we went out of our way to hand-hold the amateurs through an institutional maze of preobserving preparations and postobserving data analysis routines. Not counting volunteer staff time, the entire annual budget for the program of amateur use of Space Telescope was $8,000, a pittance compared to the multi-billion-dollar price tag for the rest of the *Hubble* project.

In return, the chosen amateur astronomers became among the best ambassadors for the *Hubble* project. Together they mounted a media briefing that shamed NASA's often staid press conferences, they toured the country speaking out positively and accurately about the *Hubble* mission, and they brought a breath of fresh air to a project that, given its multiple complexities, personality clashes, and many varied pressures, desperately needed it.

Eventually, like pulling nails out of hardwood, some parts of NASA Headquarters warmed to the notion of having *Hubble* used by, to cite its rude words, "a few street people." Once the space agency agreed that our program was running smoothly on a small budget, officials proposed, wisely, that students should also have a crack at using the telescope. We at the institute responded enthusiastically, outlining in detail how we might aid NASA in such a wonderful educational initiative. But to my astonishment, this absolutely innocent idea of involving schoolchildren in world-class science got thoroughly bogged down in the entrenched bureaucracy symptomatic of today's NASA. Scores of middle managers worried about "NASA's control of the program," about likely competition among different NASA field centers, and about on whose turf the program would operate and who would fund it.

Some two years of meetings were had, endless discussion ensued, lawyers were consulted, and bureaucratic concerns were addressed. The student-oriented program crashed when the Goddard Space Center concluded that such an endeavor—actually requiring less time, effort, and resources than our amateur program—would cost the government three-quarters of a million dollars per year to manage. The bean counters within the space agency based this cost on a nationwide contest in which NASA had asked schoolchildren to suggest names for the newest space shuttle to replace the exploded *Challenger*. The winning entry, *Endeavour*—the namesake of Captain Cook's vessel used by the Royal Society for the 1768 astronomical expedition to Tahiti to observe the transit of Venus—is a good one, but NASA had eaten up close to seven figures in dollars to conduct the contest.

So, given the early opposition that NASA expressed toward the institute's program of amateur use of the telescope, and faced with the apparent need for a small army and a large bankroll to operate its own similar program of student use of the telescope, NASA officials found it best to backtrack. They abandoned the idea of an educational program for the *Hubble* mission altogether, and again began criticizing the institute for having created our amateur program. As in nearly all such cases, I treated NASA's criticisms of our public-outreach activities with benign neglect and moved on. In fact, we subsequently broadened our amateur use of the orbiting telescope specifically to include students and their teachers. And we did so on the same $8,000 budget and an increased volunteer effort at the institute. Alas, the initiative that we took to make *Hubble* available to the amateur community, and thereafter to the educational community, was merely the opening round of a long and sordid series of jousts we had over the years with a defensive and resentful NASA concerning our educational and public-outreach activities that were embarrassingly seen by the agency as being smaller, faster, and cheaper.

Opposition to our minimal involvement of ordinary citizens in the *Hubble* odyssey is but a microcosmic preview of the twin difficulties I faced almost daily both before and after launch: dealing with a bloated NASA organization short on technical management skills yet flush with personnel insecurities, and interacting with a scientific community reluctant to share the intellectual riches of the *Hubble* project with the general public. Despite all the wonderfully positive milestones and achievements of *Hubble*'s path-breaking adventure, these troubling themes thread throughout our narrative. Sadly, they are among the foremost reasons why the Hubble Space Telescope mission has been deprived of its true greatness.

2

Jitters,
in Space and on the Ground

Some years ago, as Your Serene Highness well knows, I discovered in the heavens many things that had not been seen before our own age. The novelty of these things, as well as some consequences which followed from them in contradiction to the physical notions commonly held among academic philosophers, stirred up against me no small number of professors—as if I had placed these things in the sky with my own hands in order to upset nature and overturn the sciences. They seemed to forget that the increase of known truths stimulates the investigation, establishment, and growth of the arts; not their diminution or destruction.

Letter to the Grand Duchess Christina
GALILEO GALILEI, Florence, 1615

The first few months of Space Telescope operations were not uneventful. The badly wrapped antenna cable and repeated entry into (relatively benign) safemode were only the beginning; soon other more serious issues emerged to plague *Hubble*'s attitude, stability, pointing, and focusing. Some of these problems were expected, in the sense that we knew they might occur at some time in the mission, while others came out of the blue, so to speak, managing to surprise just about everyone in the project.

An example of an expected though serious problem occurred exactly one week after launch. On Tuesday evening, May 1, I was hosting the Science Institute's monthly lecture series, *Open Night at the Institute*. This was an ongoing program of public outreach that I had begun several years earlier, in which (usually) an institute scientist gave a semitechnical talk for the general public, after which we adjourned to the backyard, where some amateur astronomers from the Greater Baltimore and Washington areas kindly set up their home-made telescopes to view the heavens. (The Science Institute, though setting the agenda for the most powerful telescope ever built, does not own even a 6-inch portable telescope to use on the ground.)

That night, we listened to one of the grand old men of optics, physics professor Bill Fastie of Johns Hopkins, talk about the on-orbit tests then under way—to tweak the telescope's mirrors for optimum focus, to gain more experience in the use of guide stars to point the telescope accurately, and to calibrate the precise positions of each of the science-instrument apertures. Bill gave us an exceptionally fine presentation, highlighting the history of Space Telescope's mirror design and fabrication, all the while stressing that laboratory tests had demonstrated that the surface of *Hubble's* main mirror was actually smoother than stipulated by design. My journal entry for that day noted that if we had serious problems with *Hubble* it would not likely be with its mirrors. Everyone seemed supremely confident that the optical hardware onboard *Hubble* was in superb shape.

After the Open Night program had concluded and the audience dispersed, I decided to check in on *Hubble* before driving home to Annapolis. Shortly before midnight, I negotiated the usual magnetic sentry, pushed the appropriate cypher combinations, and entered the Observation Support System area. We would eventually come to expect life in OSS to be usually boring, only rarely interspersed by periods of sheer terror—and that night was clearly one of the latter. I found several people frantically talking on the squawk box to ground controllers at Goddard, breathlessly trying to keep up with the rapid flow of telemetry from *Hubble,* and generally more frazzled than I had seen anyone since the day of deployment. I gloomily conjured up the many dire things that could have happened to *Hubble:* Had an onboard battery exploded or a solar array broken off? Or maybe Space Telescope had begun somersaulting aimlessly, much as had a then recently launched intelligence spacecraft of *Hubble's* size and shape. Daring not to bother any of the OSSers so intensely involved in the task at hand, I scanned the amber monitors glowing with *Hubble's* vital signs and saw from the telemetry that the telescope seemed to be okay—its power status was nominal as were its onboard science instruments, and its positional readouts implied that the spacecraft was not tumbling out of control.

Then, one of the operational astronomers on duty bolted for the door, on his way back to consult something in his office. In a half-trot, I followed him partway down the corridor and begged for an explanation. Over his shoulder while still running, he yelled that *Hubble's* aperture door had just slammed shut! Space Telescope had assumed its deepest form of safemode and had lapsed into a coma. What's more, just as had happened several days earlier, when the door had come to an abrupt stop after swinging open, causing oscillations severe enough for *Hubble's* alarmed computer to disconnect its gyroscopes, now the door upon closing had jolted the spacecraft enough to upset its attitude and to lose its gyros once more. Clearly this was not good. We had a distressed spacecraft that had covered its Cyclops eye and was gyrating in a complex pattern of pitch and yaw.

I reminded myself that, although we had hoped *Hubble* would never actually enter this serious form of safemode, it was nonetheless a *safe* mode—a retreat designed to activate precisely and rapidly if the onboard software senses something dreadfully wrong. In this way, *Hubble* can take its health and safety into its own hands by closing its aperture door, thus preventing direct rays of the Sun from harming its sensitive optics. But short of a subsequent repair mission, what would we do if *Hubble's* aperture door refused to reopen?

As things turned out, the cause of the safemode alarm was human error. A spacecraft controller at Goddard had not properly disabled a software command on *Hubble's* flight computer during a routine test of the spaceship's gyros. For technical reasons, this effectively caused the onboard computer to think (falsely) that the telescope was heading toward direct exposure to sunlight. Although the recovery from deep safemode went smoothly by means of a carefully crafted set of uplinked commands, it also meant that when *Hubble's* aperture door opened again, it once more struck its mechanical stops, causing the vehicle to wildly oscillate, lose gyros, stress its computer further, and enter the less severe sun-point safemode yet again.

Hubble would enter safemode more than a dozen additional times during its commissioning period, mostly during the first few months of operation. All these holding patterns were of the moderately benign variety, and not until exactly a year later would *Hubble* again feel critically threatened enough to close its giant eyelid. Even so, there was mounting concern that Space Telescope's orbital verification routine—the first part of the commissioning period—was not going to be routine at all.

Another serious problem—this one quite unexpected—cropped up a few days later. The three fine-guidance sensors (FGSs) needed to lock *Hubble* onto guide stars to point it accurately in space were thrown off each time the spacecraft crossed the day/night terminator and moved into or out of Earth's shadow. Plots of the apparent positions of several stars seen in the field of view of the FGSs showed clearly that *Hubble* suddenly began oscillating by as much as 0.3 arc second (or 300 milliarc seconds) each time it crossed into orbital day. These oscillations—or "jitter," as they came to be called—were mostly, though not exclusively, in spacecraft pitch (movement up and down along *Hubble's* long axis) and lasted as long as ten minutes, gradually damping to the designed pointing stability. Forty-eight minutes later, when *Hubble* left orbital day and entered orbital night, the oscillations would begin again, although their amplitude was less and *Hubble* regained its expected stability after only about four or five minutes.

We now know that the spacecraft's sporadic oscillation is a complex, chaotic series of vibrations, much like those of a plucked violin string. The frequency of the main oscillation ("principal mode of vibration") is a well-behaved 0.1 hertz (cycle per second), meaning that *Hubble* regularly pitches up and down every ten seconds or so. But the oscillations also have several secondary modes of vibration; for example, *Hubble's* ten-second pitch was compounded by an additional wobble of nearly two seconds duration. In turn, less pronounced vehicle vibrations, in both pitch and yaw, were superposed on this complicated motion.

Orbital instabilities of 0.3 (or even 0.1) arc second can severely hamper

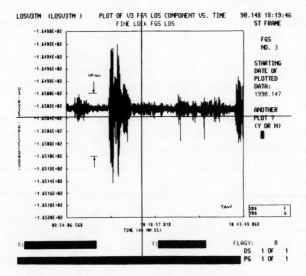

Resembling a seismograph tracing of an earthquake, this is a telemetry scan of Space Telescope's stability during a typical orbit around Earth. The vertical scale measures the telescope's aim in units of milliarc seconds and the horizontal scale is time spanning nearly a single orbit's duration of about one and a half hours. (This is actually a plot of the apparent position of a fixed star as seen by one of *Hubble's* fine-guidance sensors.) Spacecraft oscillation is clearly evident upon entering orbital day (the first series of large vertical spikes at middle) and orbital night (the lesser spikes at far right). The initial (peak-to-peak) magnitude of the "jitter" can be as large as 0.3 arc second, after which it gradually damps to the design stability of less than 0.01 arc second. Each individual, up-and-down vibration lasts about ten seconds. *[Courtesy Space Telescope Science Institute and ESA]*

Hubble's scientific agenda. The angular resolution of the wide-field and planetary camera is comparable to this, and the faint-object camera can resolve several times better in the ultraviolet part of the spectrum. If the expected leap forward in resolution is to be realized, then Space Telescope requires orbital stability of unprecedented accuracy, lest the resulting astronomical images be blurred. The design specifications called for net spacecraft stability of 0.007 arc second (or 7 milliarc seconds), and this was in fact being achieved during parts of each orbit; during the "quietest" portions of orbital night, *Hubble* is even more stable than that, often reaching a remarkable 0.005 arc second. (Technically, the 0.007 arc-second "spec" is a long-term, twenty-four-hour spacecraft stability, whereas 0.002 arc second is the expected instantaneous pointing of the telescope.) All these pointing error budgets derive from prelaunch estimates of minute spacecraft instabilities induced by the motorized flywheels and gyroscopes; the start-stop operation of the onboard tape recorders; the motion of the solar arrays, which track the Sun; the repointing of the high-gain antennas toward TDRSS; and a host of other thermal and mechanical stresses that sometimes cause *Hubble* to deviate, however slightly, from its intended target.

　　If the oscillation could not be fixed, the basic prelaunch stability requirements would not be met, amounting to a significant "hit" to the science efficiency of the *Hubble* mission. The (simplified) reasoning is straightforward: Since orbital day or night lasts only forty-eight minutes and since fine-lock acquisition of guide stars can take as much as twenty minutes and frequent passage through the South Atlantic Anomaly averages another ten minutes, if we first have to wait yet another ten minutes for *Hubble* to damp its oscillations before even beginning its search for guide stars, then we would have less than ten minutes per orbit to do science, at which time *Hubble* would begin oscillating again as it crossed the terminator.

Although for more than a month, NASA regarded these unexpected vehicle vibrations as a technical curiosity—the engineers wrongly maintaining, because of badly calibrated telemetry monitors at Goddard, that the peak-to-peak excursions of the solar arrays were some five times less than what the scientists were seeing in the star-tracking data—the institute made it clear to the space agency that this state of affairs was unacceptable for the proper conduct of the science mission. Shortly thereafter, we heard that NASA had unleashed a "tiger team"—a small group of crack, problem-solving experts empowered to prowl through the system with speed and efficiency by cutting the red tape—in order to quickly assess and fix what had become known as the "jitter problem."

What triggers vehicle oscillation, and why wasn't it anticipated in the design of Space Telescope? Early on, it seemed certain that the cause must be related to thermal effects. Telemetry from thermocouples aboard clearly shows that as *Hubble* enters sunlight, the surface temperature of some parts of it exposed directly to the blazing Sun increases rapidly—a rise of roughly 130°C (or 230°F) in about ninety seconds. And since the oscillations begin abruptly (as seen in the sudden "movement" of stars in the fine-guidance sensors), this heating is akin to *Hubble* being hit with a hammer. Engineers dub it a "thermal snap"; in the near-vacuum of outer space, any absorbed heat energy has no place to dissipate quickly other than into mechanical motion. A simple case of conservation of energy, a basic law of physics. It's a little like placing an ice cube into a hot drink; the cube quickly and often audibly snaps, producing visible cracks in the ice. And only 100°C separates ice and boiling water—a temperature change considerably less than that suffered by *Hubble* on every orbit.

Mere days after the discovery of *Hubble*'s shuttering, I attended several informal though superb briefings given by the chief European engineer on the institute's staff. A most personable and intellectually sharp fellow trained at Paris's version of MIT, the Ecole Centre, Pierre Bely had considered all manner of possibilities that might be causing *Hubble*'s jitter, including a scholarly analysis of the thermal properties of the spacecraft's body. As *Hubble* enters sunlight, one side of its hull heats dramatically while the other, in the shade, remains extremely cold, in principle distorting the hull and transforming it from a circular cross-section to a slightly oval shape.

Like a breath of fresh air, our Parisian friend was honest, revealing one afternoon that the thermal distortion of *Hubble*'s hull, while real, was much too small to explain the evident instability. The hull is wrapped in multiple layers of insulation, or thermal blankets of aluminized foil, called Mylar, designed to reflect light, and any heat absorbed would be distributed slowly, owing to the hull's large mass, and in any case would not likely explain the sudden and rapid oscillation of the telescope. Furthermore, tiny, thermostatically controlled heaters are attached to scores of spacecraft components to warm them during the eclipse phase of an orbit; throughout commissioning, *Hubble*'s thermal control system worked better than expected, freeing up about 20 percent more power for other tasks. Lowering his voice, for the European Space Agency was about to be embarrassed, Bely admitted that he and some colleagues had calculated that the fundamental bending mode of the ESA-built solar arrays (whose design Bely had had nothing to do with) was approximately 0.1 hertz—which meant that the arrays would move up and down in a time equal to the inverse of this, namely ten seconds—almost precisely what was being indirectly observed.

Alas, most fingers began pointing at the solar arrays as the cause of the problem. As noted, each array is designed to absorb rather than reflect sunlight, and is made of exceedingly flimsy material—mostly silicon wafers and metal rods amongst thin plastic blankets roughly the height of a four-story building. Many computer simulations of their behavior in space had been run before launch, but hardware tests had been virtually impossible because of Earth's 1-g environment.

Nonetheless, slight thermal expansions and contractions of solar arrays were generally known to have affected the pointing of at least two previous civilian astronomy satellites as they crossed the terminator. The U.S. X-ray spacecraft *Uhuru* (Small Astronomy Satellite–1) and the international ultraviolet observatory *Copernicus* (Orbiting Astronomical Observatory–3), both launched in the early 1970s, had experienced vehicle oscillations during passage from darkness to sunlight, and vice versa. But since those crafts' solar panels were much smaller and more rigid, and their telescopes' resolution more primitive than *Hubble*'s, such jitter did not have a negative impact on their missions.

As we at the Science Institute gradually came to know more about the design of various parts of Space Telescope, many of us began wondering how and why engineers had done certain things. We were of course eager for them to get on with the calibration of *Hubble* so that we could begin doing some of the frontier science for which the telescope was intended, but we found ourselves unpleasantly amused each day by the growing list of engineering quandaries. Why wasn't the high-gain antenna constructed better or packed into the shuttle more carefully? Where was the quality control? Given that only a few such antennas were made, not hundreds on a production line, one might expect them to have been fashioned perfectly, especially since they are crucial items—in NASAspeak, "single point failures"—on the most expensive piece of scientific equipment ever built. What about the design of the aperture door assembly? Didn't the engineers realize that when brought to a hard stop, this massive piece of metal would upset the delicately pointing, free-falling *Hubble*? And now *Hubble* was oscillating twice every orbit because of, as the European Space Agency eventually put it, a "design oversight" in its solar arrays.

Many of us scientists watched from the sidelines, foreheads furrowed and eyebrows raised, regularly questioning the engineers' moves and motives. What was the rationale for various tests then under way? Why were the telescope operators at Goddard requesting of the institute's schedulers daily revisions in *Hubble*'s carefully crafted calendar time lines, all the while seemingly scrambling to drive the spacecraft by the seat of their pants? Ours was a passive inquisition, however, for with NASA management and its engineering contractors fully in charge of the commissioning, we at the Science Institute got approximately nowhere when quizzing the space agency about its engineering intentions.

Some say that scientists and engineers live in two separate worlds, divided by a culture gap nearly as wide as C. P. Snow's celebrated gulf between technologists and humanists. Here, though, the rift seems to divide the strictly technical calling, neither scientist nor engineer truly appreciating each other or even speaking the same language, yet each crucially important to the mission— a little like oil and vinegar: both are needed in the salad, but they surely don't mix. At the institute, science and engineering offices are mostly on separate floors, our colloquia at different times, our workday hours partly out of phase (engineers often arrive around 7 A.M. and leave at 4 P.M., whereas scientists often work from about 10 to 7 or later). Even in the cafeteria, more often than not, the scientists have lunch together, and so do the engineers. Whenever we did integrate, there was sure to be teasing of each other's domains—good-natured ribbing for the most part, but ribbing nonetheless. Not that we would avoid cross-cultural discussions when confronted. To be sure, engineers were wont to lecture us celestial types on aerospace dynamics or structural issues confronting *Hubble;* all the while we scientists repeatedly reminded our pen-pocketed partners that Space Telescope was built *to do science,* not merely to play with a technically sweet gadget for the sheer engineering challenge of it.

As for the "oversight" plaguing the solar arrays, apparently the culprit was the bistems, which pull each window-shade-like panel from its cassette and thereafter frame the array with a certain amount of rigidity. These rods are made of stainless steel—a nickel-iron alloy having a "coefficient of thermal expansion" large enough to warp the arrays considerably. When sunlight strikes the arrays, the sunny sides of the rods heat dramatically (to about 50°C, or 120°F) and therefore expand, while the shaded sides remain cold (about −80°C, or −110°F). The result is that the pair of panels bend like a huge banana or longbow—a 40-foot longbow! "Thermal excitation" of this type is akin to sharply striking a tuning fork: the arrays generate a prolonged vibration at one or more of their natural frequencies, eventually quelling to a more stable state. Space Telescope must sometimes resemble a huge bird in flight, with the tips of the arrays moving up and down nearly a foot about every ten seconds, much like the flexible wings of a spaceborne pterodactyl.

To complicate matters further, the solar arrays sway out-of-sync, with the starboard forward tip up when the port forward tip is down. And since *Hubble* is a free-flyer not anchored to anything, movement of its large arrays (which are not at the vehicle's center of mass—apparently another design flaw) can in turn cause movement of the main body of the spacecraft itself. It's nothing more than action and reaction, namely, Newton's third law of motion.

Specifically, in order to conserve angular momentum—another inviolable law of physics—*Hubble*'s hull compensates for the solar-array disturbance by moving in the opposite direction to the array displacement. This keeps the net vehicle—hull plus arrays—steady in space, though because the arrays contain much less mass than the rest of the vehicle, the central optical axis of the telescope oscillates much less than a foot—in fact, only 0.002 inch. Nonetheless, even slight pitches in the hull of the spacecraft translate into a pointing problem for the scientific equipment onboard, causing frequent loss of fine-lock on guide stars—a case of the shakes bad enough to lose sight of its stellar signposts and thus make impossible even short exposures without blurring the images somewhat. In other words, for up to a fifth of each unocculted half-orbit, *Hubble* could not keep its cross-hairs effectively on the target.

Some months later, while briefing the astronaut corps at the Johnson Space Center, the STS-31 shuttle commander told me that, although he didn't notice a genuine flapping of the arrays with *Hubble* still attached to the Canadarm, part of the arrays did seem to be slightly bent. Some curvature can

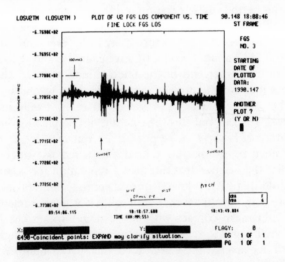

Closer analyses of myriad telemetry scans show that *Hubble* also experiences random and unpredictable vibrations—"microbursts"—at any time in its orbit. Perhaps caused by the heating or cooling of some metal hinge, a flange, other parts of the solar-array rods curled up in their casettes, or the sticking and then releasing of any number of external appendages on or near the spacecraft's hull, the result is a momentary "twang," displayed here as thin spikes. Some of them might even be triggered by miniature debris hitting the spacecraft. Since their cause is unknown and their appearance random, such twangs cannot be easily eliminated with revised software, nor can they be scheduled around. Some of these unexpected oscillations have been violent enough to alarm *Hubble*'s gyroscopes, causing the spacecraft to enter one of its safemodes.

Note how in this NASA graph only an *average* pointing inaccuracy is displayed vertically—the so-called root-mean-square, or "rms," jitter. The raw, peak-to-peak oscillation shown in the previous figure and that most affects image sharpness is actually about three times larger than the rms value. This is a small example of how the space agency, under attack from the media early in the mission, attempted to "package" its problems for public consumption, only to have the packaging later unravel, thus creating a crisis of confidence. [NASA] be seen in the IMAX photo of the just-deployed *Hubble* reproduced in the color insert, but much of that bending is probably due to the "fish-eye" view of the camera.

Flexure of this type is one of the trickiest problems to manage in space-craft design. Once a solar panel, antenna, or instrument spar of some sort begins vibrating, vehicle control can escalate from a headache to a disaster. Various support structures can reduce spacecraft oscillation, but they inevitably add mass, which adversely affects mission orbit and launch costs. For years, during the 1980s, missilemen of the former Soviet Empire experimented with huge boomlike appendages on suborbital ICBM test flights over the east-Siberian Kamchatka peninsula (all the while American intelligence monitored their performance by a number of black assets). Such tests had a military objective, but they also may have compelled the Ministry of Machine Building (the Soviet space agency) to shore up the solar arrays on its own space telescope, the 5-metric-ton *Granat,* with unflappable, inch-thick cast-iron frames that I saw with my own eyes before its December 1989 launch from Baikonur. Or perhaps the massive and highly rigid arrays merely reflected the Russian-bear philosophy of doing things.

Throughout this period of early testing, calibration, and confusion, the White House kept calling me. What's really happening? Is this morning's negative news account of spacecraft jitter correct? Why isn't *Hubble* producing wonderful pictures? I felt like a juggler on a tightrope, at one and the same time giving the man on the other end of the line accurate answers while remaining cautiously upbeat about the project's prognosis. During one such call, I suggested that the caller contact NASA for information, if only to get its viewpoint on *Hubble's* status, but this time he did not laugh. Instead, he lectured me on the nature of NASA's, as he put it, "bureaucracy, hype, glitz, and glitter." He explained to me that he wanted a balanced assessment of what was happening, that the Science Institute had a reputation for accuracy, and that he had been told that I had a knack for explaining technical issues in a reasonably understandable way. In another such call, I was given a telephone number at which I could make contact without passing through the White House switchboard. As I came to regard our tête-à-têtes as increasingly awkward, I never initiated a conversation, never used the special number. Nor, henceforth, did I report our dialogues to NASA.

While engineers pondered how they might fix the jitter problem, I was asked to give a briefing to some people from my past who normally work on "the dark side." The meeting carried a "secret" classification, was held at a secure installation near Washington, D.C., and was closed to all but several dozen officials from throughout the military-intelligence complex. Why was I asked to do this? Perhaps because they wanted to know more about how *our* space telescope was functioning, merely to be entertained by tales of a space vehicle having proportions of interest to them. Or perhaps because I was the only person (that I know of) at the Science Institute who had the requisite security clearances needed to talk shop with the attendees. To be sure, I was asked to speak in the language of technical Pentagonese: units of microradians rather than arc seconds for resolution; nanoradians for spacecraft stability; the metric system, which NASA never adopted (the Space Station is designed in inches); and other corresponding terms for those of the civilian world, all the while referring to benchmark surveillance projects by their cryptonyms—*Lacrosse, Hexagon, Crystal/Kennan, Cobra Judy.*

I note this meeting if only because when I discussed the jitter enigma we were experiencing with *Hubble,* I was astonished to see so many nodding heads. Right then and there, midway through the briefing, a rage came over me. I felt like shouting, "Damn it, why didn't you tell us!" For, apparently, these people—some of whom were *Keyhole* controllers—had years ago first noticed specifically this problem. (Perhaps the graceful flapping of solar arrays was the origin of an informal code name in the early days of the *Keyhole* project—Big Bird.)

After the briefing, several attendees approached me for further information, but two stood out. An Air Force colonel wanted me to reproject a slide showing a raw scan of *Hubble's* pointing accuracy during orbital night, at which point he approached the screen, extracted from his pocket an unfamiliar measuring device, took a few readings, and pronounced that, *Hubble's* troubles aside, he was mighty impressed with our on-orbit stability. Maybe, Colonel, but not good enough for unlocking secrets of the Universe. While exiting the briefing room, an old friend whom I hadn't seen in years puzzled me when he asked why we were "still using those old, unwieldy panels"—the implication being that clandestine space telescopes now use solar arrays quite different from those sprouting from *Hubble.* It would be another year before I fully understood his cryptic comment.

Later that evening in a lounge at the Officer's Club, I attempted to redirect casual conversation among many of the same people in hopes of learning more about the jitter problem and how to correct it. Just as it seemed that I had several Defense Intelligence Agency personnel primed to tell me what they knew, the Secret Service rather suddenly entered the lounge, scaring the hell out of me for fear that I was about to be arrested. Instead, they merely requested us to vacate the room so that they could electronically sweep it for bugs, bombs, and other bad things, after which the lounge would be sealed pending the president's arrival early the next morning. On my way out of the mostly windowless cluster of buildings—typical of those known in the trade as "the cube" or just "213," I was stopped by a serious-looking person sporting short hair, gray suit, ID leash around his neck, and absolutely radiating that woods-are-lovely-dark-and-deep demeanor. He told me the name of someone to contact at Lockheed who, he said, might be able to help us. At which point the intelligence operative did an about-face and marched away.

On my return to the Science Institute, I passed along the name of the Lockheed person to our chief engineer. Upset that the space spooks hadn't warned the *Hubble* project of this fundamental difficulty, I also made a few phone calls—into my past. I wanted to find out more about similar jitter that had apparently plagued military *Hubble*-class vehicles, of which nearly twenty have been launched since the mid-1970s. By *Hubble*-class vehicle, I mean a minimum of a 60-inch mirror feeding captured light to electro-optical detectors of the digital CCD variety. (Given the dominant technical heritage of Space Telescope, an Air Force intelligence officer once corrected me at a briefing at the Naval Academy: "Actually," he said, "your *Hubble* scope is a *Keyhole*-class satellite.")

In a nutshell, what I found out was this, none of which gives away any secrets: The jitter problem had been known for several years prior to *Hubble's* launch, but reconnaissance analysts were not bothered by it, largely because they had never needed to expose their surveillance cameras for more than a fraction of a second. Whether peeking, for example, at work in progress at the Krasnoyarsk radar site, or sensing how many infrared-emitting people inhabit a specific tent outside Tripoli, spying spacecraft need not take long exposures. They can quickly gather their data whether the spacecraft is stable or oscillating (provided it is not tumbling rapidly). Consequently, a ten-second vibration was of little consequence to them, and they probably did not pass along knowledge of it simply because it did not affect their landscape. Even so, I had always thought that the Lockheed/Perkin-Elmer team had won the contract to build *Hubble* because they could incorporate knowledge and experience gained from having built and operated predecessor big birds. However and evidently, the industrial contractors had compartmentalized their sensitive intelligence work so thoroughly that there was little or no cross-fertilization—and the civilian world was the loser.

Soon thereafter, we began hearing from NASA that a solution for the jitter problem would be forthcoming. This remedial computer program, targeted for installation by Lockheed at the end of summer, took the form of a "software patch" to the way that the pointing-control system operates. Essentially, since much of *Hubble's* oscillation is periodic and well characterized in amplitude and frequency, one should be able to compensate for it. In effect, by changing the spin of the reaction wheels and their associated gyroscopes, the spacecraft could be commanded to point ever so slightly in the opposite direction from the expected oscillation, thereby canceling or at least damping the vibration.

In reality, *Hubble's* gyros were not driven as such; rather experiments attempted to arrest the annoying jitter by "stiffening the control law" while upgrading the pointing control system. But given the nature of the dilemma, software changes to *Hubble's* ground-based computer system would not suffice. The fix would have to be accomplished by radioing up a change to the onboard memory of *Hubble's* flight computer, which at 48 kilobytes was in ridiculously short supply.

Although the engineers considered *Hubble's* "twang," as it was also called, solvable, the problem had decimated the original plans for orbital verification. Most early tests of the spacecraft and of its science instruments were judged unreliable, or were simply undoable while *Hubble* was ringing like a bell. As things turned out, the engineers were in fact unable to solve the jitter problem during the commissioning period, which eventually stretched to nineteen months, just about a year longer than expected. Throughout the remainder of 1990 and during all of 1991, numerous attempted solutions in the form of revised software were uplinked to the spacecraft, but none of them worked satisfactorily and some even worsened the jitter.

Attempts were also made to decrease the jitter by feathering the arrays at an oblique angle to the Sun each time *Hubble* entered orbital day, and this seemed to help somewhat but also caused nightmares for spacecraft schedulers. (Telescope schedules normally begin "setting into concrete" weeks before the spacecraft's latest ephemeris, or orbital-track update; thus, it is hard to predict precisely when *Hubble* will cross the twice-per-orbit day/night terminators at the time observing calendars are being built.) Of perhaps greater concern—for this was a health and safety issue for any spacecraft experiencing more than 5,000 thermal cycles per year—would metal fatigue cause *Hubble's* arrays to break apart? You can only bend a metal coat hanger so many times before it snaps.

Jitter of the spacecraft on orbit was one thing, but jitter among people on the ground was something else again. And some of the strongest jitter, or in this case vacillation, occurred regarding the early pictures expected to come from Space Telescope. The very first picture, called by astronomers "first light" and often accompanied by a bit of ceremony and fanfare, was scheduled for only a few days after *Hubble's* deployment—so early in the mission that astronaut Kathy Sullivan had wondered if I might redirect the inaugural image back up to her so she and her crewmates, then almost close enough to look through *Hubble* itself, could see the magnified star fields. Alas, Shuttle *Discovery* apparently did not at that time have onboard the latest symbol of high technology: a fax machine. And as things turned out, that first-light event would be postponed for nearly a month.

Some two years before launch, I was charged by the Hubble Science Working Group to develop a small observing program that would acquire some Space Telescope images of visually spectacular cosmic regions as quickly as possible after launch. The group was acting in accord with Hubble mission policy #5, one of a dozen or so official NASA guidelines issued in the late 1970s to govern the conduct of the mission: "Reasonable latitude may be exercised in allowing a limited number of nonproprietary scientific and public-information observations during the OV period, so long as this does not significantly disrupt the orbital verification process."

In the aftermath of the stunningly successful *Voyager* mission to Jupiter, Saturn, Uranus, and Neptune in the late 1970s and 1980s, some leading Hubble scientists wisely realized that it would be fitting—indeed, perhaps necessary—to release soon after launch some early science results from the heralded spaceborne observatory. Such a program was judged by a few of us to be especially important, given that *Hubble's* commissioning period was then expected to last the better part of a year. Certainly the press would demand to see some early images, as would educators who had prepped their students to expect great things from *Hubble*—not to mention the citizenry who were taxed the approximately $2 billion to transform the telescope from idea to reality.

In addition to the partly altruistic and partly PR reasons for acquiring a handful of early images to share with the general public, there was a good operational reason for doing so: The early images would form a "contingency sample." The first act of the *Apollo* astronauts, upon stepping out of their lunar landing craft, was to reach down and secure a handful of Moon dust and pebbles, which they promptly put into their pockets. Such a quick-and-dirty specimen gave them a small stash of lunar goods in the event they had to depart the Moon prematurely. Likewise, a handful of Hubble snapshots, taken as soon as reasonably possible after telescope deployment, would give us something to show for our efforts, should the spacecraft severely malfunction very early in the mission.

As a person reasonably well known in the astronomical community for a strong interest in education and public outreach, I welcomed the opportunity to select some celestial objects whose pictures would impress the public with *Hubble's* fine capabilities. I also knew that data from a group of representative cosmic sources (acquired for scientific and not just calibration purposes) would help my astronomer colleagues to assess the technical performance of the telescope. But I had no idea how politically charged and controversial this task would become.

The Science Working Group for the *Hubble* mission—sometimes pejoratively called "the swig," after its acroynm SWG—was a committee of little more than a dozen top-notch scientists, all of whom had a longstanding involvement in the development of Space Telescope and especially of its scientific agenda. (Although I was not a formal member, I usually attended their meetings in the years surrounding launch.) The group included all six principal investigators (called "PIs"), each of whom was in charge of one of *Hubble's* onboard instruments. It also included several "interdisciplinary" scientists and observatory astronomers-at-large who had an intimate association with one or another aspect of the project. For example, one was a specialist in data handling and spacecraft operations, another did yeoman service navigating appropriations for Space Telescope through the halls of Congress in the mid-1970s, and yet another was a mirror expert who kept an eye on the construction of the telescope's optical systems. This latter interdisciplinary group was known as the "seventh team," and some of its members felt that their status was secondary to that of the six principal investigators; occasionally this touchiness surfaced in a number of unproductive ways.

The Science Working Group represented an even larger ensemble of researchers who, because of their occasional advice to the Space Telescope project over more than a decade, were called guaranteed-time observers, or GTOs; in all, ninety GTOs from thirty-eight (mostly academic) institutions comprise the seven teams. They were "guaranteed" because NASA had assured each team some 350 hours of observing time to address whatever astronomical problems the team wished—to get first dibs on the most intriguing celestial sources of our time. This is how NASA paid them for their many years of work and advice on behalf of the project.

I first drew up a list of spectacular-looking cosmic objects by discussing my task with colleagues at the Science Institute. There, I found immediate support for the venture among key officials, especially from the institute director, Riccardo Giacconi, a rare person in the world of science who recognizes the need both to pursue scholarly research and to share with the public the fruits of that work. He had been my best ally in my attempts to create an effective educational program at the Science Institute in the years leading up to launch, and now he declared himself strongly in favor of an early release of stunning *Hubble* pictures into the public domain—"disseminate *tout le monde*," he would often say to me.

In 1978, shortly after the launch of an X-ray spacecraft called *Einstein*, he had secured four brief exposures of bright and splendid cosmic objects, processed them quickly, and shared them directly with some science journalists, who were so grateful they judged the mission favorably even when it terminated prematurely. At the time these images were colloquially called "the pretties," and on the evening of our discussion I returned home and found in an old interoffice envelope a set of the four "pretties" he had given me in Cambridge some ten years before. They included a nearby spiral galaxy, a black-hole candidate, a well-known nebula, and a mysterious quasar—heavenly bodies sure to fascinate the public. This set of *Einstein* pictures had been distributed virtually everywhere—to the news media, to schools, to civic and political officials, into seemingly every nook and cranny of American life. And they served their purpose well, for the pictures both excited and satisfied the public as well as assessed clearly for the scientific community the capabilities of the *Einstein* spacecraft. The scientists involved in the *Einstein* mission were thereafter able to take their time calibrating their spacecraft and unlocking secrets of the Universe without the media breathing down their necks; their obligation to the public had been discharged magnificently.

The "pretty picture" program of the *Einstein* project seemed an effort well suited to the *Hubble* mission. But the Science Working Group had placed on me one heavy restriction: I was informed that "the early images should release no new science." I was bemused by this stipulation, because if *Hubble* was going to be the time machine of great sensitivity and resolving power that we all expected, then wherever the telescope looked it would likely see new things. Should I choose only boring celestial regions solely to guard against the possibility of *Hubble*'s finding something new? What would the public think if the first few images came up essentially blank? What if, despite my efforts, one of the early images serendipitously contained a discovery? After all, that's the nature of discovery—an unanticipated finding. Would the project suppress it until the scholarly papers had been published? In an editorial more than a year before launch, the British journal *Nature* clearly saw the dilemma: "The institute is thus faced with the task of producing first images which are both spectacular and uninteresting, a riddle not yet solved."

For better or worse, I privately resolved to weakly humor the working group's restriction, and I drew up a list of spectacular-looking astronomical objects that were reasonably bright, in this way avoiding the need to use valuable telescope time making long exposures. I had a wonderful time doing so, calling upon a lifetime's experience in sky watching, thrilled with the opportunity to choose *Hubble*'s first meaningful science targets. I wondered aloud how Messrs. Galileo and Hubble would have gone about this task; indeed I carefully reread a new translation of the former's *Sidereus Nuncius*.

The targets I chose included stunning galaxies with sweeping spiral arms, spectacular gaseous nebulae that glow in many different colors, compact star clusters that had not yet been resolved, and a few other heavenly bodies that would yield wonderful pictures for public dissemination. In drawing up this list, I carefully avoided regions that other scientists had been officially granted time to observe with *Hubble;* I certainly did not want to "scoop" any colleagues by depriving them of the thrill of their own discoveries. After winning approval from Giacconi, I presented the list and outlined my intentions in a written communication to the six principal investigators, who also approved, some enthusiastically.

Unbeknownst to me at the time, a torpedo had been fired at my list of cosmic targets by a member of the working group's "seventh team." In retrospect, I had made a tactical error by not apprising each member of that team of my intentions—for, as noted, these at-large members of the group, mostly because they had not built any instrument aboard *Hubble,* tended to feel like second-class citizens—and thus I learned indirectly that I would be verbally attacked at the next group meeting. I didn't feel too badly about the impending ambush, for this is often the norm at such gatherings.

All Hubble Science Working Group meetings I've ever attended (with one exception to be described in Chapter 4) were regularly punctuated with disordered, rambunctious, and occasionally uncivilized behavior. Sometimes a sort of controlled anarchy prevailed, quite unlike industrial or military briefings that show a measure of decorum and adherence to the agenda. Members criticized each other at will, formed cliques, made derisive remarks, and often ganged up on speakers, all under the guise of "academic freedom" and "critical thinking." "I've never seen anything like it, never before encountered so many groups of scientists feuding with each other," noted a wide-eyed visiting astronomer who attended one such session.

To be sure, we scientists are a peculiar lot, arguably trained to be contentious, to exercise our egos regularly, and in the spirit of free inquiry seemingly wield a license to intellectually cut up one another. In an odd sort of way that greatly contrasts with the staid and studious manner of most scientists in public, behind closed doors in committee meetings we sometimes act spontaneously, often without thinking or caring, at times childishly. Of course, this Hubble Science Working Group was particularly prone to rowdy meetings largely because of the emotional anxiety accumulated during the long decades spanning concept design to flight-ready hardware. Considering the heated dialogues and the episodes of infighting, no doubt exacerbated by the frustrations of *Hubble's* troubled development history and launch delays, some of their meetings went beyond the threshold of pain. Said a friend at the Marshall Space Center who sometimes attended these meetings, "Your working group sessions are so wild, I'd be willing to pay admission to get in if I had to."

At any rate, the objection of the seventh-team member to my list of targets centered about those that had not been officially approved by the Time Allocation Committee—an international peer review board of eminent astronomers that annually recommends celestial objects to be examined by Space Telescope. (This committee, or "tac" for short, composed of the chairpersons of yet six other committees—some five dozen scientists in all—has, astropolitically and perhaps wisely, never allowed its picture to be taken.) I had specifically chosen objects outside the extensive lists of those to be observed in the early years of the *Hubble* science mission because I wanted to stay clear of any objects for which "proprietary rights" had been granted. We were, after all, talking about cosmic turf.

It is hard to believe that a celestial body or region can become the legal property of an astronomer, but that is the way NASA, in its *Hubble* mission policy #2, had designed the rules of telescope engagement: "All observations obtained by the Space Telescope on the basis of a peer reviewed and selected proposal shall be proprietary for a normal, one-year period." In fact, at an earlier meeting in 1983 entitled "Dividing the Universe," the seven teams of GTOs had convened for the purpose of carving up the cosmos, and I was most eager to avoid tampering with anyone's claim to celestial real estate. The objection, therefore, concerned my having no "legal" right to image a handful of objects that were not mine to observe.

It was against this background that I changed my approach (but not my mind) regarding early *Hubble* images. Before the next meeting of the Science Working Group at the institute in the summer of 1989, I consulted with several of the principal investigators to rearrange the list of targets completely. This time, hoping to head off trouble, I decided to choose approved objects for which telescope time had been granted specifically to members of the working group. In this way, those scientists who had worked hardest on the project would be intimately involved in acquiring and analyzing the early images, and I could be sure that they got the credit as the images entered the public domain.

The principal investigators again agreed and I drew up a new list—spiral galaxies, colorful nebulae, star clusters, giant planets, and the like. But when I projected the list formally on a big screen for all the Science Working Group to see, the same member of the "seventh team" who had objected earlier now objected again. And this time he did so violently. In the most uncouth manner I've ever experienced in my career (up to that point—more later), he screamed that a couple of his favorite cosmic objects were on my list. It was as though the sight of "his" celestial objects among the candidate early targets had triggered a proximity fuse. While I was formally addressing the group, attempting to articulate carefully the rationale used in preparing the list, and while perfectly willing to entertain deletions of specific objects, he repeatedly and heatedly interrupted, eventually blurting out loudly and with great emotion, "If you look at those objects before I do, I'll kill you."

Only slightly perturbed, I took this outburst in the spirit of a grown man whose delusions of grandeur had gone to his head—but it was a highly charged, albeit irrational, statement, one that several members of the working group found egregiously offensive. And he didn't stop there. Without my director present to defend me—"I'm tired of rolling around in the dirt with those guys"—the objector continued to rave in the most bitter tones, clearly taking out on me his *Hubble*-related frustrations that must have been pent up for years. It quickly became clear that the objector was steadfastly opposed to sharing *any* Space Telescope findings with the public; indeed he successfully filibustered to prohibit me from completing my report. As some members of the group now yelled toward the objector, "be still" and "let him continue," while a few others threw up their arms in disgust, the meeting entered a non-linear realm the likes of which I had never before witnessed, even among academics. Some who were present later called this particularly vicious broadside a "three-sigma event," a technical term used to denote a rare statistical fluctuation.

Only more recently did I realize that I had resurrected a concern that had been revisited by the working group several times over the years—a concern that sprang from what is known in some circles of the astronomical community as the "Morabito incident." It happened that, during the *Voyager* space-probe encounter with Jupiter in 1979, a young technician named Linda Morabito had noticed a plume on the limb of Jupiter's moon, Io, while she was trying to determine the precise location of the edge of Io (a task needed to navigate *Voyager* through the congested Jovian system). She realized that she had sighted a volcano in the process of erupting—a notable discovery, and one that apparently embarrassed the doctoral scientists who had overlooked it.

The Hubble Science Working Group did not want a repeat of the Morabito incident under any circumstances. A few members of the working group simply preferred not to risk someone other than themselves, and certainly not a mere technician, making an early discovery with Space Telescope. One of the most uptight of the group was later quoted candidly in a trade magazine: "If [we] immediately release that information, every Joe Blow can read the newspaper, get out his plastic ruler, and scoop you. And there are people out there like that." Added the article's author, a little disbelievingly, "In this case, of course, 'Joe' would have to have at least a graduate student's knowledge of astronomy." Flatfootedly put, a handful of scientists wanted no one else to have even the slightest chance to "skim any cream off *Hubble*'s crop."

Indeed, the working group had debated at length, at their earlier "Dividing the Universe" meeting, where *Hubble* might be initially aimed to show absolutely nothing of scientific value. "Where are the darkest places in the nighttime sky likely to yield no new science?" a former NASA associate administrator told me some of them would ask. Such foreboding realms of space—virtual blanks—represented the essence of the boring images some of the group's members wished to present to the public when *Hubble* first opened its camera shutters.

Despite the nerve I had struck, the idea of an early release of a few spectacular *Hubble* pictures was still considered a good one by a substantial majority of the working group and by key members of the Science Institute. Some argued that it was crucially important to the success of the *Hubble* mission and to the funding of subsequent space projects. The chairman of the working group, Al Boggess, who was the leading *Hubble* project scientist within NASA, also thought well of the program and quietly argued that it should be salvaged in some form. So on the second day of the meeting, a "pretty picture" committee was formed, to be chaired this time by a member of the "seventh team"—in that way intending to neutralize objections arising from the vociferous scientist.

A good part of the second day was then spent criticizing the term "pretties" that I had appropriated from the earlier *Einstein* mission to informally describe this program. I had no sexual connotations in mind nor did I worry about "politically correct" terms, but the word had rubbed a few people the wrong way. Physicists routinely use metaphorical jargon like "charm" and "beauty" to describe properties of subatomic quarks, but apparently some astronomers rebel at the thought of heavenly bodies being pretty. Or perhaps because NASA needed an acronym to make the program official, a new term was chosen to earmark the images that would be released to the public soon after launch: early-release observations, or EROs. Frankly, whether they were pretty pictures or EROs images, the irony was not lost and I didn't really care. I just wanted to see the general public, including schoolchildren and their teachers, share in the excitement and spirit of adventure upon which we were about to embark.

Over the course of the next several months, I worked closely and well with Bob O'Dell, the EROs chairman, to orchestrate a meaningful early-release observations program. Copious quantities of electronic mail were sent back and forth among committee members, and we chose an exciting list of a dozen targets that spanned the spectrum of all kinds of cosmic phenomena. I sent out detailed letters to several dozen potentially affected GTOs around the world informing them of our intent to take an early peek at "their" astronomical objects, specifically soliciting their objections and, receiving none, offering to share with them the data acquired. I prepared detailed proposals for the target observations and guided each one through the Science Institute's User Support Branch, since an operations division chief had rightfully made it clear to me, "No formal proposals, no pretty pictures."

At the next Science Working Group meeting, at the Kennedy Space Center in November 1989, our efforts won formal approval. A sensitive subject to say the least, the early-release observations program now seemed to be gaining momentum and widespread support, although a vast amount of hand-holding and politicking had to be done behind the scenes to keep the program glued together.

All continued to go well with this largely volunteer effort, centered solidly around the notion of altruism—a simple desire to share with the tax-paying public and the world at large some of the early riches returned by the Space Telescope. Some guaranteed-time observers even sent me letters and e-mail messages congratulating us on our plans and reinforcing our belief that a small but meaningful series of early images would be crucial to the long-term success of Space Telescope. Typical among the positive feedback I received was this note from Dennis Ebbets of one of the *Hubble*'s spectroscopy teams: "Science education is an important part of my professional life, and I strongly support the need for communication of HST contributions to the general public." Nearly a year later, Ebbets would play a pivotal role in helping to prove that *Hubble* could do good science.

I was further fortunate at the time to have the opportunity to consult at length with the chief scientist of the highly successful *Voyager* space mission, Ed Stone of Caltech, who was visiting the institute to deliver the prestigious *Hubble* Lecture in late 1989, a public-outreach event offered annually to share the best of science with people of the greater Baltimore and Washington areas. Recalling that he had been the driving force behind the quick release of a handful of extraordinarily popular science images as *Voyager* bypassed most of the outer planets, and admitting that he had to drag his normally cautious scientific colleagues kicking and screaming toward the idea that publicly financed big science sometimes demands immediate results, Stone urged me to stick with it. "Just show the images as nature displays itself, note this and that intriguing feature or mystery, and avoid any instant science or premature interpretations." As I drove him to the airport for his return flight to California, Stone said as clear as a bell, "Your efforts may well be crucial to the future conduct and funding of all of space science generally." His encouragement made me feel that this massive astropolitical exercise was worth pursuing after all.

And then, as we neared launch, the objections began again. Of the ninety guaranteed-time astronomers who were entitled to object to the EROs program, three began worrying that our efforts might deprive them of full credit for any *Hubble* observations they might make. One was the original objector from the seventh team, who now refused to allow us to observe any of his celestial objects; these we crossed off our list, as it had now become clear that we were not going to please him regardless of how we organized this effort.

A second objection arose from another GTO who was something of a rival to that seventh-team astronomer—the upshot being that if the original objector was going to remove all his globular clusters from the EROs list, then the second objector didn't want to let the first objector see what *Hubble* saw in any of *his* globular clusters either. Professional jealousies, normally run rampant among some publicity-hound scientists, were coming to the fore—although having a child in kindergarten at the time helped me understand. So the globular clusters—which are densely packed groups of stars thought perhaps to harbor black holes in their midst—were also deleted, and soon the entire list was ravaged when a third objection arose, embarrassingly, from a Science Institute colleague who was clearly worried that he might be robbed of discoveries in science programs that he had hoped to address several years later.

The EROs list of cosmic sources was now decimated. Besides the magnificent star clusters we had targeted, all the galaxies were gone, too, including several beautiful spirals. So were some colorful nebulae, leaving only a handful of candidates. At least, with these objections out of the way, the chairman of our EROs committee would have an easy time gaining final approval to schedule for *Hubble*'s first observations the remaining targets at the next Science Working Group meeting. Also, unlike myself, O'Dell was a longstanding member of the "seventh team," so he knew how to massage his often-temperamental colleagues on the working group. Or so I thought.

The last working-group meeting before launch was held at the Goddard Space Center in January 1990. Given that the launch was to become a reality within a few months, this was a higher-level (and a higher-pitched) gathering than most. To say that *Hubble* project managers were becoming increasingly fretful about the vehicle's expected performance on orbit is an understatement. Some officials again spoke glibly about their lives being a whole lot easier if the spacecraft "ended up in the ocean" or reached orbit "dead on arrival."

After a multitude of weighty preflight issues had been reviewed at the meeting, the time came for the chairman of the EROs committee to present what was left of the early-release targets for final approval. O'Dell and I had caucused for an hour or so in an adjacent room before he was to address the working group, and despite the list having been gored of many of its most spectacular astronomical objects, we felt satisfied that we had done a reasonable job of navigating this program through a minefield of policies and personalities. I remember thinking what a valuable experience it would be for me to watch a veteran astropolitician address such a sensitive subject.

O'Dell stepped into the well—NASA working groups usually meet at tables in the form of a U, with the speaker in the midst of the well, or U—and within three minutes he was taking crossfire. The original seventh-team objector was now protesting the entire EROs program, and this time he was aided and abetted by a senior manager from NASA Headquarters who was attending his first Science Working Group meeting in years. For these two guys, EROs had become a four-letter word. From opposite sides of the room, they let loose in a seemingly coordinated manner with a stream of strongly negative statements about using valuable *Hubble* time simply to engage the public, which they characterized in a way that would have outraged any taxpaying citizen. ("The public is illiterate about science anyway." "It's none of their business." "Do you mean to say that informing the public is an important element of this project?")

O'Dell was hardly able to get a word in edgewise and was victimized in much the same way that I had been when I broached the subject the previous summer. Much heat and little light followed, as this working-group meeting quickly entered once again one of its absurd nonlinear modes whereby civility and rationality were summarily abandoned. O'Dell gamely attempted to make his case for at least a token set of early images, but within another few minutes, rattled and exasperated, he slammed his papers and transparencies down on the overhead projector hard enough to crack its glass and stomped out of the room.

Most people present were stunned. The six principal investigators, who had originally agreed to support the EROs program, seemed emotionally drained—unquestionably intimidated by the involvement of the powerful NASA Headquarters official who controlled their grant money—and only one or two of them said anything useful. The director of the Science Institute tried to salvage some part of the EROs program but he only confused the issue with histrionics, and I too was of little help in stating meekly that such early observations would be scheduled only during orbital intervals—"dead time"—when *Hubble* was otherwise idle.

It now seemed clear that there would be no early *Hubble* pictures released to the public, perhaps none at all until after the commissioning period had ended. The idea of sharing with the world a few of *Hubble's* initial views of the cosmos seemed dead, and too many other legitimate preflight concerns meant that this hot issue would not likely be resurrected before launch. As I left the meeting to head back to Johns Hopkins to teach a course, I passed O'Dell outside the meeting room, now alone and clearly feeling betrayed by the ambush, swearing uncharacteristically at the floor.

The popular image of the scientist as a breed apart, a cold, calculating person incapable of communicating in layman's language, content to remain aloof and preoccupied with research wherever it may lead, is only true in the minority. But it is this vocal minority that often prevails in university life, as in the *Hubble* project.

NASA's apathy toward the public might well have been predicted. It had already been illustrated by another prelaunch episode—a request from the Science Institute to alter the schedule or "time line" for *Hubble's* first-light observation—the very first image that would be taken with Space Telescope. First light, as noted, had initially been planned for four days after launch. It was to be a brief exposure, taken for a single second by the wide-field and planetary camera largely for engineering purposes—an event that NASA had repeatedly warned us not to discuss with the press. But first light would be truly historic, for if Space Telescope actually proved to be the greatest leap forward in astronomy since Galileo's pioneering work, then its first image might be expected to adorn textbooks for a long time to come. Had we a copy of Galileo's very first observation it might well be still displayed in our textbooks today.

To make the observation slightly less boring than planned, I conspired with Jim Westphal, the camera's principal investigator, to formally request a change in *Hubble's* observing calendar. The original plan was to expose for one second the light from a star cluster called NGC 3532 through a single filter ("monochromatically" green, or essentially in black and white), but we knew that this stellar family is reasonably colorful; some of its stars are red and orange, along with stars of the usual yellow and white. An additional observation for a total of 1.3 more seconds through two different filters—red and blue—would enable the star cluster's true color information to be captured. Admittedly, the inconvenience (or "hit to the time line," in project vernacular) would have amounted to about ten minutes in *Hubble's* onboard calendar for that day, but Westphal and I had argued in writing to NASA that the payoff—a first-light picture in color—was well worth this small investment required to execute it. Our request was summarily rejected by NASA: "The current form and implementation of proposal 1476 [first light] is frozen."

A second episode that should have telegraphed NASA's disdain for the public occurred at the Science Institute that same January. In conjunction with Johns Hopkins University, the institute's educational and public affairs group hosted a Science Writer Workshop that attracted scores of leading educators and journalists from around the globe. Our intention was to brief them on the upcoming mission, and thereby elevate the level and accuracy of science reporting and precollege teaching. We also arranged for NASA's cable television network, NASA-Select, to videotape the workshop for rebroadcast to precollege teachers as well as to hundreds of journalists who normally tune in but could not attend in person.

The speakers were drawn from NASA and the academic community, and most of them used a common method of delivery among scientists who are generally unaccustomed to talking shop with nonscientists; they projected hand-drawn and hastily prepared transparencies overhead and spoke technically as though they were addressing their colleagues. The result was that the audience understood little of what was said; most of the educators were too polite to register a complaint, but many of the attending journalists made their unhappiness abundantly clear to me. By contrast, the four institute members who spoke reviewed the *Hubble* mission pedagogically and worked closely with an artist to have illustrations drawn especially clearly; our lectures were also carefully planned and coordinated beforehand, so that together they formed a comprehensive, educational package.

The response from the attendees to the institute presentations was flattering, not only because we took the opportunity to correct several misconceptions about the capabilities of Space Telescope— including some exaggerations that NASA had regularly made to win public support for the project— but also because we highlighted when and how it might be most appropriate for the media to cover *Hubble*'s early activities, something that NASA seemed bent on trying to hide, lest some miscue be revealed. When the NASA-Select cable network rebroadcast our Science Writer Workshop a few days later, the network was ordered by officials at NASA Headquarters to censor the four educationally oriented *Hubble* presentations. Only the technical briefings, largely useless in the public domain, were transmitted. No doubt, NASA feared that it was losing control of public relations to some people on the Science Working Group and at the Science Institute. Most of all, by our stressing education and its consequent fundamentals, the agency was torqued that we were exposing a good deal of the prelaunch hype swirling around the project.

Despite all its rhetoric about education, NASA had built history's most expensive and visible science instrument without having any associated educational program. No part of the space agency—neither NASA Headquarters nor its relevant field centers at Goddard or Marshall—had any plans to share the fruits of Space Telescope with the nation's youngsters. What's more, virtually every precollege educational initiative mounted at the Science Institute was accomplished *despite* NASA. The space agency regularly and consistently objected to the institute's teacher workshops, media workshops, teacher kits, classroom posters, and educational telecasts, among a whole host of associated activities and products designed to help teachers, students, and the public understand better the subject of space science in general and the *Hubble* mission in particular.

To take just one example, television programming illustrates an especially vexing area in which we ran into major difficulties with NASA—another genuine "hot button" in the agency's lexicon. I happen to believe that we live, for better or worse, in a video society. Rather than cursing video and how it might be corrupting our kids, my attitude has been to use video constructively in classroom settings, thus manipulating this medium to our advantage and thereby giving teachers novel ways to enthuse children about science and learning in general.

We had earlier established at the Science Institute an Astronomy Visualization Laboratory where with sophisticated computers—mostly Silicon Graphics workstations having lots of "bells and whistles"—we were able to generate broadcast quality animation of, for example, *Hubble* in orbit, its onboard instruments, and eventually its scientific findings. Well before launch, I had proposed incorporating our lab's video creations into a weekly television program as part of the NASA-Select cable network—a program that would bring the latest results of the *Hubble* mission into the classrooms of America on a regular basis. This cable network is freely broadcast via satellite across the nation to anyone having an inexpensive backyard dish antenna, but for lack of good ideas and not for lack of money, the network was idle most of the time.

Normally, NASA-Select televises shuttle flights that usually show capsule communicators on the ground staring for hours at banks of computer consoles, and occasionally a press conference featuring a lineup of bobbing heads behind a plastic dais talking in such monotone boredom that no kid would ever conclude that science is fun. Yet the space agency was having no part of our initiative that its cable channel become more educationally oriented. NASA had become so comfortably governmental that it was frightened of new and creative initiatives, even when keeping with President Bush's avowed mandate to improve our nation's schools. At the heart of the issue was the bureaucrat's greatest fear—loss of control.

Accordingly, as with our teacher workshops and educational materials that were ultimately promoted under the aegis of Johns Hopkins, I arranged for a "third party" to sponsor our television efforts. Since my education group at the institute had become involved with the Maryland Governor's Academy for Science and Math, I developed a relationship with Maryland Intec, the educational arm of PBS's Maryland Public Television. In no time at all, we were working alongside warm and knowledgeable professionals, producing a weekly television series, called *Starfinder,* geared for middle and junior high schools—the instructional level where most educators judge the need for improvement is currently greatest.

Each program was divided into three segments: one presented a new and exciting science result, usually from the *Hubble* mission, a second segment drove home a single science concept, such as magnetism or gravity, and in a third segment a person "behind the scenes" of the *Hubble* project was profiled as a role model who might enthuse children to consider careers in science and technology. Throughout prelaunch, launch, and commissioning of *Hubble,* our TV team managed to script, film, edit, and produce some thirty-two weekly programs that specifically matched an entire yearly school curriculum. And gratifyingly, before long, *Starfinder* was being used in thirty-seven states and most provinces of Canada, carried throughout the nation on The Learning Channel and The Discovery Channel cable networks, and transmitted via

Congress is to fund our programs." Today's technomanagers (and some scientists, too) just don't understand that honest arguments articulated well and with enthusiasm are often more effective than hype or the hard sell.

Here's a glaring example of one such mistake. NASA repeatedly stated as part of its prelaunch publicity that *Hubble* should be able to see seven times farther than any other telescope. This is patently false, if only because our ground-based telescopes already perceive objects, albeit dimly, close to the limits of the observable Universe. Since the Universe is reasoned by most astronomers to be about 15 billion years old, and since ground-based telescopes can detect celestial objects out around 10, 12, maybe 14 billion light-years, to claim that *Hubble* can see seven times farther implies that it would be able to probe to distances of almost 100 billion light-years. Yet if the Universe is 15 billion years old, nothing can be farther away from us than 15 billion light-years. This kind of NASA propaganda gives rise to all sorts of serious misconceptions among educators and the media, who begin thinking, teaching, and reporting that *Hubble* is designed to "see beyond the edge of the Universe," to "study the 'Big Bang' directly," to "explore creation itself," and even to "probe another Universe!"

Of course, people always ask if there was any official, written response to our principled campaign for accuracy. In truth, I received only three messages on this topic from space-agency managers as the debate reached a climax during the pressured months leading up to launch, and one of these does grant insight regarding NASA's attitudes. The chief of Goddard's Public Affairs Office (and a former school teacher) wrote me that, "Accuracy of information is a highly valued quality not commonly shared by teachers. . . . Teachers believe that life experiences that stimulate an interest are highly prized over the 'accuracy' question." Therein, perhaps, lies the disconnect—a typical response from an educational administrator, and a prime reason why today's education schools are part of the problem, not part of the solution, regarding science literacy in our nation.

Another message was a reactionary response, from a technical manager at Goddard, ordering me to write a letter to the editor of the *Baltimore Sun* to correct a minor misprint about NASA made by one of that paper's reporters—all the while ignoring my plea for honesty in NASA's statements. At least these were more substantive replies than the response from the chief of Marshall's Public Affairs Office, who sent us three bags of 2,200 lapel buttons praising "NASA's Incredible Time Machine."

All this hair-trigger tension pervading project sociology came to a head when, a couple of months preceding launch, the news media forced the issues of hype and accuracy into the public domain. A leading science journalist, Ron Cowen, who had attended either the institute's Science Writer Workshop or one of our teacher workshops, picked up on some of the incorrect assertions that NASA had trotted out to peddle Space Telescope. Shortly before launch, Cowen authored an insightful article in the superb weekly magazine, *Science News,* attempting to clarify much of NASA's extravagant claims for the *Hubble* mission. In the process and specifically regarding the blatant exaggeration of *Hubble* seeing much farther than other telescopes, he made it sound as though my clarifications of NASA's boasting meant that the space agency "was overselling" the *Hubble* project—something I had never (until now) said aloud, but do feel is true. After all, among other assertions for *Hubble* during the past decade, two successive NASA administrators, James Beggs and James Fletcher, respectively, had testified before Congress, to wit, "The Space Telescope will be the eighth wonder of the world," and with Space Telescope "we're talking about the salvation of the world." This, to me, is pure, unadulterated hype—overselling of the most egregious kind.

At any rate, Cowen's article did not initially bother too many people at NASA since few politicians read this or any other science magazine. However, since we were within a month or so of launch, many other media organizations were writing and airing major pieces about *Hubble,* not least the *New York Times.* And when the *Times* printed a marvelously comprehensive piece in its Sunday supplement, written by John Nobel Wilford, its arch-rival the *Washington Post* was caught unprepared with nothing to counter. So the *Post* reran the *Science News* article, and now the implication that NASA had oversold the *Hubble* project was in black and white for all the Washingtonians to read. The reaction from NASA management was not nice.

Goetz Oertel, the president of AURA, the parent consortium of American universities that manages the Science Institute, was summoned to NASA Headquarters, where high officials bullied him that they were tired of hearing about Space Telescope from his troops. They didn't care whether there were scientific errors in NASA's press handouts or gross exaggerations of the telescope's capabilities in its educational materials. The government men wanted no more statements of any kind that appeared critical. They especially wanted me fired, partly for saying aloud that the telescope "was built with the NASA philosophy that big is beautiful," along with another senior institute scientist who had been quoted in the *Post* article saying that, "We're launching a telescope with 1970s technology." For the record, both quotes are indeed true given the attitude and time required to design, develop, and test the huge and complex vehicle, including the years in storage during the post-*Challenger* delay.

Oertel, a most amiable and professional gentleman—and the most politically astute science manager I know—thereafter drove directly to the Science Institute where he first met with the institute director and then with me late into the night. Both Giacconi and Oertel were strongly opposed to stopping anyone from trying to elevate the degree of science understanding among the public, and they themselves flirted with the notion of resigning to protest NASA's heavy handedness—a sentiment that I found surprising at the time, but in retrospect not terribly so, given the degree to which both were fed up with the machinations behind the scenes.

The issue was clear-cut: Aside from intimidations, firings, and resignations, should we kowtow to the agency's demands and thereby enhance the hype, knowing full well that NASA was doing a disservice to the enterprise of science? Or should we hold our ground by continuing to push for technical accuracy, assuming that this check and balance was what the National Academy had in mind when recommending an institute separate from the space agency? At the suggestion of Oertel, I contacted three eminent astronomers, explaining our predicament and asking their advice. All three disappointed me with their responses, for if they held the same attitudes toward factual information in their research endeavors, they would have been discredited long ago.

One told me in a phone call that our position was technically sound, but that we should keep a low profile during the media coverage of Space Telescope's launch and deployment—"NASA is so paranoid about the telescope's potential failure, it's best to praise it at all costs." A second astronomer, visiting the institute for a committee meeting, came by my office to tell me that he applauded our "campaign for accuracy," but that he had never seen NASA so upset at the institute. Then he lectured me bluntly: "You have a lose-lose position; you are a scientist seeking truth, yet operating within a public-affairs system that regularly distorts it." When he quoted me his version of the Golden Rule—"He who has the gold rules"—I realized that he was held hostage to his NASA grant. And a third astronomer brushed off the issue, advising me in writing that science popularizations for the general public had always seemed to him laden with incorrect statements, a little like what they call in French "vulgarization": "I can well understand your chagrin concerning the various scientific errors you have found in this material, especially in view of your own efforts to correct such errors. My own experience is that misconceptions and misstatements frequently abound in popular scientific writings prepared by nonscientists." Fearing dismissal, I felt I had no recourse but to agree reluctantly not to campaign publicly against NASA's intellectual dishonesty. Nor, however, would I shirk from correcting the agency's errors when addressing educators.

If anything, the hype spiraled upward and out of control as the launch date approached. Reporters seemed intent on one-upping each other by making claims that no one at NASA was willing to correct. To cite but one example, I cringed when, while doing an interview for the *Today Show* in the frenzied weeks before launch, NBC's chief science correspondent went beyond even the "seven-times-farther" nonsense: "If it works, the Hubble Space Telescope will revolutionize astronomy, allowing us to see ten times farther into space than ever before." But what really got my goat was a subsequent remark by a Goddard public-affairs official who, knowing that the report was false, took pleasure in seeing me bite my tongue, lest I be fired: "Looks like my boss got the best of your boss on this issue," she said, smirking.

About a month prior to launch and with NASA still terrorizing the institute's burgeoning educational efforts, I received welcome backing from a powerful political ally. The senator who chairs the subcommittee on NASA appropriations was asked to participate in a press conference at Goddard—the traditional "L-30" media briefing. As Senator Barbara Mikulski entered the visitor center at Goddard, she immediately bypassed the NASA dignitaries assembled to greet her and instead went out of her way to embrace my director, Riccardo Giacconi. Without missing a beat, she then turned to the red-faced director of Goddard and asked what educational products and programs he had mounted to celebrate the launch of Space Telescope, at which point he assumed a blank stare.

And she didn't let up. During the press conference, in answer to a reporter's question about the social payoff likely to come from the *Hubble* project, Mikulski embarked on a lengthy soliloquy on the need for better math and science education and how *Hubble* should "play a major role in the education of young kids, from grade school to grad school." The senator ended rather eloquently I thought, by paraphrasing the late scientist-philosopher, Teilhard de Chardin, to the effect that, "By studying the Universe we are also learning more about the phenomenon of man." Naturally, while I was beaming, NASA officials were fuming. Even today, the space agency thinks that we arranged this lecture-like performance by a powerful senator, and maybe we did. The outcome was that it saved my skin, and more importantly it gave a genuine boost to our precollege educational programs within a governmental system that was otherwise decidedly hostile to them.

The saga of the early-release observations was far from over. The next episode occurred at the Kennedy Space Center on the day prior to launch, during the traditional "L-1" media briefing. Lennard Fisk, NASA's associate administrator for space science and applications—effectively its chief scientist—was to lead a nationally televised press conference designed to make clear some of the most profound intellectual issues that *Hubble* was about to attack. Although I had cautioned many project officials prior to the conference that the media had on their minds only two matters—*Hubble*'s cost and its early pictures— the word apparently did not filter up the ranks to Fisk. The result was a disaster of sorts—one from which the project has not yet fully recovered.

At various times leading up to launch, NASA had announced widely differing figures for the total cost of the Space Telescope project, including operations in orbit. In the end, the media had to badger Fisk into giving some of the relevant costs, for rather than clearly breaking down the bottom line of this big-science program, he seemed intent on lecturing the reporters that the *Hubble* mission is really "a small science project, because it's little observers who are the users." The truth is that approximately $1.54 billion was spent between 1977 and 1986 to design and build *Hubble*—a cost overrun of nearly triple.

The bulk of this huge R&D expenditure owes to several factors of roughly equal weight: an optical system of unprecedented intricacy, a ground system of unprecedented complexity, and a modular design of unprecedented boldness whereby some eighty devices can be repaired and replaced during shuttle servicing missions. This amount includes neither an additional $0.3 billion (that is, nearly $7 million per month) for storage and upgrade from 1986 to 1990, nor another $0.3 billion for an exhaustive series of "end-to-end" tests on the ground-computer system and the science instruments aboard the vehicle while in storage (but not on the telescope's optics). This subtotal of $2.1 billion furthermore includes neither another $0.3 billion contributed by the European Space Agency, nor an annual budget of about $250 million needed to fund operational aspects of the Space Telescope project—all of which totals close to a projected "run-out cost" of well more than $6 billion for the expected fifteen-year mission, inflation not included.

will take a picture which is part of your engineering test and then we will process that picture as we will all of the *Hubble* data in time and produce an image which can be distributed. So, the taking of the picture will occur fairly quickly, ah, the processing will, ah, you know, can take on the order of weeks in order to, ah. You know this is the first picture through the system here to produce the, ah, the image that can be distributed. I also think that, you know, if you, and I don't know how either you or your readers are, are, ah, experts in looking at astronomical pictures and saying, 'Boy that's a good picture, that's a bad picture.' Ah, I think one of the things we would like to do for you is take a picture of an object for which there are objects for which there is a ground-based plate available where this object has been seen by, say, Mount Palomar telescope. And then we will put them side by side for you, ah, the picture that is available from the ground and the picture that is available from *Hubble,* and so you can see the kind of resolution and clarity that *Hubble* achieves that we've never had before, and so that will take a certain amount of time to put this together. We're going to look for the first pictures in engineering in a few days and look for a picture to distribute, ah, to share the capability of *Hubble* as we've gone through this process."

KATHY SAWYER, senior science reporter for the *Washington Post:* "Well, John took my question. But I'll follow up on that."

FISK: "Only one question on early images."

SAWYER: "I wanted to know about the, ah, the debate over the strategy for releasing the first picture. I, I gather there's some disagreement whether it should be, ah, whether the public has the right to whatever is shown, whatever falls on the lens of the *Hubble,* ah, immediately or, ah, whether it can be, ah, negotiated or what. What is, what is the strategy and the, the legal aspect of this discussion?"

FISK: "Legal aspects! This is getting . . . [garbled]. Somehow, ah, I'm, ah, I'm, ah, I'm intrigued by the complexity of this issue. Because let's, ah, let's just think about this for a minute. I mean the, the, what we'd like to do is, is demonstrate to everyone who's interested, and there's a wide range of people that are interested, the public, the Congress, ah, ourselves, that *Hubble* is as capable as, ah, as, ah, we have, ah, advertised that it's going to be. So, ah, you know, I think, ah, we have an obligation to, as soon as this, this is available in terms of its, ah, having been taken and having been processed to put out a picture that, you know, shows that this is really going to be the fantastic device that we expect it to be. But, now, after that process, and that comes out, as we just, as I just talked to John about, ah, in, ah, over, ah, you know, ah, ah, being taken in a few days and distributed as soon as it's processed in a few weeks. Ah, now, the main thing we need to do with *Hubble* though, you know, in the next few months is spacecraft checkout and then science instrument checkout. Ah, after all this is probably the most complicated spacecraft that we've ever flown, ah, and, ah, we need to make sure that we are, we know how to use it correctly on orbit, that everything works just as we expect it to, and then we need to, ah, ah, begin to set ourselves up for the observing sequences that will occur. You know, one of the great challenges with any observatory whether it's ground-based or now space-based is to be efficient in the, ah, in the use of it. And I'm sure we've got some interesting lessons to learn in terms of efficiency, of developing a more efficient use of this thing as we go through it and all that will take place over time. And, ah, ah, so I think, ah, I guess my strategy is that I'm going to demonstrate that this is the wonder that we think it is, and then I'm going to let the people that need to learn to use this thing do what they need to do to make sure the spacecraft is safe and, ah, we know how to use it efficiently so that we get the maximum science out of it over time."

DICK RATNER, science reporter for *ABC News:* "I'm not sure you answered Kathy's question about who is going to decide what picture is going to be released and under what circumstances. Would you be a little more specific please?"

FISK: "Well, the processed picture, you know, *the* picture that is going to, ah, the, ah, of, ah, what object is it? Is that your concern? Ah, I'm not sure I know what object it is. But, ah, ah, it's ah, the ah, the picture that I think we need, ah, to put out will be released by the administrator of NASA. This is, ah, that's how it should be. This is, this is, ah, ah, you know we will, when we have this processed picture that I think will give you the comparison that you need, which is a comparison against a ground-based set of observations so you can see the capability of *Hubble,* then, ah, we'll call a press conference, stand Admiral Truly up and say: Release this picture!"

MORTON DEAN, science correspondent for *ABC News:* "To follow up on Kathy's question that was stolen by John, I, ah, I gather what you're saying is that we will not see the first picture you people see but will just see the earliest picture that you people feel will present the telescope in its best light. And, secondly, with all the planning, I have trouble believing that you don't know what the first picture is that the telescope will see."

FISK: "Well, let's put it this way. Someone's undoubtedly told me. What you're asking me is can I remember what it was. Ah, the ah, if it's, ah, I mean it wouldn't surprise me that, we can ah, you know that, ah, you know that people will be around to, ah, ah, you know while we're operating *Hubble* at both, ah, Goddard and at the institute, and of course the picture will come in there. But I think it's, ah, it's ah, I think you're overestimating what it is you're going to see when you have seen a raw picture from a spacecraft. And, ah, I mean again, you know, think of the *Voyager* analogy. Even though *Voyager* was a well-proven spacecraft, in which we had lots of history in being able to turn the little bits. And remember this spacecraft doesn't send down pictures; it sends down signals which are reconstructed into pictures. And, ah, in the case of, ah, even of, of, *Voyager,* although we'd gotten quite practiced with this thing, we were able to do in a matter of days at the Neptune encounter, ah, it takes a certain amount of time to turn, ah, to turn an image from a spacecraft into an image that you can look at and say, you know, this is, all the calibration is right and if I've got all the colors right on this thing, all the things that you need to do. And so, that's the process that we will go through to put this into, ah, a picture that I think, you know, people should look at and, ah, and judge the mission on. Ah, now, if somebody is really anxious to see the raw data as it comes down, if that's the issue, I'm sure we can probably find a way to show it. But, ah, it's, but I think you, you know, it's, it's, it is the issue that you face always is that, you know, if you look at, ah, at ah, at ah, picture that comes down, you'd like to know was that a good picture or was that a bad picture. I can't judge that. I mean, I would, ah, I'd like to have a comparison to know whether, you know, what it, what it was that we were supposed to see, that's the comparative ground-based telescope's, make sure the calibration is correct on the thing, and that's, that's the process we'll go through over the next couple of, ah, period of weeks."

LEE DYE, senior science writer of the *Los Angeles Times:* "Len, I'm going to really belabor a point. Ah, during the *Voyager* encounter, one of the things that made it so exciting for the American people is that they *did* see some of those images in real time. Some of them didn't work, some of them did. But at least they got a chance to look at them to see what they were paying for. Could I ask you specifically: Will you let us see some of these earliest images in real time, even if they aren't what you want?"

FISK: "Yes. But I, I expect, I expect, ah, ah, you to give it an interpretation that would make an astronomer proud."

MARK KRAMER, science correspondent for *CBS News:* "Dr. Fisk, one more time."

FISK: "I tell you what, this time I'll give no for the answer."

KRAMER: "Is there concern at NASA that this mission which has been pumped up to extraordinary proportions and, and probably justifiably, this is an extraordinary instrument that you're going to launch. But is there some concern that this initial data will not meet those high expectations because science normally takes time to, ah, to, ah, digest the information and, ah, determine what's been learned. Is there concern that this, ah, snapshot will be unfairly judged and the budget folks in the Congress and the American people will be disappointed?"

FISK: "No, no, absolutely not. Look. I, I mean, ah, as I say, I'm, ah, I'm really intrigued by the, ah, by the complexity of this thing. Ah, I have absolutely no concern about that whatsoever. I mean, we will, I mean, this *Hubble* is going to work, it's going, it's going to be the, ah, breakthrough in astronomy that we expect it to be. We were simply looking to, ah, to have a way in which this, ah, you know, proceeded in an orderly, orderly fashion. If someone wants to read an enormous amount into that, that's, ah, that's, ah, unwise, and so if you want to see the bits, come see the bits."

Suspecting that this was the famed and dreaded "media ambush," later that day I asked Kathy Sawyer of the *Post* if her colleagues had organized this wrenching of the NASA man, at which point she answered, heatedly, "No, we

hadn't, but the next time we're going to. We're really tired of the evasions."
Needless to say, neither NASA's astronomy head nor the influential seventh-
team member who had conspired to thwart our carefully planned program of
early picture releases for the public were present while a bloodied and sullen
Len Fisk twirled in the wind. All I could think of was Emerson's dictum:
"Nothing great was ever achieved without enthusiasm."

Viewing the debacle with several members of the Science Working Group
on a television monitor outside the media center at Kennedy, I had the sinking
sensation that the seas were rising. It was clear in "real time," while the press
conference was under way, that this was the public-affairs equivalent of a single-point failure—such a classic example of how not to give a press conference
that we later used a videotape of it as part of media training sessions at the Science Institute. We were astounded that the NASA men were being jerked
around so badly regarding a subject for which we had tried so long and hard
to plan. A leading member of the wide-field camera team captured most of
our feelings with a Yogiism: "I've never seen anyone muddy the waters so
clearly."

The media fallout was nasty indeed. Fisk had been given the chance to
reverse past animosities right then and there by embracing the EROs program
(or inventing some other one like it) while stating proudly that NASA had had
the foresight to make explicit plans to share *Hubble*'s findings with the public
at large. He could have made some magnanimous gesture about the desire on
NASA's part to contribute mightily to the cultural advances of our civilization
by granting our fellow citizens a sense of excitement generated by our nation's
premier space mission. Instead, after much browbeating, he was prodded into
promising the media that they would be allowed to witness live, and even to
televise, the first-light image from *Hubble*. Referring grudgingly and sarcasti-
cally to the digitized pieces of information that form an image, Fisk's glib, "If
you want to see the bits, come see the bits," left a lasting negative impression
on virtually all who witnessed the agency's near intransigence that day.

None of this lack of attention to detail was ameliorated when the Marshall Space Center followed up the press conference rout at Kennedy by hus-
tling out a formal press release describing *Hubble*'s first "astrological" target.
Some say it was a faux pas, but I think not. The damaging release was symptomatic of NASA's inept public-affairs operation, caused an increasingly skeptical media to lose faith in NASA's ability to describe even the most rudimen-
tary scientific aspects of the mission, and served to alienate further the Science
Institute astronomers who found this error to be egregiously offensive.

3

Hubble's First Light

Well, the passage of time has revealed to everyone the truths that I previously set forth; and, together with the truth of the facts, there has come to light the great difference in attitude between those who simply and dispassionately refused to admit the discoveries to be true, and those who combined with their incredulity some reckless passion of their own. Men who were well grounded in astronomical and physical sciences were persuaded as soon as they received my first message. There were others who denied them or remained in doubt only because of their novel and unexpected character, and because they had not yet had the opportunity to see for themselves. These men have by degrees come to be satisfied. But some, besides allegiance to their original error, possess I know not what fanciful interest in remaining hostile not so much toward the things in question as toward their discoverer. No longer being able to deny them, these men now take refuge in obstinate silence, but being more than ever exasperated by that which has pacified and quieted other men, they divert their thoughts to other fancies.

Letter to the Grand Duchess Christina
Galileo Galilei, Florence, 1615

When it finally arrived, the day of first light—Sunday, May 20, a few days short of Space Telescope's four-week anniversary on orbit—was a memorable one. At one and the same time, we were eager to get on with it yet genuinely worried about the outcome. We knew we were a part of history in the making, but we were unsure how good or bad the first astronomical observation would actually be. Significant pointing and focusing problems (described in the next chapter) had repeatedly caused postponement of first light—in all, nearly a month-long delay that in turn increased the media's suspicions and grumpiness. NASA had been stingy and often evasive with explanations for the delay, and the institute had been reordered by the agency to say nothing to the press about the growing difficulties in handling the bird. The resulting news vacuum meant that the media set the agenda. Amid numerous complaints from journalists, the public, and Capitol Hill, it was during this period that I first heard the wisecrack about the "real" meaning of the NASA acronym—Never A Straight Answer—a sentiment that would surely amplify in the coming months. The upshot was increasingly skeptical press coverage, which might have been avoided had NASA leveled with the media, involved it in the ongoing adventure, however troubled, and explained clearly the heroic human efforts then under way to tame the resistant robot.

Early that morning, I scanned the previous evening's summary reports from the orbital verification team and learned of an ambiguous analysis implying that the tweaking of *Hubble*'s optics had perhaps improved its focus during the past twenty-four hours, but that it was still likely to produce a first-light image worse than one obtainable from the ground. Large ground-based telescopes at the best sites—such as at Mauna Kea or atop the Chilean Andes—routinely yield pictures that have about 1-arc-second resolution, and occasionally twice as good (that is, about 0.5 arc second) when Earth's atmosphere is exceptionally still. This was to be compared with an expected 1–2 arc-second seeing for *Hubble,* which the fine-guidance sensors had implied during much of the previous day. Even worse, a minority of the optical experts at Goddard felt that the telescope might be way out of focus, perhaps as bad as 4-arc-second resolution; stars so seen would appear as "doughnuts," much as points of light are degraded into rings or halos when shown through a greatly unfocused slide projector.

Literally, first light had already occurred, but no pictures had been taken. As noted, to point Space Telescope with precision, a pair of guide stars need to be acquired in two of the three fine-guidance sensors. *Hubble*'s first attempt to lock onto guide stars had used the light of an anonymous star in the constellation Lyra on May 4. And although the exercise was only partly successful, one of Space Telescope's instruments had technically seen its first light from a specific cosmic target. In fact, several other tests with all three of the fine-guidance sensors had by this time collected jillions of photons (quantum bundles of light), but official first light is traditionally reserved for the first image recorded with a new telescope—if only because, for the general public and professional astronomer alike, no other kind of science data comes close to the majesty of a real picture. Thus, the honors would go to *Hubble*'s workhorse science instrument, the wide-field and planetary camera.

The wide-field and planetary camera—the acronym for which is WF/PC, pronounced "wiffpick"—is actually eight cameras in one. Four of them operate simultaneously in the wide-field camera mode, the other four in the planetary camera mode—hence the slang term for the camera in either mode: "four shooter." Both modes are designed to be sensitive to radiation ranging from 1,150 angstroms, deep in the ultraviolet, to 11,000 angstroms, in the near-infrared part of the electromagnetic spectrum.

At visible wavelengths longer than about 4,500 angstroms, each mode is more sensitive and faster than *Hubble*'s other imaging device, the faint-object camera, although here the WF/PC regularly suffers from a case of the "measles"—ugly blotches appearing on the raw images presumably because of ice condensing on the camera's window. "If the public ever saw a raw tele-

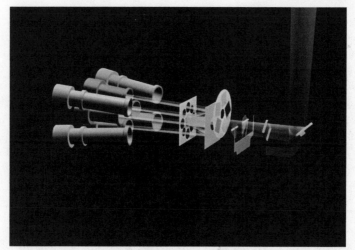

An artistic rendition of the heart of the "radially mounted" wide-field and planetary camera. Light entering *Hubble* is "picked off" (at right) by a diagonally oriented flat mirror that reflects the light through a color (filter) wheel and onto one of four faces of a pyramid-shaped mirror (at center) and eventually onto a sensitive charge-coupled detector (CCD, at left) array of 800 x 800 pixels. Operationally, this is done for all four pyramid faces simultaneously—thus the name "four shooter"—thereby yielding an image four times as large. Light not intercepted by the diagonal mirror moves onward to the other "axially mounted" science instruments. *[D. Berry]*

A note on CCD technology: A charge-coupled device is a microelectronic chip made of thin silicon wafers not unlike those in the electro-optical system of a television set or a video camera (though today's TVs in our homes have only about 500 pixels on each side of the image, somewhat smaller than *Hubble*'s WF/PC). The chips comprise numerous light-sensitive pixels, which emit electrical signals in proportion to the amount and pattern of light striking them during an exposure; each pixel has its own memory, collecting and storing electrons released when light hits the silicon. You can think of a CCD device as a forest full of millions of buckets each collecting raindrops, and for a brief shower the amount of rain collected in the array of buckets would mimic the geometrical shape of the rain cloud overhead.

In the case of Space Telescope, digital signals for each pixel are transmitted to the TDRSS network, which relays them to ground stations on Earth. The digital signals are then reconstituted into a cosmic image. Normally, CCDs detect radiation over a broad spectral range—from 4,000 to 11,000 angstroms—but the element silicon does not permit short-wavelength radiation to penetrate sufficiently to release electrons and therefore to act as an efficient ultraviolet detector. To overcome this, the CCDs onboard *Hubble* were coated with a special organic phosphor, trade-named Coronene, that converts photons of ultraviolet radiation into photons of visible light, which the miniature silicon sensors can detect. (This is not thought to be the source of the aforementioned contaminant plaguing the camera.)

To increase their sensitivity further, *Hubble*'s CCDs are electrically cooled (essentially by a refrigerator powered, like all else on the spacecraft, by the solar arrays) to −90°C (or −130°F), thus diminishing the background "noise" and allowing observations of very faint objects. The WF/PC is designed to photograph an object of the tenth magnitude in a second, the fifteenth magnitude in a minute, and the twenty-eighth magnitude in an hour. CCDs, which were pioneered by the military-intelligence community for reconnaissance purposes, are much more efficient than photographic film (70 percent versus 7 percent of all photons falling on the detector are captured), and are superior to any kind of visual detector used in the civilian world. Large ground-based telescopes equipped with CCD arrays can image celestial objects of the twenty-seventh magnitude, an improvement of two or three magnitudes (or roughly a factor of ten) over telescopes using lowly efficient photographic film; CCD-equipped telescopes are fully a hundred million times more sensitive than the human eye.

Here, then, is how a picture or "digitized image" is transmitted piece by piece from space and reconstructed on Earth: Each of the millions of pixels comprising an image is given a number in the computer memory both in the spacecraft and on the ground. In addition, a different number is assigned to many intensities of light—black and white at the extremes and hundreds of shades of gray in between. When *Hubble* views a celestial object and acquires its image, the onboard computer assigns an intensity number (the shade of gray) to each pixel (its numerical position in the image) and transmits these numbers to the Science Institute. There, computers display the intensity of each pixel on a two-dimensional grid, much like a design on a tiled bathroom wall or a child's paint-by-numbers game. Several such gray (or "monochromatic") images, taken at different wavelengths, can be combined to make color pictures. (Technically, only a few seconds are needed for signals to travel the route *Hubble*—TDRSS—White Sands—DOMSAT—Goddard—Institute, but another minute or so is required to read out an entire, 1600 x 1600 pixel array of the wide-field camera.)

The wide-field and planetary camera has an array of forty-eight filters for capturing light of different wavelength ranges that can be rotated into the light beam's path. The filter wheel also contains gratings to allow for slitless spectroscopy and polarizers for polarimetry. The latter device allows studies of the polarization of light—that is, the extent to which light waves have an aligned orientation, as do those passing through a Polaroid lens, for example. This can tell us useful information about the celestial source of the light or about the cosmic matter through which the light travels on its way toward Earth.

Both the wide-field mode and the planetary mode of the WF/PC can make use of a tiny "Baum spot" painted on the optical pyramid. This small round disk is only 1.2 arc seconds in diameter and is named after Lowell Observatory's Bill Baum, a planetary astronomer who suggested it be made a permanent part of the camera. The spot has been given low reflectance—it is essentially black—to suppress the light from a bright cosmic object in order to examine better the object's surroundings. To take advantage of this occulting device, *Hubble* needs to be positioned precisely so that the light of the bright object is focused directly onto the Baum spot. In this way, for example, dim planets can conceivably be sought near bright stars. Owing to both pointing and optics problems throughout *Hubble*'s commissioning period, the Baum spot was never used as designed—although it certainly fooled many people who noticed a little black dot in some of *Hubble*'s images.

Throughout the early commissioning, NASA had insisted on absolute and total control of events—especially regarding release of information to the public. Because of statements made by its chief scientist at the prelaunch press conference, the Goddard Space Center now had to generate a public-relations extravaganza that it had neither intended nor contemplated. Specifically, Goddard was directed to allow reporters to see *Hubble*'s first picture as it came down from space, and television crews to transmit the picture in "real time" into the homes of millions of Americans. Planning for this multimedia event became increasingly absorbing—and uptight—in the weeks leading up to first light, involving dozens of highly skilled technicians who would otherwise have been engaged in more useful tasks than organizational meetings, practice press conferences, and simulations of the proposed TV broadcast.

Magnetic tapes containing test data were repeatedly sent by Goddard to the Science Institute, where we were expected to manipulate the fake data through my *Nuncius* computer, then on to higher-powered workstations, *Paris* and *Cassandra* in our Astronomy Visualization Laboratory, and finally to the institute's photo lab, where hundreds of high-quality glossy photographs would be reproduced for the media. Several times we were ordered bluntly to redo the test because the NASA logo below the practice image was not big enough or was not centered prominently. Those in charge at the space agency seemed to care more about the credit line than about the quality or effective presentation of the results; in fact, we were specifically told not to acknowledge anywhere on the photos either the European Space Agency or the Space Telescope Science Institute.

NASA's emphasis on teamwork—a hallmark of its total-quality-management mumbo jumbo—had become just a bit hypocritical; the agency didn't understand that it gets the credit, and the blame, for virtually everything done in space. NASA simply could not conceive that its contractors might work together as a team alongside NASA, in the process helping to make the space agency look, not bad, but good. Regardless, there was to be no mention whatsoever that the first-light pictures had been processed at the institute. Every single high-resolution glossy produced at the institute had to be accounted for and immediately hand-carried back to Goddard where NASA would give the impression that all of the data processing had been done on-site. In any event, written instructions from NASA's Public Affairs Office made clear a central piece of orchestration: "If [*Hubble*'s] image looks good . . . put ground-based image also on same screen. . . . If not, no ground-based image will be shown."

Some scientists and engineers became disgusted with the importance that first light had assumed, especially given the heavy workload of many engineers at a time when Space Telescope's teething difficulties were, if anything, growing. Ever since launch, some regularly scheduled tests with the telescope had been rearranged or postponed to address the issue of first light, and key personnel were occasionally taken out of the operational trenches and told to rewire some parts of the Space Telescope Operations Control Center, at Goddard, where the television program would originate, or otherwise support numerous dry-runs of what NASA was calling "The First Light Show."

Planning for this event got so disruptive at one point that a group of engineers formally charged that the entire telescope commissioning process was jeopardized simply because NASA had failed to embrace a well-planned program of early snapshots having negligible impact on *Hubble's* observing time line, only to have its chief scientist invent on the spur of the moment a time-line-busting spectacle during the fateful Kennedy press conference. Even I, who had championed the program of early releases to the public, thought that the whole issue of first light had taken on ridiculous status.

Engineers struggling to empower Space Telescope were not the only ones distressed by the attempted media production. NASA's public-affairs personnel were greatly strained, its ranks recently thinned by firings and defections. Shortly before launch, amid arguments as to how to handle Hubble publicity, a highly skilled, veteran chief of public affairs for space science at NASA Headquarters had been dismissed (almost surely for warning in writing that prelaunch hype and postlaunch problems could cause the media to turn ugly), and the principal public-affairs contact for the *Hubble* mission at Goddard had quit the project. Now, as the pressure and confusion of the first-light event built, more Goddard public-affairs employees assigned to aid first-light activities were either fired or walked off the job, making that space center's media operations, already widely considered the least effective among all NASA organizations, even more technically uninformed and insecure.

Recognizing NASA's obsession about credit being given solely to NASA, we at the Science Institute elected to stay out of the way of the space agency publicly. Director Giacconi chose not to participate in "The First Light Show," describing it as more of a "side show," and all of our astronomers decided to view first light at the institute. Frankly, we didn't mind being invisible, for this was clearly NASA's day in the sun. Besides, owing to the faster speed of the institute's computers, we knew that we would actually witness first light before our counterparts at Goddard.

Since the first object to be imaged was part of a focus test of the telescope, the intended target was not of much interest visually or astronomically. The target name, NGC 3532, derives from entry 3,532 in the *New General Catalog* of reasonably bright celestial objects. NGC 3532 is an approximately 3-billion-year-old group of well more than a hundred stars scattered over a loosely defined region some 25 light-years on a side. Technically termed an "open" or "galactic" cluster, this family of stars is estimated to be 1,300 light-years away, in the direction of the constellation Carina (the Ship's Keel), in the southern hemisphere. The cluster was chosen because it contains several reasonably isolated, bright (up to eighth-magnitude) stars that would serve as excellent targets to help bring the telescope into progressively finer focus.

Given the historic import of the first-light event, many colleagues and the public alike have asked me why *Hubble* wasn't commanded to look at something more spectacular. Less than 3 arc degrees away from NGC 3532 is one of the most beautiful regions of star formation in the southern hemisphere, a region called Eta Carinae. I had indeed proposed this magnificent nebula as a first-light target several times in the year prior to launch—in fact, Eta Carinae had become the number-one EROs target—but I lost the argument so badly it's not worth reporting. By contrast, when the Keck Telescope atop Mauna Kea—perhaps the finest observing site on Earth's surface, some 14,000 feet above sea level on the island of Hawaii—took its inaugural peek at the heavens several months later, its first-light target was NGC 1232, a majestic spiral galaxy and one of the most visually stunning objects in the sky.

Most of Sunday morning I spent making several star charts of the NGC 3532 area, celestial maps that would help me mentally navigate the region amid many other stars and well-known objects in Carina. Since we were still unsure how accurately Space Telescope was pointing, I wanted to be prepared with a variety of charts in order to identify individual stars in the first-light image. Matching expected star fields with an actual observation—especially in a relatively boring region where few of us had worked the skies before—is not as easy as one might think. And if *Hubble* could indeed sense the cosmos as deeply as expected, then the sight of bunches of previously unseen stars would make even well-known parts of the sky differ considerably from what we were used to. Mostly, though, I was killing time, anxious to see first light yet worried that the image might be so poor that the media would declare, as the *Washington Post* had implied that morning, that Space Telescope was a lemon.

The computer commands programmed the day before by schedulers at the Science Institute had been uplinked to *Hubble* at about 6 A.M., local time, and there had been nothing anomalous about that transmission. Those com-

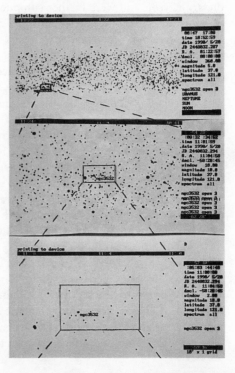

Star charts of the NGC 3532 star cluster in the constellation Carina. Prepared on a personal computer, this progressive series of "zooms" shows, from top to bottom, the field of stars surrounding NGC 3532.

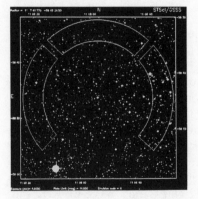

Another star chart of the NGC 3532 area, this one extracted from the Science Institute's *Guide Star Catalog,* showing the three fields of view (or "pickles") of the fine-guidance sensors astride the target star cluster. The "first-light" area imaged with the wide-field camera is centered on the small cross at the middle of the photo and is only a few times larger than the size of that cross. The bright star closest to the cross is an eighth-magnitude star with the catalog name HD96755. *[Space Telescope Science Institute]*

mands included some additional, rather large adjustments of the onboard mirrors in an attempt to better focus the telescope. The optical engineers at Goddard who were in charge of optimizing Space Telescope had been having a devil of a time aligning the secondary mirror, not least because they had been forced to work only with the fine-guidance sensors, which sample stars near the edge of *Hubble's* "eyesight"—almost like trying to use peripheral vision to determine the view directly in front of you. To do so, they had to resort to a "bootstrap" technique whereby the focus was "guesstimated" by sending uplink commands on one day, checking the progress by studying downlinked data on the next day, adjusting the focus some more, followed by more checking, more vehicle cross-talk, more data analysis, and so forth. All these focusing adjustments were made by incrementally changing the position of the secondary mirror, and were further complicated by the gradual and expected loss of moisture in (and therefore shrinkage of) the telescope's superstructure upon exposure to outer space.

Everyone knew that a more intelligent way to achieve focus would have been to open the shutter on the wide-field camera, which looks down the barrel of the telescope, but NASA had by then elevated the act of first light to such circus proportions that the engineers were prohibited from doing so prior to "The First Light Show." (To use the European-made faint-object camera to aid focusing and thereby achieve first light with a non-American telescope would have appalled NASA Headquarters.) As things turned out, these most recent and overnight focusing adjustments, made largely in the blind, saved the day—and probably a few people's jobs.

The shutter on the wide-field camera was due to open shortly after 11 A.M., but the image would not be immediately relayed to us. Neither of the TDRSs that *Hubble* was using at the time was available for a real-time transmission of first light to the ground; the inaugural picture would be stored onboard one of *Hubble's* tape recorders for a few hours. Some of us wondered whether the TDRS system had been preempted to relay images taken by reconnaissance spacecraft of the upheavals then beginning to break apart the Soviet Union—a not unlikely supposition, given that the use of this vital communications network is governed mainly by the Defense Department.

Actually, there would be two shutter openings that morning. One, as

steadily cooled to the ambient temperature of outer space, but its onboard coolers had not yet been activated to further chill the CCDs, so the image was "noisier" than it would be in normal science operation. Some stars could be seen, but they were hard to match quickly with the stars on the charts I had prepared.

Again, there was no emotion, just a studied curiosity at what we were actually seeing. Although the data had gone through Goddard on their way to the institute, we could see from our closed-circuit television that the first-light image had not yet been displayed in the Goddard control room, making a tense situation there seem overly strained, the people on camera awkward at best. We felt badly for our counterparts at Goddard, whose natural uneasiness while waiting for the first picture was recorded on live TV. And the bright, hot, "tormentor" TV lights wouldn't help anybody to see the image on the monitors once it did appear, a factor that hadn't been considered in all those dry runs.

I spread out all my star charts and several of us began trying to identify the specific part of NGC 3532 electronically captured by *Hubble*. This open star cluster extends over nearly a full arc degree, so the wide-field camera had sampled only about one percent of it. Given the fact that *Hubble* was continuing to experience faulty aiming, further complicated that day by an uncertainty in the roll angle of the spacecraft, it was something of a challenge to match the image with the now instantaneously outdated star charts. But it was fun, and we were almost doing astronomy.

Using the institute-built data-analysis system that allows the astronomer

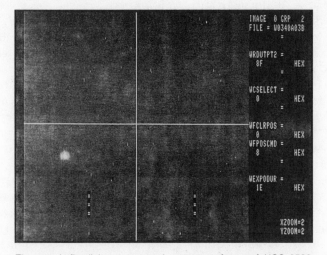

The actual, first-light, one-second exposure of part of NGC 3532 taken with Space Telescope—raw, untouched, uncalibrated, unmassaged, and just as we saw it in real time at the Science Institute. The image captured light having a wavelength of 5,550 angstroms, using a filter set before *Hubble*'s launch and which had not jostled loose during the vibrations of liftoff. Little can be seen here as computer processing had yet to be done. The bright, obvious speck of light would, minutes later in a longer exposure, be recognized as a resolved, twelfth-magnitude double star. As with all the images taken by either of *Hubble*'s cameras, this is not really a picture. Rather, it is a photographic reproduction of a digital electronic image. [NASA]

to interact with and manipulate the data, we zoomed in on a small segment of the image in order to make a quick estimate of telescope focus. Al Holm, the chief of the OSS area, deftly used a track ball (much like a computer mouse or joystick) to move around within small portions of the camera's field of view. And now we began hearing some encouraging murmurs from the assembled group. Light from the one prominent star visible on the one-second exposure seemed to spread across only 5 or 6 pixels. Since the wide-field camera's angular view is about 0.1 arc second per pixel, this implied that the telescope's resolving power was on the order of 0.5 arc second—much better than anyone had expected. Apparently, the mirror adjustments—based on a combination of reason and guesswork—telemetered up to *Hubble* earlier that morning had greatly sharpened the telescope's focus.

Suddenly and for the first time that day, there was some emotion and excitement. Several of us exclaimed in unison, as though we had practiced it, "Hey that's pretty good!" Hardly were the words out of our mouths when someone noticed that *Hubble*'s tape recorder was dumping the thirty-second image to the ground. Goddard was still struggling to display the shorter exposure on its monitors, but we immediately switched to display the longer exposure. Now we saw many stars, and the sense of astonishment grew. People drew closer to the computer monitors, effectively surrounding the new image, and a rapid-fire collective commentary began: "Look at this feature! . . . It resembles a double star. . . . Where exactly are we pointed? . . . Who cares, we have real data. . . . Has anyone identified the roll angle of the spacecraft yet? . . . Back off, I'm trying to figure where the hell in 3532 we've landed! . . . Just what are we looking at? . . . It doesn't matter, *Hubble* works!"

Space Telescope had indeed opened its eye and had seemingly demonstrated some excellent seeing—in fact, resolution apparently already better than that routinely achievable at any of the world's ground-based telescopes. We were ecstatic! We had crossed a frontier, and an adventure with unknown consequences had begun, one destined to be a roller-coaster ride of emotional highs and depressing lows. It would have been a wonderful moment for the world's media, had they (or at least a few of their representatives) been there to record astronomers concerned, puzzled, and then confident while at work—unlike at Goddard where the distant scene was staged to the point of looking plastic. Coverage at the institute also would have doubtless shown scientists in the act of sharing champagne from a single bottle that a deputy from the institute's directorate had thoughtfully brought along—a spontaneous, human reaction to a historic cosmic event that, I suppose, would have driven stuffy NASA officials directly up the wall.

Later that afternoon at Goddard, several "wiffpickers"—members of the instrument development team that designed and built the wide-field and planetary camera—computer-corrected the thirty-second exposure to remove some of the excess noise corrupting the fidelity of the raw image. This was done by subtracting a closed-shutter exposure (called a "dark-current," which contained a representative sample of the camera's noise) from the open-shutter exposure, thus revealing a cleaner image of the stars clustered within NGC 3532. The image was then hand-carried on magnetic tape to the Science Institute, electronically processed through our computers and photo laboratory, and made into hundreds of glossies for release to the media (but only after the final product was shipped back to Goddard).

Upon seeing the comparison images side by side, Jim Westphal, the principal investigator for the camera team, announced for the TV audience in his usual bubbly tone, "I'm just pleased as punch at this point. Focus is better than we thought it might be and that's certainly a good thing."

Unfortunately, that evening's telecasts and the next day's news stories carried poorly reproduced images photographed with a hand-held camera by NASA personnel directly from Goddard's computer screens. The American press, seeking instant gratification, couldn't wait the few extra hours for the institute's pipeline to return our high-resolution copies of NGC 3532 to Goddard, nor did NASA allow the media to obtain directly from us the higher-quality institute results—a clear case of the agency's public relations again gone afoul, for it was to NASA's advantage that *Hubble*'s first picture, however boring, be reproduced with pristine clarity in the U.S. press. "Prints of the images distributed by NASA lacked some of the detail seen on television screens and some of the stars were not clearly pictured," noted the *New York Times*. Submitted a *Washington Post* op-ed piece: "The captions said that the *Hubble* photos were clearer. To me it seemed they were fuzzier. But how could I argue with NASA? I probably needed a new eyeglass prescription." Actually the writer's eyeglasses were fine, yet NASA's obsession with getting sole credit for everything continued to chip away at that same credibility. Ironically, the European and Canadian press printed better pictures of first

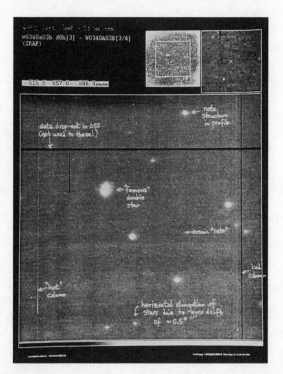

The somewhat cleaned, thirty-second, first-light exposure of part of NGC 3532 taken with the Hubble Space Telescope at visible wavelengths. Handwritten annotations of an astronomer at the Science Institute drew attention to some features shortly after the image appeared on our monitors, including long black lines that are momentary transmission failures, "cosmic rats" or hot (white) pixels peppering the image because the camera had not yet been cooled, and an astute comment (made May 21, the day following first light) regarding structure in the profile or spread of light surrounding the stars, all of which are members of the Milky Way Galaxy. *[Space Telescope Science Institute]*

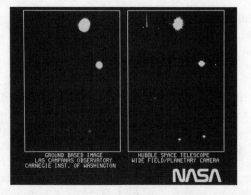

A comparison of part of the previous *Hubble* image (right, courtesy of NASA) and a high-quality ground-based photo taken of the same region at the DuPont Observatory at Las Campanas, Chile (left, courtesy of Carnegie Institution), equipped with a visual filter (approximately 5,500 angstroms) and a charged-couple device (CCD) much the same as that used on *Hubble*. *Hubble's* view is notably sharper, displaying about 0.5 arc-second resolution; the Las Campanas data are resolved to about 1 arc second, resulting in a fuzzier picture. The stars appear to have slightly different orientations in each frame because the roll angle of the orbiting telescope differed slightly from the tilt of the ground-based telescope; otherwise there is a direct star-for-star correspondence in the two images. The area shown, extracted from one of the wide-field camera's CCD chips, is about 11 x 14 arc seconds in size, which is only a small part of the camera's full field of view; it does not include the centrally targeted star, HD96755.

light, since we at the institute felt free to digitally ship the computer-enhanced version to them by electronic means.

Several interesting features were immediately noticeable in this first set of *Hubble* images—one telling, another intriguing, and a third disturbing. First, the brightest source of light did not arise from a single star. Even in the Las Campanas photo, this blob of light was not spherical, suggesting to astronomers that probably more than one star was present. Detailed photographic analysis (including spectroscopic observations) had earlier implied the presence of two stars nearly overlapping, although they cannot be individually resolved from the ground. By contrast, the *Hubble* image showed the two stars separated by about one arc second. (Despite what was reported in the press, these two stars do not comprise a binary system; they are unrelated, and merely lie along similar lines of sight. And despite what was widely announced at the time by NASA, *Hubble's* splitting of the oblong blob into two stars was not a "great discovery.")

Second, although the *Hubble* inaugural picture was sharper than the ground-based version, it still did not accurately represent Space Telescope's focus at the time of first light, because the image was taken in an "aim and shoot" mode, with the spacecraft's attitude fixed only by means of its gyroscopes. No guide stars were used to precisely aim the craft in space, since we had not yet learned how to finely lock the telescope onto guide stars for "rock-solid" stability. Careful examination of a pair of dim stars toward the bottom of *Hubble's* thirty-second exposure showed a horizontal shift of the stars' images. Space Telescope had drifted a few tenths of an arc second during the exposure, blurring the image somewhat. Other spacecraft with lesser resolving power can easily tolerate this kind of slight movement, but *Hubble* cannot. Only with the gyroscopic drift removed could engineers make an estimate of the telescope's true focus at the time. The easiest way to do this was to examine the one-second exposure, during which the spacecraft movement was less. And this image in fact showed each star dramatically separated. Apparently, some of the light resolved by Space Telescope was already as good as about 0.2 arc second, better than any telescope can see stars of this (approximately twelfth) magnitude from the ground.

Most puzzling of all was the shape of the individual images of every star in both of *Hubble's* first-light pictures. The pattern of starlight surrounding each object seemed odd to several people on that historic Sunday afternoon, and some of us in fact made passing remarks about it. Within a day or two, after the first-light hoopla had subsided, a handful of astronomers at the Science Institute and on the wide-field and planetary camera team began closely examining the individual stellar images. In particular, they noted numerous "hairy tendrils" and spiderlike tentacles emanating like spokes from the center of each of the bright stars. These tendrils can be clearly seen in the thirty-second exposure and were the topic of much debate at the Science Institute during the week following first light. What's more, the symmetrical pattern of light in and around each of the stars, called the "point-spread function," was strange, certainly not what we had expected.

A cross-sectional cut through the center of any of the stellar images disclosed that the light within the image was composed of a bright central peak sitting atop a wide, dim base. In other words, when the light in any of the stars' images was dissected in minute detail, we could see that part of it was concentrated into a very small point, with the rest spread throughout in a surrounding disk or halo. The central spike measured about 0.2 arc second across, nearly as good as specified for a properly focused *Hubble* telescope, whereas the blurry disk extended out to well more than 1 arc second. This kind of twofold pattern of starlight had never been seen before with any properly operating ground-based telescope, nor had we expected to see it with *Hubble*.

Ground-based telescopes always show the light from any star as a bell-shaped (or "Gaussian") distribution, the smearing of which is largely caused by the "twinkling" of starlight seen through Earth's atmosphere. The expectation was that *Hubble* would display a similar bell-shaped pattern, albeit narrower and highly peaked because of the superb optical clarity expected from a telescope operating above our atmosphere.

I initially took this bizarre point-spread function to be merely an indication that *Hubble* was not yet focusing starlight well, compounded by a blur induced by the spacecraft's drifting gyroscopes and jittery pointing. But throughout the intense weeks following first light, a handful of instrument experts at the Science Institute were talking in guarded tones about more

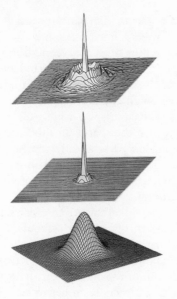

Top: This three-dimensional plot of a star's image recorded on the thirty-second first-light exposure shows the distribution of light energy, or "point-spread function," extending outward from the core of the star's image. As can be seen, part of the recorded light is concentrated within a small, approximately 0.1-arc-second interval, and the rest is scattered throughout a much wider range extending beyond 1 arc second. The central spike rises from a turbulent base, resembling a mountain peak overlooking a series of foothills. Eventually, scientists would come to realize that within the hills and valleys at the base of the peak lay the crux of *Hubble's* biggest problem.

Center: Prelaunch design of *Hubble* specified that most of the acquired light be concentrated within a narrow spike with very little spilling over into a surrounding halo. The spike would rise from a nearly flat base because most of the captured light would be focused at a virtual point.

Bottom: A typical "point-spread function" for a star observed with a large ground-based telescope shows a bell-shaped curve having a breadth (full width at half maximum intensity) of about 1 arc second.

basic worries. They could not understand how even a grossly unfocused telescope—and especially one with a slightly wild aim—could yield such a symmetrical, spoked, twofold pattern of light from every star in an image.

Additional wide-field and planetary camera images were acquired some ten days later, and they, too, showed the disturbing twofold light pattern, including the tendrils extending from each of the stellar cores. Since Director Giaconni and I were soon to leave on separate speaking trips to apprise astronomers elsewhere in the country of our progress, we were given a frank and thorough briefing by the institute's telescope and instrument branch on the increasingly tense discussions and debates then under way within the institute. Ominously, we were told that the observed light pattern was likely indicative of a fundamental problem: either we were seeing the scattering effects of "microroughness"—dust, paint flakes, or small-scale imperfections on the surface of the mirror—or there was a near-fatal flaw in the telescope's construction. Either possibility, we knew, could become a show stopper for some of the most exciting and profound science expected to be done with the Space Telescope.

The European Space Agency has had over the years a love-hate relationship with NASA. Leading up to the launch of Space Telescope, ESA officials had often expressed disappointment that NASA's most important international partner was being neglected by its bigger brother. The thirteen western European countries that comprise ESA had collectively contributed a substantial amount of money and expertise to the *Hubble* project. (The member states are Austria, Belgium, Denmark, France, Germany, Ireland, Italy, the Netherlands, Norway, Spain, Sweden, Switzerland, and the United Kingdom; Finland is an associate member and Canada a cooperating state. Each contributes to ESA in proportion to its respective gross national product.)

ESA had provided the solar arrays that power the spacecraft and had designed and built one of the cameras onboard *Hubble;* its member countries also supply about 15 percent of the scientific staff at the Science Institute. But NASA regarded this contribution as, variously, not enough or too much, or in any case that ESA was stealing its thunder. For its part, ESA regularly complained that it was being treated unprofessionally and often ignored by NASA. Still smarting from the United States' having backed out of its share of the *Ulysses* mission to study the Sun, frequent changes in the space-shuttle manifest to favor domestic payloads, and repeated revamping without consultation of the international space-station design, shortly before *Hubble*'s launch ESA once again became irked by what it felt was U.S. high-handedness.

At another unfortunate televised press conference at the Kennedy Space Center on a Sunday morning preceding the historic blastoff, NASA trotted out all sorts of U.S. officials, mission experts, and contractors. The agenda listed ESA's contributions to the Space Telescope project last, its people scheduled to mount the dais to address the mass media at about noon. As the morning droned on with NASA managers repeating their often exaggerated claims for *Hubble*'s science capabilities (and one leading scientist making off-color remarks about God engaging in foreplay before the Universe began), it became evident that the assembled journalists were not warming up to the occasion—even those not offended by sacrilegious remarks on the Sabbath.

At about 1 P.M. and with the press conference running late, the ESA representatives finally took center stage, yet a NASA official simultaneously announced that a free buffet lunch was being served in the next building. Not surprisingly, the press room emptied almost instantaneously, leaving our distinguished European counterparts facing a cavern-like hall devoid of nearly everyone except a few captive cameramen and a sound man apparently wearing defective earphones. No *Hubble* scientists or managers remained, and certainly no reporters. The ESA contingent was fuming, but gallantly went through with their prepared remarks for the benefit of the television cameras, after which there were no questions from a media throng that was imbibing elsewhere with U.S. officials. That evening, while running along Cocoa Beach, ESA's chief scientist, the likable Parisian Roger Bonnet, told me in no uncertain terms that he was going to file a formal protest with NASA, which I think he did.

This lack of sensitivity among other obviously political mania—such as the big NASA logo appearing without any mention of ESA on *Hubble*'s first-light image—fueled the notion that Europe was again being slighted and made perhaps inevitable the need for a "second first light." Since the Europeans had built the more powerful of *Hubble*'s two cameras, they, too, wished to release their first photograph with some public fanfare. This event occurred a few weeks after the images released by the wide-field camera.

The faint-object camera (FOC)—built for about $105 million mainly by two European companies, Dornier Deutsche Aerospace and Matra Marconi Space—is designed to detect stars as dim as the twenty-eighth magnitude, and perhaps considerably fainter, by means of long exposures. This camera is more sensitive and faster than the wide-field and planetary camera in the blue and ultraviolet wavelengths shortward of about 4,500 angstroms. It also has a higher angular resolution than the WF/PC at all wavelengths, but, like a tele- photo lens, the trade-off is a narrower field of view. The FOC is the only onboard instrument to utilize the full angular resolving power of *Hubble,* namely, about 0.02 arc second deep in the ultraviolet—all of which means that the faint-object camera should be able to study, for example, Cepheid variable stars, which are benchmarks in determining the size and scale of the Universe, out to distances some four to five times greater than that presently achieved with ground-based telescopes.

A dual detector system comprises the heart of the faint-object camera. Each detector uses a series of optical relays and an image intensifier tube (or amplifier) to focus a highly magnified electronic image onto a phosphor screen, so that the photons detected become some 200,000 times brighter than the light falling on *Hubble*'s main mirror. And it does so without altering the pattern of information intercepted by the mirror. In principle, these devices work much like an amplifier in a home stereo system, transforming the weak signal produced by the cartridge in the turntable stylus into one strong enough to rattle your neighbors' windows. The instrument's phosphor image is then scanned by a miniature television camera and the individual photons counted.

As with the WF/PC, the product is a two-dimensional grid of digital data, which is transmitted to Earth and reconstructed into a picture. Also like the WF/PC, the faint-object camera has a variety of prisms, filters, and polar-

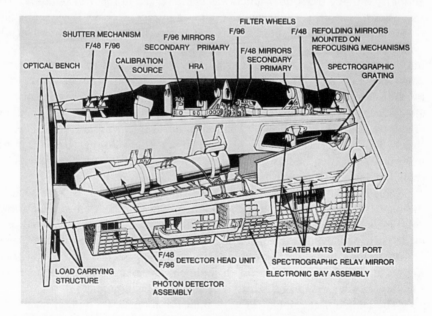

This cutaway drawing of the faint-object camera aboard the Space Telescope illustrates some important aspects of *Hubble*'s most sensitive camera. The camera measures 3 x 3 x 7.2 feet, weighs nearly 700 pounds on Earth, and resembles a large refrigerator. *[ESA]*

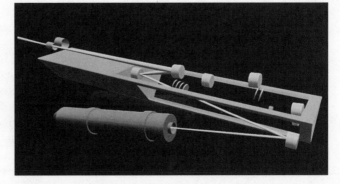

This artist's conception illustrates the two pulse-counting detectors of the faint-object camera in the plane containing the principal optical axis of the Hubble Space Telescope. *[D. Berry]*

izers that rotate in and out of the light beam. These allow, respectively, an incoming light wave to be split into its component wavelengths, sampled at a specific wavelength, and analyzed for any preferred plane of oscillation. And it has two "occulting fingers," one 0.4 and the other 0.8 arc second wide, designed to eliminate the glare of a bright object (such as a star) so that nearby fainter objects (such as planets, perhaps) may be seen. As an analogy, an occultation device could serve to block the headlights of an oncoming car, making it feasible to read its dim license plate. This camera system is so sensitive to faint light sources (hence its name) that objects brighter than about the twenty-first magnitude must be dimmed by neutral-density filters, to avoid saturating and thus harming the camera's silicon innards. By contrast, the Palomar Sky Survey photographs used to build *Hubble's Guide Star Catalog* display stars no fainter than the twenty-first magnitude.

For the technical aficionados, the faint-object camera offers two focal ratios, f/48 and f/96, on a standard television format of 512 x 512 pixels. An f/48 FOC image spans 22 x 22 arc seconds and yields a pixel size of 0.04 arc second, while the f/96 mode halves these values. A special f/288 option is also available by utilizing a miniature telescope *inside* this camera; although its field of view is only a tiny 4 arc seconds across, this is the observing mode that matches the so-called diffraction limit of Space Telescope (~0.02 arc second in the ultraviolet), the clearest conceivable view attainable with a 2.4-meter mirror.

The principal investigator for the faint-object camera, Duccio Macchetto, who is the leading European scientist in residence at the Science Institute, not only was eager to demonstrate that his team's camera worked, but also was concerned that he not be upstaged by another European group, which was soon to release high-resolution photographs taken with a new telescope built for the European Southern Observatory. Called the New Technology Telescope (NTT), this novel ground-based, 140-inch (3.5-meter) instrument is perched atop the 8,000-foot-high Andean mountain of La Silla, in the Atacama Desert about 400 miles north of Santiago, and was constructed using "active optics," a technology unavailable in the 1970s, when *Hubble* was designed. This "new technology" uses a thin (10-inch), flexible main mirror, whose shape can change slightly several times an hour by the action of scores of computer-controlled motorized pistons in order to counteract gravity's distorting effects on mirror shape as the telescope mount rotates to compensate for Earth's spin. The result—for bright cosmic objects—is a substantial improvement in focusing, and thus in resolution, over that normally achieved by ground-based telescopes with rigid mirrors.

In the months prior to the launch of *Hubble*, the La Silla astronomers had waited night after night for the stillest of atmospheric conditions, and had eventually succeeded in using active-optics techniques to photograph some relatively bright star fields with a spectacular resolution of about 0.3 arc second—the best civilian, optical results ever accomplished from the ground. The La Silla telescope cost only $13 million—a bargain, by all accounts, in today's world of high-powered astronomy. Alternatively stated, if only for emphasis and not to be judgmental, one could buy well more than a hundred NTT-class telescopes for the price of a *Hubble*.

Several of us at the Science Institute had seen a draft of a press release claiming that the La Silla group could do practically as well from the ground as we could from orbit: "The telescope therefore functions as if it were situated in space . . . the technique is therefore entirely complementary to the Hubble Space Telescope concept." Subsequent accounts in the press were misleading, for we knew that this claim was an exaggeration; they failed to mention that no ground-based telescope can operate in the ultraviolet, where *Hubble* would do much of its work, nor was the European-based group forthcoming about the fact that sub-arc-second seeing from La Silla was possible only for nights when the air was exceptionally calm, and only toward relatively bright objects (no fainter than about the fifteenth magnitude) for which exposures could be accurately guided. Even so, there was a bit of a race between ESA and ESO for the sharpest images of the sky.

First light for the FOC was scheduled for mid-June, but even by then, nearly two months after launch, we could not reliably maneuver or point Space Telescope. NASA managers were reluctant to command *Hubble* to move much around the sky, fearing that control of the spacecraft might be lost altogether; it had become politically safer to leave the bird's attitude fixed than to risk breaking it further. For what seemed an endless time, we stayed in the area of the Carina constellation, imaging and reimaging NGC 3532 and especially a bright calibration star, Iota Carina, with the WF/PC. Guide-star acquisitions were especially problematic, causing any exposures longer than a minute to blur. Fortunately, the first-light target for the faint-object camera—a star cluster called NGC 188—had some bright stars within the chosen field of view, and therefore exposure settings could be kept short. In fact, a neutral-density filter had to be used in order to diminish the captured starlight by six magnitudes, so that the camera's delicate sensors would not be harmed.

Like NGC 3532, NGC 188 contains hundreds of stars, the majority fainter than the thirteenth magnitude. Approximately 5,000 light-years away, it is the oldest known open cluster, having an estimated age of about 12 billion years—approximately the age of the Milky Way Galaxy. The tricky part of this observation was that NGC 188 is about 4° from Polaris, the North Star; *Hubble* would then have to slew to the opposite part of the sky in the northern hemisphere—a maneuver that terrified some program managers.

A somewhat inconvenient target, NGC 188 was chosen not because it would make a "pretty picture" but for engineering purposes; the positions and magnitudes of the cluster's stars were accurately known, and thus *Hubble*'s second camera would provide an independent check on the confusing focusing tests, then continuing on the telescope. But it was also chosen because it was an "open" cluster—one of those near-voids in the nighttime sky unlikely to yield a "premature" discovery. (Earlier appeals to have the FOC capture the image of a companion of the well-known North Star—a target to which the public might better relate and for which *Hubble*'s handlers would still get their engineering data—fell on deaf ears.) Because the widest aperture of the faint-object camera was only some 22 arc seconds across, a decision was made to take many short exposures at slightly different positions, in order to cover a wider field of view. Although this first-light event was not to be televised (nor would NASA allow any reporters to be present), there was nonetheless a fear at Goddard that if only one narrow-field picture were taken it might, given *Hubble*'s pointing problems, come up blank by falling in between the scattered stars of the cluster—the target region was that boring!

On early Sunday morning, June 17, the first couple of attempts to take a picture with the faint-object camera failed, because *Hubble* was unable to acquire guide stars properly. Finally, some seven images were exposed and telemetered to the ground, where a small "liquid celebration" was held to acknowledge the camera's inaugural operation. These images were taken in visible light; no radiation was sampled in the ultraviolet part of the spectrum, where the FOC would do most of its pioneering work. Regardless, the camera team was thrilled. Exclaimed one member half out of breath, "The images not only have stars in them, but they are the correct stars!"

All was not well, however, and we knew it. Like the first-light images taken with the wide-field camera, starlight captured with the faint-object camera seemed strangely unfocused. In principle, this was not overly surprising, for *Hubble*'s optical system was still degassing, and was probably not yet optimally aligned. But disturbingly, the new images showed the same kind of irksome halo seen earlier, with the same spidery tendrils and other odd radial structure surrounding each star's pointlike cusp of light. Up close, these stars certainly did not look like the "jewels of the night" we had anticipated seeing with Space Telescope. Instead, every star in each of the new images displayed

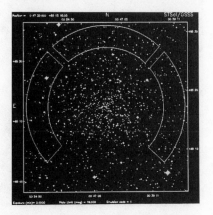

This wide-angle, ground-based finding chart of the NGC 188 region was used to navigate Space Telescope in the Cepheus constellation, and to help identify which stars' light had been exposed to the faint-object camera. The cluster is composed mainly of some 150 stars loosely scattered over about 20 arc minutes in diameter, which is roughly the angular size of this print. Like all the others in this book that have the three semicircular fields of view (called "pickles") of *Hubble*'s fine-guidance sensors, this photo-chart was generated by the Science Institute's Guide Star Selection System. *[Space Telescope Science Institute]*

the same twofold point-spread function: most of each star's light was spread over a wide, fuzzy area well more than an arc second across, an imaging quality worse than that obtainable with even a small ground-based telescope.

Even so, astropolitically, it was important to the Europeans that they get an FOC picture into the public domain. Uncertain what to make of the faint halo of light surrounding each star in the images, the European team essentially disregarded it. This they took to be a kind of "image enhancement"—some colleagues called it "a creative use of gray scale"—and they noted the diffuse halo only parenthetically in their press release. The remaining 15 percent of the light was focused exquisitely within an approximate 0.1-arc-second spot, making the two stars they elected to show the public indeed resemble diamonds on a dark velvet background. They then compared *Hubble*'s view with a ground-based image of the same two stars, but they didn't use one taken with typically good seeing conditions.

Largely because they wanted to present a comparison image taken from a European observatory (but, surely, not their competitor's facility at La Silla), they used a photograph of part of NGC 188 acquired with the 100-inch (2.5-meter) Nordic Optical Telescope on the Canary Island of La Palma—a picture that displayed relatively poor resolution, of nearly 2 arc seconds. Naturally, when the comparison was made—a less-than-optimal ground-based photograph with a doctored Space Telescope image—*Hubble*'s view of the Universe seemed spectacular, vastly outperforming conventional techniques. I suggested to various members of the camera's team, aloud but privately, that this was unfair and misleading. A Swiss astronomer on the institute staff agreed, claiming to an unsympathetic Dutch project scientist that this was tantamount to cheating, yet managed, he told me later, "only in making an enemy for life." NASA, through which every *Hubble* press release must pass before being issued publicly, had no comment—other than their complaint that the NASA logo was printed too small below the image.

The first-light images for each of nine positions in NGC 188, taken with *Hubble*'s faint-object camera. The stars are shown mostly as small points surrounded by corrupting noise, including the unfocused halos around the stars. Exposures at most positions were for ten minutes, though a few were cut short owing to faulty telescope guidance. The two frames at the bottom right are essentially blank since the camera's shutter was commanded to close when *Hubble* was unable to acquire guide stars. Each frame measures 22 arc seconds on a side. These first-light images were not released to the public. *[ESA and NASA]*

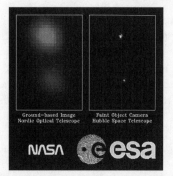

A small (5 x 6 arc-second) portion of one of the faint-object camera's first-light images showing two of NGC 188's stars (right, courtesy ESA and NASA). The pair of stars shown here separated by about 3 arc seconds are the same two stars at the top of the middle, top frame of the previous figure, but with the halos subtracted. This is compared with a poor-quality ground-based image of the same two stars taken with the Nordic Telescope at La Palma (left, courtesy of Observatorio del Roque de los Muchachos). When presented in this way, a dramatic improvement in spatial resolution is apparent. This is the "first-light" image that was released to the public, including the "big" ESA logo, which only fueled the conflict between the two space agencies.

In the end, it didn't matter how the data were manipulated and packaged, for hardly any newspapers ran a copy of the faint-object camera's first-light image. And those that did used dot-screening techniques that did an injustice to both the new image and its comparison photograph. The fuzzy ground-based stellar photo was washed out almost entirely, and the sharp points of light in the *Hubble* image looked like hardly more than a single white dot on a black piece of paper. Alas, the published image very nearly *was* blank—which the public, by this time registering increasing complaints at the Science Institute and elsewhere, obviously saw as disappointing. A scientist at the Jet Propulsion Lab sent me a terse note on e-mail: "We are anxious to see more results than the lousy image in the *LA Times*." The *Baltimore Sun* editorial page was not as patient, remarking that *Hubble*'s pictures thus far "looked as though they had been taken by a chimp and developed at Fotomat."

Since guide stars had eventually been acquired and had locked the telescope into position during the faint-object camera exposures, we could no longer blame the peculiar spread of light on gyro drift. Nor was it likely that the mirror was, after all, rougher or dustier than expected. In the weeks since first light, we had exhaustively reviewed prelaunch tests that convincingly proved that the smoothness of *Hubble*'s mirrors had exceeded the design specification; although we knew at the time of launch that about 2 percent of the primary mirror was obscured by dust, it could not have caused the type of symmetrical distortion we were seeing. Instead, these observations bolstered the truly dreadful notion that the orbiting observatory was suffering from a major optical flaw.

*   *   *

What about active optics, adaptive optics, and other new electro-optical technologies now coming on line? Can they compete with *Hubble*'s superb angular resolution? Do they make expensive orbiting observatories obsolete? Like smart weapons, "smart mirrors" have been pioneered by the U.S. military for tracking and surveillance purposes, and the smartest among them can impressively compensate for the twinkling of starlight caused by the churning of Earth's atmosphere. This technology, currently only partly in the public domain, comprises not just "active optics" that uses slow mirror bending (or simple tip-tilt corrections) to counteract gravitational and mechanical distortion, but also "adaptive optics" that uses rapid mirror warping to vary the focus across the mirror and thus to see through Earth's turbulent atmosphere with a kind of crystal clarity.

As with the New Technology Telescope just mentioned, most new scopes now being constructed on Earth—including the world's largest, the Keck Telescope atop Hawaii's Mauna Kea—incorporate some sort of active optics, thus achieving results better than that possible with a fixed, rigid mirror. All such systems depend critically on tracking guide stars. They expose light from an arbitrarily chosen reference (or guide) star near the edge of the field of view, after which a detector compares the image with what is expected from a true point source. Commands are then sent via computer to pistonlike actuators on the support struts of the mirror, causing the mirror, which can be subtly warped by gravity, wind, and heat, to "shape up" for better focus. This is analogous to squinting, which causes the eyeball to deform, the focal path through its lens to change, and a blurred image to become better focused. (Incidentally, the main mirror at the New Technology Telescope at La Silla was found at its first light to have a large error in its shape, similar in form—and probably with similar cause—to *Hubble*'s error, but the NTT's support actuators are sufficiently robust to strongly warp its main mirror into the desired shape, thus removing the error.) Such active-optics systems often produce superb seeing on the order of 0.5 arc second, although the observed stars still "twinkle" and are therefore fuzzier than they could be. Furthermore, the technique is currently practiced only toward celestial objects brighter than about the fifteenth magnitude.

A particularly promising scheme that does remove much of the adverse twinkle of stars was developed in the 1980s for the Strategic Defense Initiative ("Star Wars") Organization, although it was first proposed theoretically by an astronomer in 1953. This is an adaptive-optics system, which uses sophisticated technology originally designed to help laser beams slice through the atmosphere and destroy incoming enemy missiles. Operational years ago, where I first encountered it atop Hawaii's Haleakala Mountain, selected features of this new instrument were recently declassified, allowing astronomers to explore its usefulness for civilian purposes. The heart of the device is an ultra-thin (0.08-inch, or 2-millimeter) glass mirror that, as customary, is the recipient of light entering the telescope. Unlike a conventional telescope, however, the pattern of this light is constantly measured while striking an array of small lenses, each outfitted with a detector (called a "wavefront sensor").

As the precise position of the starlight falling on the mirror dances

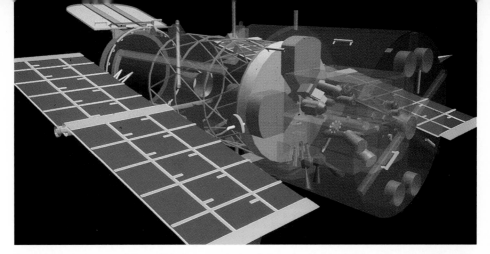

The major components of Space Telescope are depicted above in this simplified transparent diagram. The primary mirror is encased within a central ringlike structure that allows the primary and secondary mirrors (light blue disks) to be mounted inside the body of the spacecraft. Starlight enters the telescope at left and is focused in the aft-bay at right where onboard scientific instruments (whose essential components are shown within as red, green, and blue boxes) are used to analyze it in a wide variety of ways. [D. Berry]

Below is one of the earliest images taken with the improved *Hubble* telescope after the repair mission in late 1993. The object shown is M100, a spiral galaxy member of the Virgo Cluster, approximately 50 million light-years away. The image is rendered in full color, the result of combining 30-minute exposures at red, green, and blue wavelengths with the new wide-field and planetary camera 2. This camera, WF/PC-2, has corrective optics built into it to compensate for the main mirror's aberration, and the result is indeed more sensitive seeing of fainter light—such as that in the outlying spiral "arms" in this image.

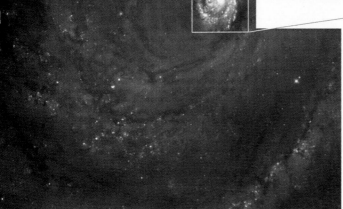

But one trade-off of this improvement is that the camera's field of view has been diminished, resulting in this strange, chevron-shaped image (a mosaic of the three chips of the wide-field CCDs and the one chip of the planetary camera CCD). The inner, brighter part of the galaxy, caused by vigorous star formation and seen here through the planetary camera, is magnified in the box at upper right. Resolution is about 0.1 arc second, just about the same as achievable with the old WF/PC-1 camera before the repair mission. [NASA and J. Trauger]

Space Shuttle *Discovery* (at left), mated to its ruddy bullet fuel tank flanked by two rockets resembling thin Roman candles, and with the Hubble Space Telescope safely tucked inside its cargo bay, leaves Launchpad 39B at the Kennedy Space Center on the morning of April 24, 1990. Except for the five astronauts in the crew compartment at the top of the winged spaceship, no humans were this close to the shuttle at launch; the photograph was made by a remote camera whose shutter was triggered by the noise of the rocket engines. *[NASA]*

Below, the act of *Hubble*'s deployment was caught by a remote-control IMAX movie camera located in the cargo bay of *Discovery*. This single frame, isolated from the camera's extra-wide (70-millimeter) film, is among the most dramatic images of the Hubble Space Telescope in orbit. As shown here, the Canadarm had just moments before released *Hubble*, making the telescope a free-flying unmanned vehicle. Notice the bent solar array at the top, partly due to the fish-eye lens, but also perhaps indicative of a regular "flapping" of the arrays as the vehicle experiences uneven heating while passing through orbital sunrise and sunset every three-quarters of an hour. *[NASA, IMAX Corporation, and Smithsonian Institution]*

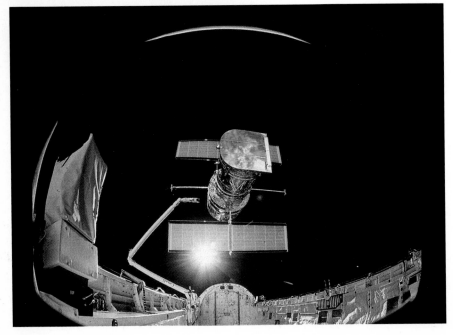

After *Hubble*'s commissioning period ended, members (mostly) of the wide-field and planetary camera team returned to Arp 220 and successfully imaged this peculiar object, thought by some astronomers to be two galaxies in collision. This Space Telescope view furthers the idea and also suggests that giant clusters of stars are produced at a furious rate from the gas and dust supplied by the interaction of the two galaxies.

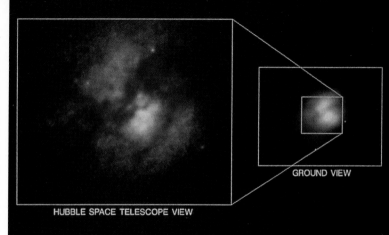

HUBBLE SPACE TELESCOPE VIEW

GROUND VIEW

Until *Hubble*'s planetary camera took this multicolor image of Arp 220's central, double-lobed structure (about 10 arc seconds across), such a "starburst" galaxy had never been seen in such detail. It is in that central region where, amidst a dark dust lane bisecting the object, *Hubble* has revealed several gigantic young star clusters some ten times larger than groups of stars previously observed anywhere. Six prominent clusters are visible in this image as bright blobs in and around the dust lane, all within an arc second or about 2,000 light-years from the object's nucleus. Since massive stars are known to proceed through their evolutionary paces most rapidly, the core of Arp 220 has become a unique laboratory for studying the late evolution of massive stars, especially perhaps frequent supernovae detonating like a string of firecrackers popping off several times per year. *[NASA, U.S. Naval Observatory]*

Below is an artificially colored image of supernova SN1987A, exposed at 5,007 angstroms and not computer enhanced. It shows an elliptical ring of matter around the remnant of the supernova itself, flanked by two prominent stars. Since this wavelength, corresponding to the emission of doubly ionized oxygen atoms, is seen as yellow-green light, we gave the ring that color. The reddishness of the supernova at the center of the ring reflects the true color of its expelled and now-cooling debris, and the bright, hot blue stars astride the ringed supernova show their approximately true colors as well. For scale, the apparent diameter of the ring averages 1.5 arc seconds, while the thickness of the matter in the ring itself is marginally unresolved at approximately 0.1 arc second. For reference, at the distance of the Magellanic Cloud, 1 arc second equals about 1 light-year. *[NASA and ESA]*

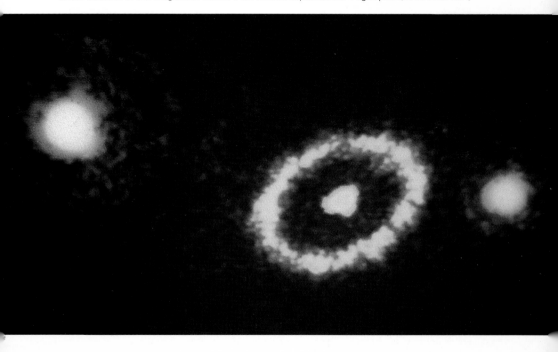

These three frames show computer-enhanced *Hubble* images of Saturn through each of the red, green, and blue filters used. Specifically, the observations were centered at 7,180, 5,470, and 4,390 angstroms, respectively. Here, the field of view is the nearly complete 80 arc seconds of a single wide-field camera CCD chip; resolution is about 0.1 arc second. Each image was deconvolved separately and shipped electronically to the Science Institute's astronomy visualization laboratory where they were combined to form a true-color image. *[NASA and S. Ewald]*

Toward the end of commissioning, a remarkable image was obtained of the globular cluster 47 Tucanae, an extraordinarily dense "beehive swarm" of upward of a million stars roughly 15,000 light-years distant. At left is a ground-based photograph, taken in blue light (4,400 angstroms) with a large telescope in Chile; seeing is about 1 arc second. At right is a *Hubble* ultraviolet image (2,200 angstroms) of a small part of the star cluster near its core; the scale of the space-based image is some 8 arc seconds or 0.5 light-years across, and the seeing a superb 0.08 arc second.

Made with the European faint-object camera, the new *Hubble* image shows no evidence for a bright cusp of light, or telltale sign of a central black hole. Apparently, as with the M15 globular cluster, there is no heretofore expected giant black hole in the cluster's heart. Instead, the 47 Tuc image reveals a surprisingly high concentration of a unique class of ultraviolet-bright stars called "blue-stragglers." Such star systems might actually be evolving "in reverse," from old age back to a hotter and brighter youth, their new lease on life made possible as stars in the cluster's core experience close encounters, and occasionally even capture each other, thus forming tight binary systems. Their blueness results from their renewed burning as one star in each binary siphons fresh hydrogen from its companion, and the term *straggler* suggests they have been left behind by all the other normally aging stars in the cluster.

Blue stragglers probably play a critical role in the dynamical evolution of the cluster's core, perhaps even explaining why the cores of globulars are not even more densely packed with stars. Such cores might in fact collapse but then rebound due to the presence of the blue straggler binary-star systems, which would serve as "egg beaters," stirring up the motions of thousands of other stars in the cluster and thus granting the cluster some buoyancy that prevents further collapse, and which might even cause the cluster to expand a bit. [ESO, ESA, NASA, and F. Paresce]

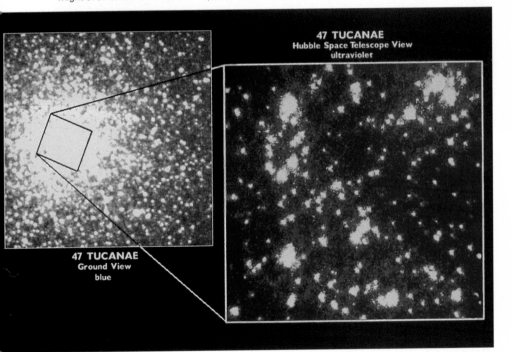

47 TUCANAE
Hubble Space Telescope View
ultraviolet

47 TUCANAE
Ground View
blue

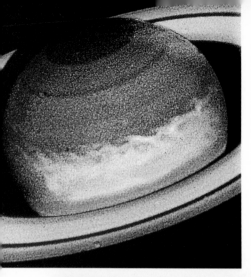

This computer-enhanced image combines the light from Saturn in the two colors observed—blue and infrared—to display novel structure never before seen across the face of the planet. Acquired in early November, more than a month after the discovery of the planet's White Spot, this (false-color) psychedelic cloudscape clearly shows that the broad white spot had by this time dispersed around much of one hemisphere. By combining the two colors in this way, it is also possible to study the vertical growth of the clouds, which are thought to be mainly ammonia (NH_3) ice crystals; hence, the spot's white cirrus cloudlike features. These storm clouds in fact seem to reach 50 miles altitude, whipped by nearly 1,000 mile-per-hour winds. Image resolution is about 440 miles (700 kilometers), compared to 75,000 miles (120,000 kilometers) for the diameter of Saturn itself, which at the time was about 870 million miles (or 1.4 billion kilometers) from Earth. [NASA, J. Westphal, and W. Baum]

In the late stages of commissioning, *Hubble* returned to study the archetypical jet of plasma within the prominent giant elliptical galaxy M87, known to dominate the center of the Virgo Cluster some 50 million light-years away. In this image acquired by *Hubble*'s planetary camera, the full extent of M87's 4,000-light-year-long jet is captured mainly in near-infrared radiation (approximately 7,850 angstroms). The jet appears as a string of emitting knots within a widening cone extending from the core of this gargantuan galaxy, thought to contain 300 billion solar masses, about 1 percent of which is probably wrapped up in a huge black hole that likely powers the jet. This image, as well as an even higher-resolution ultraviolet image of M87 taken with *Hubble*'s faint-object camera, reveal unprecedented detail at the 0.1-arc-second level, resolving some features as small as 10 light-years across—such as the compact sources (which are globular clusters) scattered around M87's core.

This new image, spanning 32 x 24 arc seconds, exposed for twenty-three minutes and heavily deconvolved, also shows the pointlike cusp of light at the heart of the galaxy (at left). This central cluster of stars is thousands of times richer than the spread of stars in the neighborhood of our Sun and at least hundreds of times denser than expected for a normal giant elliptical galaxy—"at least" because the ultimate central density of stars and gas in M87 might be even higher, but its measurement is beyond even *Hubble*'s exquisite resolving power. Is the spike of light at the galaxy's center indicative (paradoxically) of a black hole? No one yet knows for sure. As one of the principal investigators so aptly put it: "It looks like a duck and it acts like a duck, but we haven't yet heard it quack like a duck." It seems that only *Hubble*'s ailing spectrographs can pin down the case for a genuine black hole, a task possible only if the spaceborne observatory can be fixed. [NASA, T. Lauer, and J. Westphal]

All four charged-couple detectors (CCDs) abut here to display a full wide-field-camera mosaic of a1-light-year square area of the Orion Nebula. The thin black cross running through the midst of the image delineates the edges of the sensors, which together span 130 arc seconds. In this new *Hubble* view, which has been deconvolved to correct partially for the telescope's flaw, red outlines light emission from the element sulfur, and those in blue and green highlight oxygen and hydrogen, respectively. Although the hydrogen and oxygen emissions seem distributed rather smoothly, the sulfur emission is seen to fragment into filamentary and clumpy structures with sizes down to the limit of *Hubble's* visibility—here, about 0.1 arc second, or 6 light-*hours*, a scale comparable to the radius of our solar system. Also note toward the left side of the image, among many other fascinating patterns, the curved "snorkel" or "elephant trunk" (whose length is about thirty times the size of our solar system) at the 7 o'clock position, the jetlike feature (in length about 100 solar systems, or 0.1 light-year) at the 9 o'clock position, and the small orange ring (which might be an instrumental artifact) at the 10 o'clock position. The black dot to the lower left of the field center is not real; named the "Baum spot," the dot is painted directly on the camera optics to aid other occultation observations. [NASA and J. Hester]

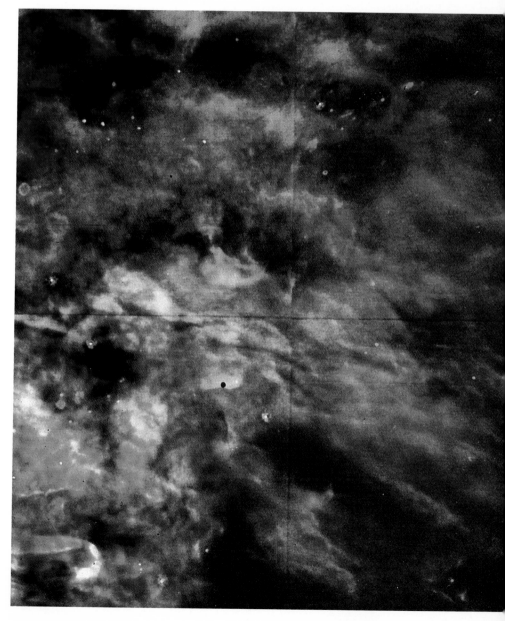

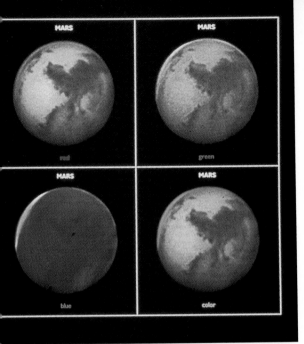

These black-and-white images of Mars were taken with *Hubble*'s planetary camera through red, green, and blue filters, at 8,900, 6,730, and 4,400 angstroms, respectively. Each image, exposed to Mars' reflected light for less than a second, was computer enhanced by subjecting it to a deconvolution algorithm about fifty times. Note that the red light is reflected (upper left) from the Martian surface, since this longer-wavelength radiation can penetrate the atmosphere in much the same way that red light can pierce fog on Earth. By contrast, the shorter-wavelength blue light (lower left) is reflected mostly from the upper cloud deck in the Martian atmosphere. The bottom-right image is a true color accumulation of the red, green, and blue images. [NASA and P. James]

The first true-color image of Jupiter was made by combining red, green, and blue exposures using *Hubble*'s wide-field and planetary camera. All the features seen here are cloud formations in the Jovian atmosphere, whose banded colors are thought to be caused by trace amounts of carbon, sulfur, and phosphorus and ammonia-ice crystals interacting within an extremely cold (about –280° F) environment. At lower right, the planet's Red Spot swirls—now no longer so red, for unknown reasons. Above the spot, we see a peculiar tentlike structure on the edge of Jupiter's south equatorial belt; below the spot, a so-called White Oval, one of several hurricanelike structures that formed about fifty years ago. The intent is to image Jupiter periodically with *Hubble*, not only in visible light, as here, but also in the ultraviolet and infrared, showing how clouds form at different heights in the atmosphere—thereby systematically monitoring the planet's weather over long periods of time. [NASA and J. Westphal]

4

Babel Revisited

And here it is fitting that all who intend to turn their attention to observations of this kind should receive certain cautions. For, in the first place, it is absolutely necessary for them to prepare a most perfect telescope, one which will show very bright objects distinct and free from any mistiness, and will magnify them at least four hundred times, for then it will show them as if only one-twentieth of their distance off. For unless the instrument be of such power, it will be in vain to attempt to view all the things which have been seen by me in the heavens, or which will be enumerated hereafter.

Sidereus Nuncius
GALILEO GALILEI, Venice, 1610

At a price of some $450 million dollars, the Perkin-Elmer Corporation of Danbury, Connecticut, delivered to NASA in 1984 the Optical Telescope Assembly, containing not only the mirrors used to capture cosmic light and direct it to Space Telescope's science instruments, but also the telescope's three fine-guidance sensors. The originally estimated cost, on the basis of which Perkin-Elmer, one of the world's premier optical shops, won the contract, was slightly less than $70 million, roughly one-sixth of the eventual bill that the government paid.

Touring the Science Institute shortly after launch, a retired vice-president of the Eastman Kodak Company, which had competed more than a decade earlier to build *Hubble*'s mirrors, described succinctly the flagrant underbidding by Perkin-Elmer: "Their low-balling was sinful." A Kodak-Itek team had bid $100 million, and this estimate included a thorough "end-to-end" test of mirror quality; this basic overall systems check was not part of Perkin-Elmer's proposal, and was not done—a circumstance that ultimately contributed to Space Telescope's myopia in orbit. "A very, very sad story," added the Kodak executive.

The design and manufacture of the complete optical package took some seven years and more than 4 million man-hours of effort. That's four and a half centuries of one person's work, an almost unimaginably larger effort than the few weeks needed by Galileo to fashion his pioneering ocular. Although *Hubble's* primary and secondary mirrors were modest in size compared with those of the largest ground-based telescopes, they were built to more exacting tolerances than for any comparably sized optical system ever made. The intent was to render a performance superior to any other astronomical instrument, to achieve for the first time in the history of sustained deep-space astronomy the "diffraction limit"—namely, the finest resolution possible with a mirror of given size, within the bounds of the known laws of physics. (In late 1989, Perkin-Elmer sold the division that had, during the years 1978–1982, designed, fabricated, and tested *Hubble's* optics, to Hughes Danbury Optical System, a subsidiary of Hughes Aircraft. Although the firm is known at the Science Institute as "h-doss," after its capital letters, in this book I shall hereafter refer to it as Hughes Danbury.)

In order to make Space Telescope as compact as possible—an essential ingredient given the need to deliver it to orbit packed within the cargo bay of the space shuttle—a Cassegrain design was chosen. Named after a French cleric who devised the basic idea in the seventeenth century, this type of optical system can be "folded" into a relatively small space. *Hubble's* focal length—the distance between the main mirror and the point at which it focuses light—is 189 feet (nearly 58 meters), or about three times as long as the shuttle, but because the light is additionally reflected by a secondary mirror back through a hole in the main mirror, a long focus can be attained within the confines of the telescope.

To be technically correct, *Hubble* is a Ritchey-Chretien telescope, a Cassegrain variant named after a pair of early twentieth-century opticians who proved (by modifying a 1637 treatise by the French mathematician René Descartes) that optical distortions throughout the field of view are minimized if both mirrors are shaped as hyperboloids, not as paraboloids or portions of spheres. At any rate, Space Telescope's large focal ratio of 24—its focal length

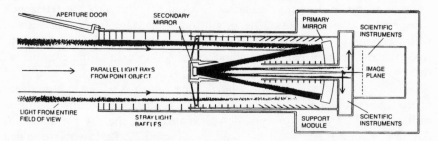

This simplified diagram of the light path within the Hubble Space Telescope shows the rays of a distant star entering the main telescope tube, past baffles equipped with fins designed to minimize unwanted stray light of bright objects such as the Sun, Moon, and nearby Earth from reaching the primary mirror. Reflecting from the concave front surface of this 94.5-inch (2.4-meter)-diameter mirror, the light returns some 16 feet (4.9 meters) up the tube to the smaller, 12-inch (0.3-meter)-diameter convex secondary mirror. The light is then re-reflected through the hole in the primary mirror to the telescope focus, some 5 feet (1.5 meters) behind the primary mirror. *(See also color insert section.)* *[Adapted from Perkin-Elmer documents]*

divided by the diameter of its main mirror, or "f/24"—derives from a common two-mirror design used during the last seventy-five years to build many
large ground-based telescopes. *Hubble's* basic optical architecture is not at all
novel or even special. It was the manufacture of such mirrors to an unprecedentedly high tolerance that was unique. (For comparison, the 2-foot hole at
the center of *Hubble's* main mirror is larger than the entire mirror aboard the
previously largest civilian space telescope—the 18-inch or 0.5-meter *International Ultraviolet Explorer* satellite, launched in 1978 and still operating in
geosynchronous orbit—a success story of the "old" NASA.)

The Optical Telescope Assembly uses a system of subsidiary mirrors to
feed light into the suite of science instruments, including the fine-guidance
sensors. Unlike most other space probes in which various instruments are
arrayed on a turntable that rotates them one at a time into the light path, the
science instruments and guidance sensors on Space Telescope are fed light
simultaneously. They are said to share the focal plane, requiring none of these
massive refrigerator-size instruments to move inside the spacecraft. Light from
the very center of the telescope's field of view is intercepted by a small "pick-
off" mirror, which reflects it to the side and into the wide-field and planetary
camera. The four other science instruments each receive the light in a middle
quadrant of the telescope's circular field of view. Around the outside of this
"science" field is the "guidance" field, which is divided among the three fine-
guidance sensors by an additional set of pick-off mirrors.

The primary-mirror blank, or block of glass, was furnished by Corning
Glass Works for about a million dollars and is made of their trademarked
Ultra-Low-Expansion silica glass, which is doped with titanium dioxide.
This material was chosen for its almost zero coefficient of expansion, so that
the telescope optics are minimally sensitive to thermal changes as *Hubble*
moves in and out of the sunshine every three-quarters of an hour. If it were
a solid slab of glass, sufficiently thick to remain stable during launch, *Hubble's* 8-foot primary mirror would have weighed some 4 tons. Instead, to
meet the shuttle's payload weight requirements, the mirror is a sandwich-
like construction of two 2-inch-thick faceplates separated by and fused to a
10-inch-thick core of glass ribs arranged in a rectangular grid pattern (much

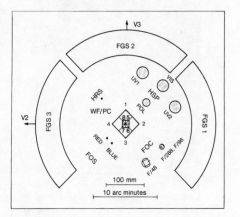

Space Telescope's field of view and focal
plane layout, as seen from the instrument
side, spans 28 arc minutes in diameter.
WF/PC denotes the wide-field (covering sectors 1–4) and planetary camera (covering
sectors 5–8). FOC denotes the faint-object
camera (with its two apertures). FOS
denotes the faint-object spectrograph, HRS
the high-resolution spectrograph, and HSP
the high-speed photometer (with its four
apertures, one for visible light, two for ultraviolet, and one for polarimetry). FGS denotes
the fine-guidance sensors. V2 and V3 are
principal axes of the spacecraft, referring to
pitch and yaw respectively. *[Adapted from
Perkin-Elmer documents]*

like a honeycomb or egg crate), resulting in a structure weighing only 1,800 pounds.

The 8-foot slab was first roughly shaved on a large, conventional grinding machine at Perkin-Elmer, transforming it from a flat and round transparent waffle into a bucketlike concave shape. In this two-year process, a disk periodically pushed down in precise strokes on a slurry of abrasive material spread on the glass, thereby removing nearly 200 pounds of glass mostly from the middle of the blank. As the huge chunk of glass eventually approached the desired hyperbolic shape, it was transferred to a sophisticated, computer-controlled polishing machine that employed finer abrasives, like jeweler's rouge, to refine it to an unprecedented surface quality; all the while, the mirror was lying on a novel fixture containing 134 counterweighted titanium levers (called a "bed of nails"), each designed to push up with a force equal to the weight of the glass above it, and thus to simulate the gravity-free environment of space.

At virtually every step of the manufacturing process—a total of some twenty times over eight months—the figure or shape of the mirror-to-be was measured with precision equipment thought to be capable of detecting the slightest blemish in its contours. The resulting largest deviation (microscopic bump or valley) anywhere on the surface is less than a half-millionth of an inch (or 0.01 micron). Subsequent testing of numerous small, isolated parts of the finished mirror showed its smoothness to have exceeded the desired specification by some 20 percent. *Hubble's* mirror is, quite simply, the smoothest large surface ever made.

Once polished, the glassy surface was coated with an exceedingly thin layer of aluminum to make it reflective. This was accomplished in a specially designed stainless-steel chamber that can release fine vapors in an ultra-high

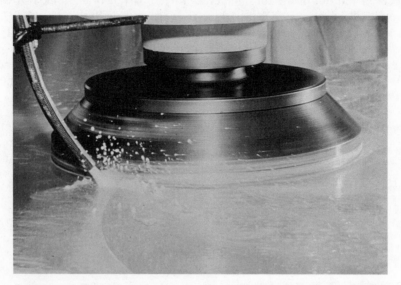

To fabricate "the perfect mirror," opticians and technicians of the Perkin-Elmer Corporation used a small, spinning pad (shown here sloshing around a slurry abrasive) whose grinding speed, pressure, and direction were controlled by a computer. *[Courtesy Perkin-Elmer]*

The rib structure of the lightweight core of *Hubble*'s primary mirror is clearly visible here, just before thinly aluminizing it to make it reflective. *[Courtesy Perkin-Elmer]*

vacuum. Although some three years were required to fabricate, grind, and polish the glass, only three minutes were needed to make it a mirror, coating it with a highly reflective layer of aluminum hardly three-millionths of an inch (0.065 micron) thick. Nevertheless, the procedure was so demanding that Perkin-Elmer engineers rehearsed it for nearly a year.

The thin metal surface was then covered with an even thinner layer (0.025 micron) of magnesium fluoride, which serves to protect the aluminum from oxidation and also enhances its reflectance in the ultraviolet part of the spectrum. The combined layers are so thin that if the coating could be removed and lifted into the air it would float for days, like a fine mist. The finished mirror has a total reflectance of about 85 percent at visible wavelengths and about 70 percent in the ultraviolet.

Hubble's finished primary mirror, here undergoing testing in 1981, is mounted in a titanium ring, the structural backbone that integrates the Optical Telescope Assembly into the spacecraft. *[Courtesy Perkin-Elmer]*

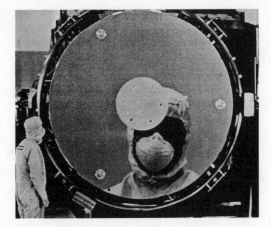

Behind the primary mirror and attached to its main structural ring is a support bulkhead, or "reaction plate," that spans the diameter of the mirror. Made of magnesium for light weight yet stiffness, this structure has two important tasks. First, the bulkhead carries an array of miniature heaters, which enable ground controllers to maintain the mirror within a degree of room temperature, 21°C (70°F). These heaters—of which there are about 800 onboard—thermostatically adjust the environment throughout the Optical Telescope Assembly, minimizing distortions in its mirror contours from uneven solar heating.

The bulkhead also supports a set of twenty-four "actuators" or motorized screws, connected to the back of the mirror. These screws, arrayed in two concentric rings of a dozen each, act like pistons; each is capable of exerting about 10 pounds of force on the mirror, and can be commanded to make small corrections to the shape of the mirror surface. However, the actuators can deshape the mirror by only about 0.01 micron and thus are meant for extremely fine-scale adjustments. (*Hubble*'s primary mirror distorts a great deal more than this as it transitions from the "1-g" environment of Earth's normal gravity to the "zero-g" conditions of outer space; the exact amount of distortion caused by orbital insertion is apparently either classified by the military or deemed proprietary information by two or three of America's best optical firms that have, together with the military, pioneered the fabrication and deployment of large spaceborne mirrors.)

The secondary-mirror assembly cantilevers off the front face of the main structural ring surrounding the primary mirror. For proper focus, the position of the secondary mirror, some 16 feet (4.9 meters) in front of the primary mirror, must be accurate to within one-ten thousandth of an inch (or 2.5 microns, which is one-thirtieth the thickness of this page), whether the telescope is operating in sunshine or the cold darkness of Earth's shadow. This precision has been achieved by designing the secondary mirror's principal structural element, called the "metering truss," to take full advantage of the unique properties of a special composite material made largely of graphite fiber reinforced with epoxy. (Again, like many structual elements of Space Telescope, graphite/epoxy was chosen for its high stiffness and light weight.) The metering truss was customized to ensure that the overall expansivity of the graphite/epoxy structure was close to zero. No mechanical fasteners are used to hold the truss together; it is completely self-bonding. The entire support structure is also wrapped in layers of aluminized Mylar, to keep its temperature as uniform as possible.

The secondary mirror itself is a convex hyperboloid approximately 12 inches (0.3 meter) in diameter and made of another kind of low-expansion specialty glass ceramic known as Zerodur^{TM}. It has the same type of "mag fluoride" coating atop an aluminum surface as does the primary mirror, and has similar reflecting qualities. Although much smaller than the primary mirror, the secondary mirror was almost as difficult to make, because of its steeply convex surface. Tests have shown that the final surface smoothness of the secondary mirror is even better than that of the primary mirror.

To help achieve the precise offset from the primary mirror, the secondary mirror is mounted on three pairs of force actuators, which control its position and orientation. These actuators are commanded from the ground to align the secondary mirror in just the right position to provide the best image quality. They can also be used to slide the secondary mirror back and forth by as much as 3 millimeters (or 3,000 microns) along the long axis of the telescope, and they can tilt the secondary relative to that axis; but they cannot be used, as with the primary, to change the shape, or figure, of the secondary mirror.

This entire optical system is mounted within a cylindrical telescope tube extending more than 10 feet beyond the secondary mirror in order to shield the optics from stray light coming from the Sun, Earth, and Moon. An extensive system of concentric baffles inside the front end of the tube provides additional shielding. And much of all of this tube and baffle combination is treated with Martin Black, which is not (as some thought) the name of a person but a code name for a classified, fragile, stealthlike paint made by the Martin-Marietta Corporation. The blackest of all materials on Earth, the porous Martin-Black substance ensures the purity of *Hubble*'s data by "absorbing" all light except that coming directly from the telescope's targets. Like almost everything else onboard the spacecraft, this paint became a cause for concern as launch neared. The telescope was mounted in the cargo bay aft-down, and some of the paint, having been applied nearly ten years before, could well have flaked off onto *Hubble*'s uncovered main mirror while *Discovery* vibrated during powered flight. No one knows how much of it actually did.

Not surprisingly perhaps, given Space Telescope's long and drawn out R&D effort, its two major contractors seemed at times to be trying to make the project as perverse as possible. A small example concerning the secondary mirror will suffice. Nearly a year before launch, in July 1989, with *Hubble* mounted vertically in its clean room at Lockheed, word began to circulate that no one had certified that the secondary mirror had been properly bolted and glued to the metering truss. Technicians then employed at Perkin-Elmer could not recall completing this task nearly a decade earlier and no one inside all of NASAdom had signed off on it. No corporate memory, no paper trail. So the engineers at Marshall went to work with their counterparts at Lockheed to concoct what must have been the world's most expensive diving board. The spacecraft was lowered to the horizontal position; a (small) man crouched at the end of the board with a bag of tools; the board was gingerly inserted into the telescope tube and past the razor-sharp baffles; and the man made his inspection. The mirror was fine. "P-E did its job okay, but NASA QC failed to follow through," read a report I saw.

Meanwhile, we at the Science Institute were cringing at the thought of a human being with a wrench and a bottle of glue so close to *Hubble*'s precious mirrors. "Humans break things" was one of our rallying cries. And, indeed, when the clean-room personnel reoriented the telescope to its dormant vertical state, they somehow managed to drive *Hubble*'s stern into the wall, destroying one of its low-gain antennas. The twin bills for the "diving board" and a new antenna were promptly sent to NASA, which given its spineless procurement and accounting system duly paid up.

*   *   *

None of the orbiting observatory's science instruments would be able to take advantage of its superb resolution if it were not for the fine-guidance sensors, which help to point *Hubble* and keep it steady while long exposures are taken. Each of these three devices, mounted radially around three of the four quadrants of the main ring of the primary mirror and fed light from that same mirror, houses a wave-front sensor and a guidance sensor. The former helps to align an array of small, rotatable mirrors inside each FGS for peak optical performance; the latter work in tandem to keep a cosmic object held firmly within *Hubble's* tiny sights. Each FGS, packed with microelectronics and yet more optics, is about the size of a large suitcase (1.6 x 3.3 x 5.3 feet) and on Earth weighs nearly 500 pounds.

All told, *Hubble's* various navigational sensors comprising its pioneering pointing-control system work essentially this way: First, the gyroscopes slew the vehicle with rough accuracy toward a preselected part of the sky. Next, *Hubble's* three fixed-head star trackers—miniature telescopes amidship, each with a large, 5-arc-degree field of view (some ten times the diameter of the full Moon)—aim the telescope with an accuracy of about an arc minute. Then, control of Space Telescope is automatically handed over to the FGSs, whose sensors are supposed to be able to discriminate among millions of stars, to "lock" onto a specified star's image, and to measure any apparent motion of that image with a much greater accuracy of 0.002 arc second—roughly equivalent to the width of a human hair at a range of 5 miles!

For those technically inclined readers who can't get enough of this minutia, here is a simplified account of how a bus-size spaceship locks onto a star light-years away. An intended guide star can be anywhere within the 4 x 16 arc-minute, pickle-shaped field of view of one of the fine-guidance sensors (there is always some uncertainty about the exact position both of the cataloged guide star and of the spacecraft itself), so the FGS must search everywhere in its own sight to find the chosen guide star. It does so by means of a pair of servomechanisms, acting much like optical gimbals, that steer small visible-light detectors (5 arc-second-squared view) in north-south and east-west directions to any position within the field of view. Encoders within each servomechanical system measure the exact coordinates at the center of the detector field. Usually, the detectors execute a spiral search pattern, starting at the expected star's position and spiraling outward until the detector finds the intended star.

While the search is nearing completion, Space Telescope enjoys a reasonably steady attitude, known as "coarse track," which is in fact better than guidance systems on any other civilian spacecraft. But not good enough, if *Hubble* is to achieve its finest resolution. So, once the appropriate guide star is found, the FGS hovers closely around its position in space and enters a condition known as "fine lock," holding the star's image exactly centered in its field of view. Fine lock can be maintained throughout half an orbit (that is, until Earth occults the cosmic target), because the encoders periodically send data ("error signals") on the star's position relative to the telescope axes to the pointing-control system—basically, the gyroscopes. This allows *Hubble* constantly to adjust its position ever so slightly and thus to remain stationary vis-à-vis a specified line of sight. Although *Hubble* cannot pitch or yaw much

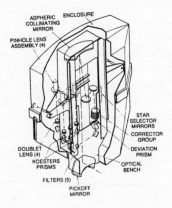
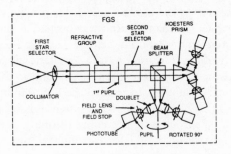

Left: A layout of the essential optics within one of the fine-guidance sensors aboard Space Telescope. The detector range of the onboard optics is strictly in the visible domain, from 4,670 to 7,000 angstroms, and is designed to work only for relatively bright stars, between the fourth and seventeenth magnitudes. Note how the optical path from *Hubble's* main mirror to the FGS detectors must be repeatedly "folded" in order to fit all the needed optics and electronics into each suitcase-size fine-guidance sensor. To help stabilize the optics themselves, the entire optical bench is made of graphite/epoxy and is temperature controlled with miniature heaters.

Right: This simplified schematic of one of the fine-guidance sensors shows how the detectors themselves form a pair of interferometers (called Koesters' prisms) fed by a beam splitter and coupled to photomultiplier tubes—one detector for each of two orthogonal axes. Operating on the arriving wave front from the distant guide star, the interferometers compare the phase at one edge of the telescope entrance aperture with the phase at the opposite edge. When the phases are equal, the star is exactly centered in the detectors' field of view; any phase difference shows a pointing error that must be corrected. *[Hughes Danbury]*

from a single guide star once fine-locked in this way, it could still roll around the position of that guide star. Thus, a second guide star is needed, within the field of view of another FGS, in order to keep the spacecraft steadfast in each of the two dimensions of the sky. The upshot is that Space Telescope—when not otherwise vibrating in the aftermath of a day/night terminator crossing—can equivalently track the motion of an aircraft's landing lights over San Francisco from Boston, or hold firmly on a dime perched atop the Washington Monument from an observing platform in Manhattan.

With any two of the three FGSs sufficient to provide the needed guidance signals to the onboard pointing-control system, the third FGS then becomes a science instrument for purposes of astrometry, the assiduous measurement of the positions of stars on the sky. Here astrometrists intend to use Space Telescope to refine our understanding of distances in the Universe. Using the method of parallax, which charts the positions of nearby stars at six-month intervals (with Earth on opposite sides of the Sun), a team led by Bill Jefferys of the University of Texas has planned to determine, with a superb 10-percent accuracy, the distances to all known stars within a few thousand light-years. This will enable us not only to build an intricate road map of our cosmic neighborhood, but also to better anchor our knowledge of the geometrical scale of realms far, far beyond our own.

A second, important science objective addressable by the fine-guidance sensors is the indirect search for planets beyond our solar system—an idea for which Giordano Bruno was burned at the stake in 1600. Stars with planetary companions should exhibit back-and-forth positional changes or periodic "wobbles," caused by the gravitational attraction of any unseen companion, some of which these sensors should be able to detect. Space Telescope seemed well positioned to resolve the longstanding debate of whether planets similar to those in our own solar system inhabit neighboring parts of the Milky Way. Alas, throughout the commissioning period, the fine-guidance sensors were mostly "on hold" for astrometry studies, owing to a plethora of problems from pointing jitter to fuzzy vision.

Throughout the tense days and weeks following first light, we at the Science Institute were distressed to hear of two distinctly different kinds of operations reports. One came from our own operations people who were often relating in closed briefings to the institute directorate that "*Hubble*'s pointing problems persist," that "the star trackers are working badly," and that "panic has set in while attempting to find the telescope's true focus." Much of these frank descriptions were elicited in response to Giacconi's continuing plea, "Be technically truthful, not Pollyannish."

Early on, guide-star acquisitions were especially problematic, even when searching for well-known, bright guide stars. The spiral search patterns of the fine-guidance sensors would often extend over relatively large areas of more than 120 arc seconds in diameter, whereas the stars' positions are generally known to a little better than one arc second. Even the "old-fashioned" fixed-head star trackers, of a type used on many previous space missions, should have gotten *Hubble* to within at least 60 arc seconds of where we wanted it to be pointing, but they themselves were apparently misaligned.

Of course, it didn't help that the star trackers' subcontractor had messed up: all three star trackers were supposed to have square fields of view, but one of them had been delivered to Perkin-Elmer with a round viewing field, accompanied by an apologetic note from the engineer who built it—a rather important piece of information, which had been misplaced and forgotten in the ten-year interval leading up to launch. Consequently, the odd star tracker was missing stars in the corner of its expected field of view, because it *had* no corners. All three of these little telescope trackers also turned out to be more sensitive than anticipated, and thus confused by stray glints of light, which they took to be stars. The result was that *Hubble* would drift ominously for hours at a time. Worse, some of the pattern matches between what the FGSs were seeing and the star fields expected from ground-based observations—not pictures per se, but merely plots of star positions on a piece of paper—were puzzling. Yet more troubling, much of what was seen in the sensors couldn't even be recognized from the existing star maps.

Many of these early pointing headaches were caused by a lack of synchronization—timing problems—between the hardware (the star trackers) and the software (the computer programs commanding them). But some were also, again, the result of human error. For example, a programming mistake made several years before launch by a Perkin-Elmer analyst, who had inserted a plus instead of a minus sign in a computer code meant to correct for the precession of stars, now puzzled other analysts struggling to decipher the first few pattern-matching tests. Then, too, the telescope's pointing instructions had been based on a 1954 sky chart, and the sky looked different nearly forty years later, because of the natural, millennia-long wobble of Earth on its axis.

Furthermore, the catalog of bright reference stars (sixth magnitude and brighter) used by the operations crews at Goddard to select the celestial targets of the star trackers was incomplete and laden with mistakes, yet the engineers there refused to allow the institute astronomers, who have much expertise in this area, to become involved. Worse still, from time to time and without warning, *Hubble*'s navigational sensors would observe "rogue" stars apparently not even listed in the star catalogs prepared by NASA's orbital verification team. The old adage, "garbage in, garbage out," applies not just to bug-ridden computer programs, but also to poorly compiled star catalogs.

Ironically, the institute's own *Guide Star Catalog* was of no use with the "ballpark" star trackers since it was meant to provide information only to the precision FGSs, and thus classifies only stars dimmer than the sixth magnitude. It would be nearly a year before a revised star-tracker catalog was ready; once it became available, telescope pointing noticeably improved midway through commissioning. Dismayingly, we learned that ground controllers had struggled with the same pointing anomalies on two previous NASA missions, at least one of which—the *Solar Maximum Mission* of the

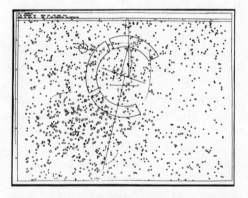

This is a copy of the first successful pattern match between stars detected by *Hubble*'s horseshoe-shaped fine-guidance sensors and those expected from ground-based catalogs. Not an image or picture, this plot of hundreds of open circles displays potential navigational (guide) stars of various brightnesses in the Northern Hemisphere star cluster, NGC 188. Since automated (computer-driven) matches repeatedly failed due to error-prone sky catalogs used by NASA, this observed pattern was pirated by astronomers at the Science Institute, manually marked up, and the stars within it successfully identified, after which it was transmitted back to ground controllers at Goddard who eventually adjusted the telescope pointing-control system accordingly. *[Space Telescope Science Institute]*

1980s—had used identical star trackers and had encountered identical problems.

Aerospace lessons not having been learned, or at least not passed along from one group of engineers to another, meant that *Hubble* was pointing off target by as much as a half-degree (1,800 arc seconds), which is the diameter of the full Moon. The result of all these pointing anomalies was a spate of increasingly sarcastic stories in the press, witness an Associated Press news wire that led off a mid-May story entitled "Hubble Space Telescope Draws a Blank": "NASA engineers struggled unsuccessfully yesterday to teach the Hubble Space Telescope exactly what in the heavens it is looking at. The telescope was told to look for stars of a certain brightness in what scientists thought was a rich field. It found none."

Aside from a host of hardware/software glitches and an embarrassing number of human blunders, both types of which were rapidly corrected during *Hubble*'s initial checkout, why were we persistently perplexed when attempting to acquire guide stars? Coarse track had usually aimed the telescope nicely once the appropriate stars had been acquired, but fine lock—*Hubble*'s normal operating mode—had not yet been achieved more than a month after launch. Worse, the fickle robot would occasionally just start drifting aimlessly again, as though completely baffled about which stars to focus on—causing more of those terror-filled moments when our eyeballs fully widened. All these efforts, among other more subtle and technical tests, were part of ongoing trials to collimate and focus Space Telescope by aligning its secondary mirror relative to its primary—a little like tuning a piano without touching it and while moving at nearly 18,000 miles per hour.

The project seemed confronted with a classic Catch-22 dilemma: accurate locks on guide-star pairs were required to focus the telescope's mirrors, but it was necessary to focus the mirrors before achieving a good solid lock onto guide stars. (In fact, it would take another eighteen months of testing, checking, and arguing before *Hubble* was declared "collimated as well as could be" to nearly everyone's satisfaction.) One of *Hubble*'s principal investigators, Bob Bless, wisely and repeatedly called into question the issue of holistic engineering: "It's a complicated system. Everything depends on everything else."

Orbital verification—the first part of the commissioning agenda, scheduled to last only a couple of months—had hardly begun, and at this rate would extend for more than half a year well into the fall. The telescope had in fact been idle (the euphemism for which is "station-keeping") much of the many weeks since launch, often fed "health and safety loads" that mostly kept it out of safemode; all the while an increasingly puzzled commissioning team based at Goddard tried to devise new pointing and focus diagnostics tests. "Actually, the FGSs work fine," said a frustrated tracking expert at the institute. "It's the people operating them who are troubled." Whatever, the bottom line was that, nearly two months into the mission, we didn't know how to reliably maneuver or point Space Telescope.

Growing lists of other, nagging difficulties were also piling up and a certain resignation set in; it had become abundantly clear that this space mission

that item within an intricately coded telescope time line—our troops were stuck in a seemingly endless state of planning and replanning, essentially in a continuous "target-of-opportunity" mode that was expected to occur very seldom in response to a rare astronomical event that would cause us to disrupt already built telescope schedules.

The project had slipped into a crisis mode of operations, which would undoubtedly increase the risk of commanding errors, which in turn would adversely affect the spacecraft. "Something aboard *Hubble* is going to break," complained an operations-crew member—an "ops jockey"—bridging the unhealthy interfaces among the institute schedulers, the Goddard commandos, and the Marshall contingent ostensibly in charge at Goddard. NASA's daily flight-status report in early June departed from its usual engineering jargon to note that, "All seems to be going rather well [with scheduling of *Hubble*] despite the fact that the content of the SMSs seems to vary on much less than a twenty-four-hour time scale. The Science Institute has been able to keep up so far, but we would guess that there must be some very overworked people up there. SMS generation is not very easy in the 'dynamic' working conditions they are working under."

To be sure, morale was beginning to sink, as the pressure and stress of the jobs in the operations areas would not let up. We could see it daily on the strained faces of our colleagues in the institute's OSS area, and especially in the planning and scheduling branch, where, to keep the orbiting beast fed, the unremitting concentration, shift after shift for weeks on end, would have severely taxed the country's busiest air-traffic controllers. It would only be a matter of time before someone in that pressure-cooker blew his brains out. Rodger Doxsey, one of our division chiefs and an astronomer himself, rightfully wailed, in a mid-June director's meeting at the institute, that neither NASA Marshall nor Hughes Danbury had any long-range, well-conceived, rational plan to analyze the telescope's enigmatic eyesight: "We are regularly receiving hit-and-miss requests for SMS calendar generation. No one seems in charge. A group meets daily [at Goddard] and decides by the seat of their pants what to do the next day. There is absolutely no method to their madness!"

In contrast to this troubling state of affairs as we saw it at the institute, a second kind of operations report was more sanitized, if not Pollyannish. These were the *Hubble* management reports to the upper echelon at NASA Headquarters—which otherwise was trying to keep its distance from the growing mess, all the while refusing to step in and lead, or at least to manage the project. Typical of the program status reports filed during this period, and which probably lulled Administrator Richard Truly and his lieutenants at the NASA summit into a false sense of security, was the following mid-June summary assessment:

The fundamental problem with the ground software that was preventing reliable acquisitions of guide stars has been resolved. Nearly all recently scheduled guide-star acquisitions have been fully successful. The previously addressed Pointing Control System oscillations are almost eliminated. Only the thermal (solar array) excitation remains, and that has been reduced significantly. Since solving the guide-star ground software problem, significant progress in the orbital verification activities is being achieved for both the optics and the scientific instruments.

About a week later, this rosy picture was followed by another positive summary to upper NASA management, which began, "The commissioning period is progressing steadily and well, with reliable achievement of test objectives. Coarse wave-front measurements were completed and the secondary mirror was moved 24 microns toward the primary mirror as part of the ongoing mirror focus exercises. By the end of June the fine focus activities should be completed."

In point of fact, little of this was true, nor was the only (to my knowledge) formal press release issued to the world at large by the Goddard Space Center during all of orbital verification: "At this time the telescope systems are healthy, in proper working order, with only occasional, minor anomalies to report." This, despite NASA's internal flight-status report listing at the time fourteen "significant unresolved anomalies" whose status was under investigation. "Hell," exaggerated astronomer Bob Bless, upon reading such grossly distorted reports, "everything that can go wrong has gone wrong. Mr. Murphy has been very busy." I began to realize why the White House had been calling me and perhaps other non-NASA *Hubble* insiders, earnestly seeking a more balanced assessment of *Hubble*'s status that was not altogether devoid of reality.

Regardless of the overconfidence of the agency's spin doctors, we all knew that the better part of a year is often needed to check out and to learn to operate a new ground-based telescope. Medieval cathedrals, too, took a lot longer to complete, and their construction had the advantage of being managed by authoritarian leaders. We also conceded that *Hubble*'s teething pains were exacerbated by the vehicle's great complexity and inaccessibility.

Although most scientists counseled patience, all the while embracing guarded, if not nervous, optimism about *Hubble*'s prospects, an occasional engineer was forthcoming about the difficulties ahead. Noting that you can't just reach over and twist a knob, Gene Oliver, one of the chief engineers of Marshall's orbital verification team, correctly cautioned: "Making complicated changes in the computer programs and the spacecraft 380 miles above the Earth is kind of like changing a sparkplug while driving down the road. We must be extremely careful in maintaining control of the vehicle and not causing more damage." But he also publicly admitted to a meeting of astronomers, for which he took flak the next day from uptight managers at NASA Headquarters, that "without resolution [of these problems], we do not have a Space Telescope."

What really worried the astronomers was the detailed appearance of the stars; they were so unlike what any of us expected from Space Telescope. If *Hubble*'s optics were optimally focused, about 70 percent of the light from any star would have been concentrated within the intended range of 0.1 arc second. But the early pictures with both cameras consistently showed that only about

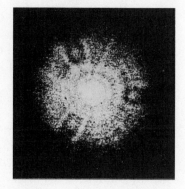

The devil's footprint: Looking like a point of light viewed through a frosted window, this detailed image of a single, second-magnitude test star, Iota Carina, as seen with *Hubble*'s planetary camera, shows that some of the starlight is concentrated within a small core, but much light is also spread out in a surrounding disk or halo. The scale of the entire boxed frame is about 4 arc seconds across. Some of the radial spikes and tendrils in the halo are caused by light diffracting off (or bending around) support structures inside the orbiting telescope, especially the four-vane spider carrying the secondary mirror. The three small circular features toward the extremities of the pattern and at the points of an equilateral triangle are images of the pads on the surface of the primary mirror fastening it to its support bulkhead. [NASA]

15 percent of each star's light was so concentrated. The remaining light energy was in the halo. Something was wrong. If most of the light from a star could not be precisely focused into a virtual point equal to or less that 0.1 arc second across, then *Hubble*'s exquisite resolution would be compromised and many of its scientific objectives unachievable.

"Level-1 requirement" is the terminology used by NASA to denote a critical specification that must be minimally met in order to fulfill a space-science mission. The *Hubble* mission has ninety-four such stringent "specs," most of which were achieved. Here, for the record, in technical but clear English, is the Level-1 requirement for the optical performance of Space Telescope, which was decidedly *not* met:

> The optical image, including effects of optical wave-front error, pointing stability, and alignment of the scientific instruments to the Optical Telescope Assembly, should satisfy the following on-axis requirements at 6,328 angstroms and be a design goal at ultraviolet wavelengths: Image resolution using the Rayleigh criterion for contrast of 0.10 second of arc. A full-width half-intensity diameter of 0.10 second of arc, 70 percent of the total energy of a stellar image must be contained within a radius of 0.10 second of arc. After correction for astigmatism, these specifications shall apply to the image quality over the entire usable field of Space Telescope.

Throughout the calamitous month following first light, and especially amidst strained "discussions" that often produced more heat than light, it was easiest for most project personnel to attribute the peculiar shape of a star's light energy to a simple, but correctable, lack of focus. The Hughes Danbury optical engineers argued that most of the light from a given star was not yet converging to a virtual point because they had been unable to locate the optimum separation (focus) and tilt (collimation, or alignment) between the primary and secondary mirrors.

The Hughes contingent in consort with the Marshall managers who were directing *Hubble*'s commissioning, still then centered at Goddard, were looking at the problem from the "front end"—attempting to assess the telescope's impaired vision by examining graphs of engineering telemetry radioed from *Hubble* via a set of onboard electro-optical devices (the wave-front sensors, discussed earlier). Working in concert with the fine-guidance sensors, these so-called "smart" gadgets give indirect indications of *Hubble's* focus by means of graphs and plots drawn up on the basis of data they collect. The orbital verification team had no pictures to look at, no images to analyze—except for the first-light images, which they largely ignored. Day after day, the engineers clustered at Goddard repeatedly tasked the institute to prepare specialized, unscheduled uplink transmissions in order to command *Hubble's* secondary mirror to move in, then out, then tilt, and so on, all the while apparently searching for the true focus by means that seemed to us astronomers like a brute-force method of repeated guesswork.

By contrast, at the Science Institute and among the camera-development teams, analyses were under way from the back end. Astronomers painstakingly dissected in intricate detail each stellar image itself—data acquired right down the barrel of the huge looking glass—attempting to solve the source of the focusing problem by unraveling the starlight captured by *Hubble's* cameras. Distressingly, these scientific analyses were suggesting serious trouble. We began hearing, in small, low-keyed briefings by our own optical experts, that one or more of *Hubble's* mirrors probably has a fundamental flaw—either "aberration," "coma," "astigmatism," or some combination of these imperfections. Some such optical distortions—especially coma and astigmatism—could be partly corrected from the ground by a judicious choice of mirror positions or by activating the small motorized screws mounted on the back of the primary mirror. But aberration was the worst of the optical flaws, if only because it would not likely be correctable from the ground. And it was aberration that a handful of institute scientists suspected was the main source of *Hubble's* blurriness.

As early as May 30, only ten days after first light, the institute's lead telescope scientist, Chris Burrows, a tall gangling British physicist on long-term loan from the European Space Agency, came to my office looking tired and serious, closed the door behind him, and showed me a thorough numerical analysis he had completed of the WF/PC first-light images. He had written a detailed, nineteen-page memorandum outlining the technical case for a "spherical aberration," an optical flaw that invariably blurs part of a star's light energy over a larger area than expected—indeed, a flaw so severe as to prohibit, even in principle, the achievement of true focus—and a condition that could be caused only by, as he put it, a "symmetrically deviant circular mirror," in short, a mirror of the wrong shape.

No slouch, Burrows is said to be the youngest person since Isaac Newton to earn a doctorate in mathematical physics from Cambridge University, and his memo was not a casual one. Given its dire implications, the memo needed to be argued closely, and it was. In places it read like an optometric treatise, and it was supported with charts, tables, and several computer simulations. Although I knew he had held this position with increasing conviction throughout the past week, I merely stared at him and then at his memo. The case for a severely flawed telescope still didn't seem to register. How could an eagle transform into a bat?

Burrows was reluctant to release his memo just yet. It was one thing to make oral arguments in the institute corridors or during private briefings behind closed doors, yet quite another to go out on a limb with such a catastrophic interpretation in writing. He needed some sort of confirmation—the hallmark of any good scientist. He asked me and a handful of others at the institute to join him in the OSS area that evening, when *Hubble's* "second light" would be transmitted to Earth. Although the engineers at Goddard had grabbed back the telescope after its inaugural 30-second exposure on May 20, we had arranged to get another shot of a real star ten days later, this time 100 seconds of exposure with the planetary camera. We didn't care which star— virtually any of the billions in the Milky Way would do nicely. But with the deeper exposure and the higher-resolution mode of the WF/PC, this new image should show us the strange fabric of starlight in greater detail.

The air that night in the OSS area seemed rarefied, the mood tense, although only a few people there knew that we were literally looking for trouble. The time approached, the telescope again failed to lock finely onto its guide stars, the exposure occurred anyway, and in minutes we saw the raw image. It indeed radiated trouble. Only a few stars could be clearly seen, and they unmistakably displayed all the earmarks of aberration. Up close to the imaging monitor in the OSS, these jewels of the night looked more like lumpy balls of cotton. I turned around to consult with Burrows, who was standing at a distance, his arms folded, his forehead deeply furrowed. Now he looked not just tired but also angry; he was living an optician's nightmare. No words passed between us as he intently stared at the monitor. There was no discussion of the star's light pattern, no instant analysis. I sensed a total deflation in my gut, like when all the air rushes out of a balloon. Burrows said only that he would now send the memo to NASA, and immediately left the room. I remained sitting quietly in the OSS for some time, a kind of listlessness paralyzing me, the case for a crippled telescope having finally sunk into my head.

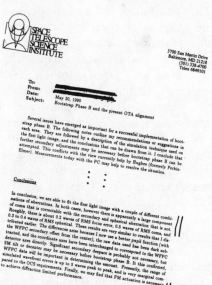

These first and last paragraphs of a pivotal memo, written by Chris Burrows, the lead telescope scientist at the Science Institute, sandwiched the technical case for *Hubble's* optical flaw. In English, it says essentially that, based on a detailed numerical analysis of the first-light image, Space Telescope suffers from a large amount of uncorrectable spherical aberration. *[Space Telescope Science Institute]*

What should I do now? As spokesman for the institute, whom should I tell? Of more immediate concern, what should I say later that evening when I was scheduled to deliver a *Hubble* talk following a private dinner attended by several senators?

I called Goddard to inquire what they thought of the latest image. Incredibly, most project personnel there had apparently gone home. As is the case at most government facilities, the parking lot empties at 5 o'clock, pivotal events notwithstanding. I asked instead to talk to the highest-ranking on-duty operations staffer. "What do you make of the latest image?" I asked.

"What image?" an engineering manager responded.

"The one just downlinked! Haven't you seen it?"

"Look," he said impatiently, "we don't have time to play with pictures in your science sandbox. We are working too many other issues, not least of which is why this jittery beast doesn't want to focus."

"Yes, I understand," I said quietly, and hung up.

Naturally, the Hughes engineers did not want to hear this kind of dour analysis. That *Hubble*'s big eye could be so critically impaired—indeed, uncorrectable from the ground—was simply unacceptable to the opticians privileged to have worked on this celebrated project. Nor was any of this comprehensible to NASA management, which was becoming increasingly frazzled in its search to understand what was really going on. In the opinion of NASA, which has historically rejected people bearing bad news until hit squarely in its face, Chris Burrows was to be ignored. Denial was a lot easier than thinking about the implications of such a major-league screw-up.

A week after the Burrows memo had been sent, the industrial and governmental opticians struck back with their own analysis. NASA's daily flight-status report for June 6 summarized their position: "At a meeting yesterday a brief report was given on the current image quality based on wave-front measurements by Hughes Danbury Optical Systems people here at Goddard. They concluded: focus ~1 arc second, astigmatism ~0.1 arc second, coma ~2 arc seconds, spherical aberration ~0.1 arc second." The last sentence meant that *Hubble* was substantially unfocused and uncollimated, but experiencing negligible astigmatism and aberration.

Thus was staged another classic confrontation between scientists and engineers. During the several months of orbital verification, the engineering teams from Hughes, Lockheed, and the Marshall Space Center were in charge of commissioning *Hubble*. They were responsible for bringing the telescope "on line" by aligning and calibrating its bulk optical systems, after which, in the science verification period, the scientists would gradually take charge of *Hubble,* first to tune up its onboard science instruments and ultimately to unlock secrets of the Universe.

For astronomers to be meddling in the work of engineers during the initial checkout of *Hubble* was anathema to most engineers on the project. Furthermore, the uneasy relationship between NASA and the Science Institute did not help matters, for, aside from our calendar building and telescope scheduling of the engineers' tests, we at the institute were to be minimally involved in orbital verification. Astronomers were the "end users" of *Hubble,* and we were essentially told to stay out of the way until NASA had fully steered the robotic observatory through its space trials—all of which, indeed, were mounting daily.

Meanwhile, NASA's public-information machine had ground to a halt. Marshall's occasional press briefings had stopped, Goddard was in a strictly defensive posture, and NASA Headquarters' public-affairs people were doing nothing at all on behalf of the *Hubble* project. Also ended were occasional two-way telephone conferences between agency officials and scores of journalists—telecons aimed at staving off the press, in hindsight famous for their stonewalling and evasiveness—which the media in any case had come to call "NASA's amateur hour," for they must have smelled the rat. In the resulting news vacuum, the press just retold and often amplified stories of earlier telescope problems.

All media inquiries directed to the institute had to be redirected to Goddard, which was in no position to field them, having in previous weeks lost several of its key public-information personnel. Nonetheless, we were instructed several times in writing by the Goddard project office not to talk to the press about spacecraft ills, and not to allow any of its representatives into our operations areas; take this June e-mail directive from Goddard to the institute: "[The project office] considered whether to lift its isolation of STScI and STOCC operations areas from the media at this time, and has decided NOT to do so. . . . We expect your continued compliance with this restriction."

As the scientist charged with overseeing the institute's Educational and Public Affairs Office, I found this media ban something of a dilemma. Even though we knew the answers to relevant media questions, NASA's gag order required us to pass the questions on to public-affairs people elsewhere in the project, only to see in the next day's papers and television programs a distorted and incomplete view of the truth. It was never clear to me whether the inaccuracies and omissions occurred because NASA personnel were simply lacking in knowledge of Space Telescope and its status on any given day, or because the media had by this time become increasingly convinced that all was not well with the project. I suspect it was a combination of both factors.

At one point in mid-June, the media did file accurate reports of *Hubble*'s mounting disorders, but the facts came mainly from scientists who had convened in Albuquerque for the summer meeting of the American Astronomical Society. There, some of the world's best science writers had gathered to hear about *Hubble*'s early science results subsequent to first light. Instead, they encountered a number of astronomers privy to the telescope's condition, and the word began to circulate that there were problems aplenty. On June 15, John Noble Wilford put things clearly in the *New York Times* when he stated, under the headline, "Troubles Continue to Plague Orbiting Hubble Telescope": "Seven weeks after its launching, the Hubble space telescope is giving controllers and engineers fits as they struggle to overcome difficulties with antennas, guidance sensors, star trackers and the spacecraft's bad case of the jitters. . . . It is uncertain whether the Hubble will fulfill expectations." And on the same day, Lee Dye orchestrated the front-page headline in the *Los Angeles Times,* "Hubble's Hopes Clouded," leading correctly with the statement, "The problems plaguing the Hubble Space Telescope are far more complex than the nagging troubles it initially suffered, and if not resolved will undermine the effectiveness of the instrument." None of these news accounts, however, revealed what Giacconi, during a formal briefing at Albuquerque, did in fact mention, albeit only in passing: the telescope was suffering from aberrated vision, like one of his bad dreams come true.

Feeling that their astronomical expertise was being summarily ignored, senior personnel at the institute, who by now had become almost certain that a serious defect must be plaguing *Hubble,* had little recourse but to recommend that more pictures be taken with many different focus settings, to be adjusted by moving and tilting the telescope's secondary mirror. Nearly day after day throughout the first few weeks in June, hundreds of stellar images of the standard target, Iota Carina, were obtained, especially with the planetary camera. Iota Carina is a bright (second-magnitude), isolated supergiant star some 750 light-years distant, and not far on the sky from the first-light object, NGC 3532. Working very long hours and appearing increasingly haggard with the passage of each day, Chris Burrows mounted an irrefutable case for spherical aberration. His numerical simulations—computer models of the specific routes taken by light waves multiply reflected inside a flawed telescope—reproduced with great precision the observed fuzzy images of the Carina test star. One of *Hubble's* mirrors, I remember him telling me somberly, must have a gross design or manufacturing error.

At one and the same time, NASA was hearing—and fearing—the institute's viewpoint. Not until the second half of June did the Marshall/Hughes contingent of the orbital verification team obliquely admit that it was bamboozled. It had tried in vain virtually every option to tweak the telescope; no one quite knew what to do next. A NASA daily flight-status report (none of which were made public) haplessly broached the growing guesswork: "Follow-up planetary camera pictures show little change in the image quality. Reexamination of the . . . data suggests that the reason for this lack of change in the image may be that [last night's secondary] mirror adjustment moved through focus to a position about the same distance from focus as they were before the adjustment." A few days later, the flight-status report had this to say, in part: "The images taken after the tilt and decentering

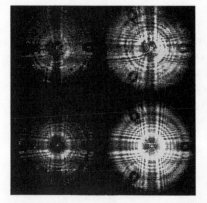

Comparisons between observed (top) and simulated (bottom) stellar images for different positions of *Hubble's* secondary mirror. The observed images were made with the planetary camera near 5,000 angstroms. Note how the simulations at the bottom, done at the Science Institute, mimic astonishingly well, almost feature for feature, the observed spread of light around the test star, Iota Carina, at the top. These numerical simulations took specific and detailed account of the optical path inside Space Telescope and include here the effects of mirror deformity; the circular, concentric features are called "diffraction rings" and the spokes "radial tendrils," the latter caused mostly by the secondary-mirror support structures inside the telescope tube. The close correspondence between the computer models and the observed images comprises the single most irrefutable factor in the case for aberration. *[Space Telescope Science Institute]*

movement of the secondary show essentially no significant difference between the before and after images. Resolution of the puzzle must await further wave-front measurements." The opticians were literally hunting in the dark.

Days later, NASA officials began cautioning me not to use publicly the "s word"—by which they meant, I guess, spherical aberration. It was the agency's first acknowledgment that Burrows might be right. They especially ordered me to make no mention of *Hubble*'s growing pains to a group of some 300 high-school teachers who were attending a nationwide teacher workshop that I was hosting at Johns Hopkins, yet another of our outreach programs that I had transferred across the street to relieve it from the Goddard micromanagers who hassled us over such educational activities. It irked me then, and still does now, that these leading educators from throughout the country were deprived of a technically honest assessment of the *Hubble* mission, and of how the scientific method really works, including all of its subjective baggage that scientists struggle with every day to puzzle through real, nontextbook problems.

Trying to respond to White House inquiries during this period was also proving tricky. All I could do was to appeal, in layman's—though accurate— terms, to the technical subtleties of this complex mission, while placating NASA by trying to keep a lid on the worst of *Hubble*'s troubles. On each occasion, I gave an honest appraisal of our ups and downs, admittedly colored by our view at the institute, and hoped that the caller wouldn't leak anything to the press, which he didn't.

Cosmic images of test stars were now rapidly reaching our laboratories. The faint-object camera was also now online, contributing additional views of a blurry Universe. We were suddenly awash in data, real images of real stars at numerous focus settings. Most of these tests were conducted with *Hubble* aimed at the same target, Iota Carina. I had never imagined that anyone would look at a single star so closely; indeed, it is probably fair to say that Iota Carina's booming image has now been studied more exhaustively than that of any other star (save the Sun) in the history of astronomy.

I also again urged that the telescope be reoriented a mere few degrees west to take a quick snapshot of Eta Carinae, a spectacular region of colorfully glowing gas that had been the number-one priority in the now defunct program of early-release observations. Such an image, I argued, would give us a quick assessment of how well or badly we could do science with a more representative cosmic object, even if that object was somewhat unfocused—an observation that was done months later with great success. But at the time, no one at NASA was listening.

During the third week of June, the institute's imaging gurus managed to convince the skeptical wide-field camera team that *Hubble*'s optics were suffering from an aberration. This led to a direct and unpleasant confrontation with the engineers from Hughes and Marshall, now ever more isolated. A series of uptight meetings at Goddard and teleconferences back to Hughes Danbury thus began the first formal and substantive discussions between astronomers and engineers regarding optical quality. Few could agree on what the insti-

A typical, reworked exposure log sheet prepared in early June at the Science Institute for uplinking to the spacecraft. Its accompanying justification specified that a sequence of observations of the test star, Iota Carina, be taken at several wavelengths and exposure settings, all in an attempt to "derive the aberration of Hubble's Optical Telescope Assembly . . . and will enable further investigation of the 'fingers' [radial tendrils] already observed." [Space Telescope Science Institute]

tute's image analyses meant, or even what the central problem was.

Some Hughes opticians argued that *Hubble* was still way out of focus and could be improved; others admitted that the telescope was focused about as well as it ever was going to be. The dreaded term "spherical aberration" was thrown around liberally, but with much dispute over how severely *Hubble* was affected by it. The arguments predictably ran along technocultural lines; the scientists were largely convinced that *Hubble* was the victim of a serious case of bad optics, while the engineers, obviously on the defensive at these meetings, maintained that the data were inconclusive, ambiguous, and contradictory. More diagnostic data were needed, said the orbital verification team. Thus commenced a marathon session of observations requiring extraordinary efforts by the calendar builders at the Science Institute, virtual open-channel communications to *Hubble* via the TDRSS network, and much real-time commanding of the telescope. The situation was desperate: more tilting and moving of the secondary mirror back and forth toward the primary, more sensor data acquired, dozens more images taken of Iota Carina, more meetings, more analysis, more round-the-clock arguing. More exhaustion.

By Friday, June 22, most people were convinced that one of the onboard mirrors was indeed defective. Which mirror was faulty was unclear, although it was the institute's position that Burrows' latest numerical simulations had shown the most likely culprit to be the main 8-foot mirror at the very core of the telescope. What could be done about the flaw was also unclear, but it was certainly not correctable from the ground; the error was just too large. It also seemed improbable that *Hubble* would achieve perhaps its most important science objective of both sharp resolution and deep sensitivity; its astronomy mission was clearly jeopardized.

Some of the Marshall engineers, who only recently had come to realize the enormous implications of *Hubble's* flaw, were extremely despondent. A few became physically or emotionally sick, others just dazed by sensory over-load; another compared his checkmate to a death in the family. The project's chief engineer at Goddard expressed concern that there might be suicides. The Hughes staff seemed numb but guarded; they were doubtless worried about their jobs, their careers. Their parent company assigned several dozen more engineers to work on the problem, still trying to discover a mistake in the analysis implicating the flaw or at least a way around it.

Those institute scientists, who knew of the consensus now reached, were of course both devastated and disgusted. The principal investigator for the faint-object camera, an unusually level-headed Italian astronomer, Duccio Macchetto, summed it up for many of us when he suddenly blurted out one evening at the institute, "I just want to shoot someone." (Later, he amended his gut reaction: "You have to realize even that's not going to buy us a new telescope. But I've spent fifteen years of my life on this, and it is really terribly disappointing.")

Still, the media (and thus the public) had no inkling of *Hubble's* sternest affliction, nor did members of the nation's scientific community, who had been fed increasingly infrequent and ambiguous flight-status reports from NASA—reports that gave little hint of the telescope's true problems. And then, during the third week in June, the space agency's daily status reports stopped altogether and without explanation, whereupon astronomers from around the world began increasingly pulsing the institute via e-mail to find out what was going on.

At Hughes Danbury's headquarters in Connecticut—according to astronomer Mark Stier, one of my former Harvard teaching assistants who was working there—the air was thick with grief. "It's like the living dead here," he said. The news from the *Hubble* trenches was sickening, impossible—a feeling of sheer disbelief. Depression was widespread, embarrassment overwhelming. "I can never again attend an astronomy meeting, or even show my face among astronomers," confessed Stier, an extraordinarily skilled technologist, like most of the other Hughes employees who had nothing to do with *Hubble's* mirror fabrication.

Again, there was concern at the plant that someone might do something foolish to hurt themselves, worsening what was already a tragedy. Skilled opti-cians elsewhere who managed to find out about the developing fiasco called the institute, such as a dark-side associate at Lincoln Lab experienced in space-craft commissioning, and who had earlier worked at Perkin-Elmer on *Hubble's* optics, urging us to check this or that. We had. And have you done that or this? Yes, every conceivable option had been pursued. There seemed no way to escape the logic, no way to squirm out of acknowledging one of the great technical debacles of the twentieth century. Aberration couldn't be kept secret forever.

All of *Hubble's* earlier drawbacks—many of them expected on a project as complex and technically demanding as this one—now paled in comparison. The long pole in the tent had been identified. The startling significance of a fundamentally deformed device at the heart of the multi-billion-dollar tele-scope was driven home for all the principal parties to hear, to see, and to internalize. Said a fazed Terry Facie, the chief scientist for *Hubble* at Hughes, "It's the very foundation stone of the telescope that's flawed. That's the sad part."

Following yet another series of failed focus attempts on the weekend of June 23–24—these managing to convince even the diehard unconvinced—project officials elected to tell the NASA hierarchy the truth that the *Hubble* mission had experienced an insurmountable setback: in NASAspeak, a "failure to meet a Level-1 requirement"—in other words, the dreaded show-stopper. "Black Monday" was the day when everyone finally agreed. In a somber meeting at NASA Headquarters that only federal employees attended, Lennard Fisk offered his opinion that *Hubble*'s failure was the space-science equivalent of the *Challenger* disaster. He also reportedly cried that evening. At the time and despite *Hubble*'s shortcomings, I thought this was an overreaction, and now I know it was. But it was symptomatic of the kind of exaggeration that continued to haunt the mission, owing largely to the way in which the pendulum-like NASA management issued its public statements.

NASA's press and educational materials had first hyped the telescope's capabilities, then engineers reluctantly confessed to *Hubble*'s early operating snafus, then the sanitized flight-status reports to headquarters suggested that all was fine, and now the agency's chief scientist was making dramatic, pessimistic claims about the effects of aberration—a bombshell that could only precipitate another public crisis of confidence in the U.S. space program.

What is spherical aberration? And why does it prohibit achieving a true focus with the Hubble Space Telescope? Concave mirrors act like "light buckets," capturing radiation from distant cosmic objects and reflecting it to a focus. In the simplest kind of telescope, a Newtonian reflector much like the one first made in 1668 by Sir Isaac Newton, all the light captured by a concave mirror converges to a point halfway across the radius of curvature of the mirror, as shown in one of the nearby illustrations. But it is not sufficient to grind the mirror into a concave shape, or "figure," that is simply a section of a sphere. Light collected across the face of such a "perfect" spherical curve does not all arrive at the same focus, hence the term "spherical aberration." Also, if the mirror is overly curved, the outer edge of the mirror focuses light closer to the mirror than the halfway point ("marginal focus") than does the center of the mirror ("paraxial focus"). Likewise and for the converse reasons, a mirror whose figure is too shallow is also unable to bring all of its captured light to the same "true" focus.

Thus, spherical aberration is an undesired optical condition induced by a circular mirror having a deformed figure—that is, a curvature of the wrong "prescription." It has nothing to do with the smoothness of the mirror, but depends only, yet critically, on the shape of the mirror. Unable to focus all the light to a single point, such a misshapen mirror actually displays a range of foci. Alignment of the optics can emphasize the light captured either by the outer or by the inner parts of the mirror, but, as illustrated, the resulting image, regardless of these attempts, is a bright point surrounded by a disk of diffuse light.

Because *Hubble* is a variation of a Cassegrain telescope, its captured light undergoes reflection from a secondary mirror, which itself can be flawed, thus contributing to "higher-order aberrations." Furthermore, if either mirror of the telescope assembly has some tilt relative to the incoming light rays—an "uncollimated telescope"—the resulting aberration can be additionally complicated, causing the shape of a star's blurry underlying disk to have a tail not unlike the image of a comet. This effect is called "coma," an undesirable condition that some of us at the Science Institute thought we noticed in the early *Hubble* images. What's more, if sources of light are located far off the central axis of the telescope (defined by the line OQ in the accompanying figure), yet another form of aberration, called "astigmatism," can be observed, distorting point sources (such as stars) into false patterns such as circles or straight lines.

Working backward by matching computer computations of theoretical images, modeled using a variety of possible *Hubble* mirror anatomies, to actual images taken of early test stars with both the wide-field and planetary camera and the faint-object camera, Science Institute personnel proved that the culprit is the main mirror, not the secondary. Additional numerical simulations showed that the type of aberration is indeed spherical—in fact, astonishingly symmetrical, a "textbook" case. The mirror is too flat; the outer edge deviates from the expected curvature by a minute 0.0001 inch, or about one-twenty-fifth of the diameter of a single strand of human hair. And since the

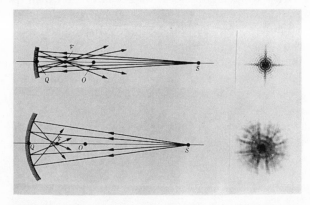

Top: A perfectly shaped concave mirror brings all of its captured light to a single sharp point, called the focus. Here, in opticians' symbols, the center of the mirror is designated Q, the center of curvature of the mirror O. Light rays of a distant point source at S (considered for all practical purposes to be at infinity) then converge to a true focus at F, precisely halfway between Q and O. The result of such a perfectly focused mirror is a bright point with faint (diffraction) rings around it, such as the negative image of a star shown at the right.

Bottom: The blurriness of spherical aberration stems from an irregularity in the shape of the mirror that does not permit all the captured light to collect at a single point at F. Instead, a distorted spot of diffuse illumination appears with a bright point at its center, since different parts of the mirror focus light to positions in and around, but not all directly on, point F. Such an ugly image, shown at right to same scale as above, is inconsistent with that expected from a telescope that is merely unfocused or uncollimated (misaligned). Rather the excessive blurring can be understood only if a substantial amount of spherical aberration has been effectively ground into the mirror. *[Space Telescope Science Institute]*

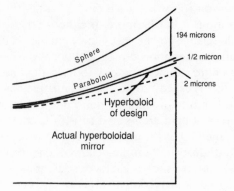

Hubble's main mirror is, in reality (dashed line), about 0.0001 inch (or 2 microns) flatter than it should have been (solid line, labeled "hyperboloid of design"). Too much glass was removed from near its edge during the polishing process. Although this error sounds small in our everyday world, 2 microns (which is about forty times thinner than this piece of paper) is actually a gross imperfection in the world of precision optical instruments. The result is a fundamental inability to concentrate even a majority of the mirror's captured light into a single point. (Not drawn to scale.)

outer half of any circle contains 90 percent of its area, the mirror is misshapen where it captures the majority of its light.

(Technically, there is one-half wave of spherical aberration at 5,500 angstroms, compared to the design specification of one-fiftieth of the wavelength of light, whereas Perkin-Elmer had officially reported to NASA, and NASA had accepted in writing, that the mirror is accurate and smooth to one-sixty-fifth of such a wavelength—some 20 percent better than spec. This half-wave of aberration is an rms, or root-mean-square, value for yellow-green light, which translates into a total figuring error of about 4 waves, or 2 microns, on the surface of the primary mirror.)

Such is modern science, big science. A microscopic imperfection only a few percent of the width of a human hair can nearly cripple a $2-billion piece of scientific apparatus.

Throughout the second half of June, with most of the world still ignorant of the time bomb that was *Hubble,* a cottage industry was born at the Science

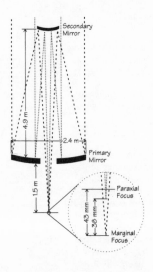

Hubble's ill-made optics caused light rays from different radii on the main mirror to focus at different distances from that mirror. The marginal focus formed by light (long dashed lines) collected by the outer parts of the mirror is 1.5 inches (or 38 millimeters) beyond the focus of the innermost rays (short dashed lines) that graze the secondary mirror. Technically, for a doughnut-shaped mirror there is no paraxial focus formed by the light normally collected by the center of the mirror. The adopted "best focus" for *Hubble* is about 0.5 inch beyond the (imaginary) paraxial focus position. This drawing is exaggerated to illustrate the point—or lack of it!—more clearly. *[Space Telescope Science Institute]*

Institute. Without any instruction from the institute's directorate or its division heads, several groups of astronomers informally began exploring how much of the intended resolution could be recovered by computer analyzing the blurred images. In principle, if you merely subtracted the "noisy" halos surrounding each source of light in a raw image, you could "clean" it for greater clarity—a technique known in the trade as "deconvolution," a fifty-dollar word for image reconstruction, or computer enhancement.

Admittedly, we would lose some sensitivity, since most of the acquired radiation (namely, that in the halos) would necessarily be discarded—the price to pay to restore the image. Such sophisticated algorithms have been used for years by intelligence experts to identify foreign assets for arms-control compliance and to examine ground terrain for cruise missile maps, among other reconnaissance tasks; radio astronomers have also successfully used a similar method—specifically, a computer code called Clean—to sharpen cosmic radio images (or "radiographs") made with the Very Large Array Radio Telescope in New Mexico. "Computer magic" of this sort also sharpened *Voyager*'s stunning views of the solar system's outer planets during the 1980s. What's more, we had fully expected that a few years after launch we would begin using such image-enhancement techniques in order to extract every morsel from the *Hubble* data, even if they had been rendered with nearly perfect telescope optics.

Two factors limit our accuracy here. First, if the exact shape of the halo is not known, it cannot be subtracted perfectly. This requires us to measure the point-spread function for the specific focus setting of *Hubble*'s secondary mirror and for the specific wavelength filter used. Ideally, we also need to know how the point-spread function varies from the center to the edge of any of *Hubble*'s fields of view. As scientists gradually derive improved point-spread functions, through repeated observations with many camera modes and filter settings, the halos should be characterized much more accurately, which in turn should allow for significantly better restorations.

A second, more fundamental limit is set by the "noise" in the observed images. Astronomers measure starlight by counting individual particles of light, or photons, that are collected by Space Telescope. For the faintest stars, only a few photons are counted, and there is consequently an inherent uncertainty in the brightness measured for those stars. This uncertainty makes it impossible to subtract halos around even the brightest stars perfectly. In fact, even to detect the presence of faint stars amidst the glare created by many overlapping halos becomes a virtually impossible proposition. This is the greatest loss to the *Hubble* mission, at least until the telescope's faulty optics are corrected: the faintest cosmic objects simply cannot be seen through the haze of crowded fields.

We knew we would have to exercise great caution when using computer-enhancement programs, for these two limiting factors can sometimes produce false features in the images—artifacts implying structures that aren't really there. The more the data are subjected to any deconvolution algorithm—computer iterations—the greater the chance that the process of deconvolution itself will amplify noise peaks—that is, add noise in a coherent way that then looks like a real cosmic feature.

Accordingly (and not having the benefit of knowledge gained by years of image-enhancement evaluation by the military-intelligence community), astronomers have spent a great deal of time and effort trying to arrive at a best compromise between limited enhancement of sharp features and excessive amplification of background noise. This has become a bit of an art as much as a science, and for some images that means ten iterations are acceptable, for others perhaps twenty or as many as fifty iterations; in only one case, that of the nebula Eta Carinae, to be discussed toward the end of Chapter 7, were there well more than a hundred deconvolution iterations applied to the raw Hubble data. As it turned out, a small minority of astronomers refused to massage their images at all, feeling that deconvolution was "simply too much magic."

To address the oft-asked question—Why can't similar deconvolution techniques be applied to ground-based images in order to improve their resolution, too?—the answer is that in a few special cases they can, but in general it is very difficult. Stellar images recorded by ground-based telescopes lack the well-defined cores of those seen with *Hubble;* hence an underlying halo cannot be cleanly identified and then subtracted. Furthermore, atmospheric instabilities continuously change the apparent position of a star ("twinkling"), making it tricky at best to determine a fixed stellar profile.

One evening at the institute, amidst intense disputes regarding what we could do about the aberration fiasco, Pat Seitzer, one of the scientists spearheading these computer-enhancement efforts, came by my office with some very encouraging results. He had programmed his Sun Sparcstation to suppress the unwanted halos of all the simulated stars in a hypothetical, dense, crowded star cluster—a realistic, tough challenge for the impaired *Hubble*. No sophisticated computer algorithms were used—just the simplest method available for cleaning an image, much like heightening the contrast on a television monitor. He then compared the result to a simulation of the finest imagery that could be acquired toward the same star field, assuming the best seeing conditions on the ground. The comparison was an eye-opener. Even the cursorily cleaned *Hubble* simulation dramatically showed that much of the superb resolution expected of Space Telescope was preserved and that *Hubble* would still be able to see cosmic objects a good deal more clearly than ever before.

Hubble's loss, we all now agreed, was to sensitivity, not to resolution, an important point that the media refused to get straight for months thereafter. The orbiting observatory would surely be unable to study unprecedentedly faint objects in the Universe, but it would indeed be able to see already known objects with spectacularly improved clarity. Aberration would not be the absolute show-stopper that some had feared. (Even today, articles in some leading newspapers and weekly news magazines get this distinction wrong, and often reversed—and sometimes just mixed up, like the Associated Press that had a penchant for reporting that *Hubble* lost two magnitudes of resolution.)

Considering the importance of these two intertwined terms—sensitivity and resolution—and given not only that they point up the greatest disadvantage to the aberrated *Hubble* but also that they summarize the single most pressing need to repair the telescope, let us state the issue flatfootedly: Space Telescope suffered from lack of sensitivity, not lack of resolution—it could not see truly faint objects, but those it could see, it could see sharply. Because about 15 percent of the acquired light was concentrated into a tight core, and aided by deconvolution, *Hubble* was still able to realize its expected leap forward in angular resolution. As advertised before launch, it could definitely discriminate between two similarly intense sources of light only 0.1 arc second apart. But the trade-off was that deconvolution discarded—literally threw away—the 85 percent of the light scattered into an object's halo.

Accordingly, the telescope could not detect faint objects by the light contained within their imaged core; most of their light was hopelessly unfocused within a larger disk. Depending on the science instrument and the aperture and filter used, the impaired *Hubble* was a factor of four to ten times less sensitive than expected, which amounted to a loss of about two or three magnitudes of sensitivity. An analogy is that we would only be able to see the tip of the iceberg, whereas *Hubble* was built to allow us to map the entire iceberg. It was in this sense that astronomers referred to *Hubble* as a "photon-starved" telescope. In popular terms, the disabled *Hubble* granted us not blurred vision as much as vision through dirty glasses.

We passed on to NASA this promising development of restoring much of *Hubble's* intended clarity of view, expecting to interest one or another of the agency's middle managers or technical officers that the fine art of computer enhancement might allow us to identify a way around part of the problem. We were also seeking to persuade the space agency to let us acquire some *Hubble* images of cosmic regions more interesting than isolated test stars. The fact that we were playing around with simulated, rather than actual, images of

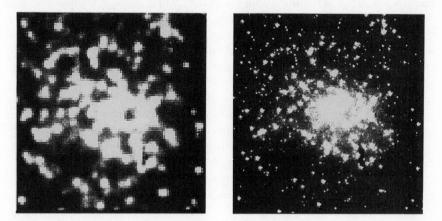

This pair of computer simulations shows how well the purblinded *Hubble* can see in a dense, crowded, hypothetical star cluster containing some 10,000 stars. The simulated image on the left is representative of the very best that can be done from the ground; seeing was taken to be an excellent 0.5 arc second. The image on the right is a simulation of a slightly enhanced or "cleaned" *Hubble* image of the same star cluster in which some but not all of the unwanted halos, caused by aberration, have been surpressed; seeing here is 0.1 arc second. Clearly, the image on the right is sharper, proving that the impaired spacecraft can still likely make some very fine astronomical observations with a high degree of clarity. More sophisticated computer algorithms would later sharpen *Hubble* images even more than this exploratory simulation. *[Space Telescope Science Institute and P. Seitzer]*

star clusters two months into the mission made even more obvious the need for real data from typical astronomical targets. But our arguments fell on deaf ears, partly because none of the telescope's handlers were working scientists having a direct stake in the outcome, and partly, I think, because it was the Science Institute making the suggestion.

The response from the Goddard project office was predictable, and as before: "Don't even mention the 's' word." Distressed and alienated, this was the point at which we at the institute decided to distance ourselves somewhat from NASA. We resolved neither to attack the agency nor to support it, yet we did feel that it was heading for a very big fall indeed.

On June 27 and 28, at the Goddard Space Center, the dozen or so *Hubble* principals of the Science Working Group, who had guided the science aspects of the *Hubble* project from its inception, were joined by another dozen or so astronomers of a newly formed Users' Committee, who were to advise NASA and the institute on the telescope's use during its first few years of operation. This meeting was declared closed to all but persons having a direct involvement in the *Hubble* program. And a remarkable meeting it was, for these were the darkest days to date in the *Hubble* project.

As I was uncertain where the meeting would be held among the mass of randomly numbered buildings at Goddard, I arrived around dawn, hoping to have a few hours before the meeting began to think through possible worst-case scenarios for the day's events, yet to unfold. But I was not the first to discover the cramped, stuffy meeting room and what I found early that morning resembled a wake, or at least a hospital waiting room. The place reeked of depression. In *Hubble's* case the prognosis seemed bad: the precious gem of American astronomy was partially blind, wobbling uncontrollably, and slow to react. Some scientists stood about awkwardly, their shoulders slumped, their heads down, talking in low, somber tones. Others just sat and stared into the air. Key engineers were also present and armed with reams of test data, now seemingly ready to offer their maxima culpas. Notably missing were the jovial greetings and strained banter that traditionally cause such meetings to begin late.

Some of the people who had been deeply involved in the attempts to focus *Hubble* during the past several weeks looked dangerously exhausted. Doug Broome, *Hubble's* program manager, just having arrived from Washington, kindly went out of his way to shake hands with me, "my pen pal" as he called me, for we had recently exchanged letters declaring a truce in an ongoing feud concerning public-affairs activities. His hands were sweaty and trembling; he was desperately seeking friends. He would survive commissioning, but not much more.

Although several astronomers at the Science Institute had been intimately immersed in the battle for true focus, several leading Hubble scientists had learned of the telescope's woes only a day or two before. These players had elected to remain at their home institutions throughout the previous month, and now were in a state of shock. Others would first learn of the problem as the morning's discussions proceeded. Astronomer Bob Kirshner of Harvard, a member of the new Users' Committee, was largely unaware of *Hubble's* fuzzy vision before the meeting began—NASA's "blackout" of focus information to the science community had been that thorough—and later recalled how the dire implications of the aberration problem did not dawn on him until halfway through the morning session. Sitting in his chair silently listening to the cold, stark facts, he was mentally blown away when he realized what he was hearing.

This was quite unlike any Science Working Group meeting I have ever attended. Contrasting with those often loud and boisterous affairs, virtually all of the two-score attendees were unusually quiet. On many occasions, whoever was talking had to be asked to speak up. I had the feeling that the briefers were speaking softly because they didn't want to say what they had to say, or at least didn't want others to hear.

The Marshall engineering contingent spoke first, admitting now that we had a serious problem with the image quality of Space Telescope—as they put it, "a textbook case of an optical system that has a significant amount of spherical aberration"—and following with an unemotional analysis of the focusing problems. Hughes' Terry Facie gave a military-style briefing—the kind of crisp, basic, technical tutorial that often puts academic talks to shame—but the result was still disappointing: Space Telescope had imperfect vision and he didn't quite know why. Furthermore, the primary-mirror actuators—the screws that impinge on the mirror's back side—would not help much.

A few astronomers—including the institute's Chris Burrows, who had originally identified the problem, and Sandra Faber, a prominent researcher on the wide-field camera team—contributed conclusive evidence that *Hubble* suffered from a classical example of aberration—just like we had all studied in our freshman physics courses in college. *Hubble's* optics were indeed perfect— "perfectly wrong," added Gene Oliver, Marshall's chief engineering analyst on the orbital verification team, "as though the error was done on purpose." His tone was not accusatory, but the comment sent a chill up my spine.

It was time for triage. Each of the six principal investigators who had headed a team building *Hubble's* science instruments spoke in turn, giving their assessment of what could and could not now be done with their respective equipment aboard the crippled telescope. The room was extremely still when Jim Westphal, the usually cheerful and upbeat lead scientist for the wide-field and planetary camera, began talking. Without even leaving his seat, he said glumly and in a sum total of less than a minute, that the discovery of aberration "has essentially taken the WF/PC out of action." He had no desire to pursue his science goals, if only because his camera worked best in the optical domain where, he claimed, ground-based telescopes could do virtually as well. "This wipes out our optical-science program," he said, adding that his camera "would not be competitive with the faint-object camera in the ultraviolet." In Westphal's view, the huge advance in resolution would not be realized for the wide-field camera, nor would its advertised sensitivity, and he just didn't want to even attempt his team's grand science objectives. As though seconding a motion, Faber concurred, saying, "Our scientific program is now fully compromised—devastated."

The interminable pause that followed was the moment of deepest despair. The project had bottomed out. The wide-field and planetary camera is the workhorse of *Hubble*'s science instruments, scheduled to be used in nearly half the mission's science agenda. And now it seemed lost.

I glanced discreetly around the room. A skilled academic optician, who had kept an eye on the manufacturing of *Hubble*'s optics, was ashen. Directly in front of me, the project's chief engineer was hunched over, elbows on his knees, head hanging low. Across the room, an eminent astrophysicist sat slumped and immobile with mouth agape, his mind obviously spinning furiously but mutely. A prominent former NASA project scientist sitting next to me dropped his pen on the floor but didn't move a muscle to retrieve it. The room was absolutely silent, motionless, for a full minute—a stretch of time that seemed like eternity. Here was one of the highest densities of scientific and technological brainpower ever assembled on the planet, yet no one seemed to know what to say or do. The deafening silence was broken when a NASA official gagged, got up awkwardly, and trotted out of the room.

Westphal's comments would come to be seen by a number of his colleagues as the opening moves to mortgage the telescope—to keep it for many years in the hands of the principal investigators, like bankers, until such time that *Hubble* might be repaired. He knew, as we all did, that a replacement for his camera had been planned well before launch, scheduled for installation aboard Space Telescope by astronauts during a space-shuttle mission within a few years. Rather than using his guaranteed time observing the cosmos with an impaired camera that yielded disappointing images, he understandably preferred to wait for the installation of the new WF/PC, which might be altered in such a way as to correct *Hubble*'s aberration. To Westphal, *Hubble*'s myopia was much the equivalent of another four-year launch delay. But, in the meantime, he was not happy with the idea of other scientists using his camera on astronomical targets that he and his team regarded as exclusively theirs. Proprietary issues, it seemed, were again coming to the fore, even at a time when one might have hoped that individual concerns would have been overshadowed by the need to sustain the mission.

Duccio Macchetto, the principal investigator for the faint-object camera, was more encouraging: first, because his camera was designed to work best at ultraviolet wavelengths, where ground-based telescopes are useless, so any imaging data acquired in that mode would be new, however blurred; and second, because his camera's first-light images, even with primitive computer enhancement, had already demonstrated that some very fine science could likely be achieved at high resolution. In fact, he stressed that the computer algorithms being developed at the Science Institute could diminish much of the noise introduced by the aberration, thus preserving a fair amount of his team's original science objectives. He bolstered this argument by projecting an example of a simulated "cleaned" *Hubble* image, with much of the stellar halos computer-subtracted. Even so, he admitted that about 30 percent of his team's planned observations—most notably the faint-light objectives, such as the search for extrasolar planets and the imaging of regions surrounding quasars—would be largely impossible. Unfortunately, most people in the room were still reeling from the previous, highly negative report, and few comprehended this more encouraging presentation.

The two spectroscopy teams also gave relatively upbeat assessments of their programs, stating that they would probably be able to undertake well more than half of their intended observations. With longer exposure times, enough radiation could be collected to address some important scientific objectives—and to do so with sensitivity hundreds of times better than that achievable with any other spacecraft or ground-based observatory. Still, some of the most exotic programs would be lost, including spectra of gas in the immediate vicinity of suspected black holes and in the hearts of some of the most violent galaxies known.

As if to put a mask on the many pained faces, a couple of people tried to lighten up the discussion. "At least *Hubble* is on-orbit," piped up Dave Leckrone, the deputy project scientist at Goddard. "The first spacecraft I worked on ended up in the Indian Ocean." But no one laughed. Pete Stockman, the deputy director of the Science Institute, dryly added that the increased exposure times per celestial object needed to partly compensate for aberration was one way to lower the overhead of the spacecraft and thus increase the science efficiency of the telescope. But people only glared at him. This was not a day for jokes or wisecracks of any kind.

The remaining two instruments—*Hubble*'s photometer and its astrometric devices (the FGSs)—were judged by their respective principal investigators to be able to acquire reasonably good data despite the faulty optics. The prospects for the astrometric program were judged especially encouraging—provided *Hubble*'s jitter could be damped—since positional work with the fine-guidance sensors would more likely be adversely affected by telescope pointing gone awry than by *Hubble*'s fuzzy vision. In any event, these instruments were together slated to utilize only a few percent of the total telescope time in the *Hubble* mission.

The six principal investigators having given their salvage reports, the "seventh team" would not be denied. One of its members, a scientist of some considerable reputation though not the team leader, rose to present a devastatingly negative analysis of where the project stood. Obviously distraught, he repeatedly wrote on Vu-Graphs in huge block letters, "BAD," beside many items on a long list of *Hubble* imaging prospects. By "bad," he said he meant that he "would not elect to spend time working on these projects, given the press of life's other duties." He did acknowledge that *Hubble* might be suitable for spectroscopy of bright, isolated objects, and suggested that the Hubble Space Telescope's initials, HST, should now stand for the Hubble Spectroscopic Telescope. He was speaking for himself and not for the seventh team, and it later became evident that he, too, wished to postpone observations of "his" cosmic objects until after the telescope was repaired—and to keep others from examining them in the meantime. His discouraging assessment reminded the gathering of the negative WF/PC report earlier that morning, and everyone left for lunch thoroughly depressed. In fact, he turned out to be quite wrong about what *Hubble* could and could not do.

In the Goddard cafeteria, a few of us, especially from the Science Institute, tried to talk up the science that could still be done rather than dwelling on those parts of the mission that had to be sacrificed or deferred. I pestered anyone who would listen that now was the time, more than ever, to take a few images of cosmic targets of scientific significance—to see by trial and error how well or badly *Hubble* could do science. During the luncheon politicking, I cringed when we learned that NASA intended to mount a major televised press conference late that afternoon. *Hubble*'s problems had to be publicly acknowledged, of course, but I had zero confidence that NASA would explain them either accurately or well.

After our meeting reconvened, the mood became more encouraging—as we say in the trenches, "the derivative was becoming positive." This was especially true during briefings by the scientists who led the development teams for *Hubble*'s three second-generation instruments—an advanced spectrograph and an infrared camera, in addition to a new WF/PC—scheduled to replace existing instruments in the mid-1990s. NASA had ambitious plans to service Space Telescope by arranging for a space shuttle to rendezvous with *Hubble* every three or four years. These so-called maintenance and refurbishment missions would replace any of eighty or so orbital replacement units that had become faulty and change out one or more of the science instruments with newer, state-of-the-art versions.

Although neither the primary nor the secondary mirror could be replaced in orbit, since each is near the virtually inaccessible heart of the spacecraft, all three second-generation-instrument briefers agreed that *Hubble*'s nearsightedness could be largely corrected by making slight changes in the new instruments, each of which was then in the planning stage. This would essentially be done by intentionally causing several of the nickel-size mirrors of those instruments to have an aberration of equal but opposite sign to that induced by *Hubble*'s main mirror.

By bizarrely reshaping the newly machined oculars, a little like those distorting our own human image in a fun-house mirror at a circus, the effects of aberration could be nearly canceled—proving that, as applied to light waves anyway, two wrongs can make a right. In each case, the compensating device was claimed to add but a small cost to the multi-million-dollar, second-generation instruments. And provided that the magnitude and extent of *Hubble*'s aberration could be precisely measured, the refurbished telescope would be able to focus nearly 70 percent of its captured radiation with 0.1-arc-second resolution—close to the bona-fide Level-1 specification. Of special import, John Trauger of the Jet Propulsion Laboratory, a designer for the new WF/PC (also known as the "wiffpick clone"), had worked throughout the night before the meeting to complete an encouraging analysis showing that a carefully shaped correcting mirror retrofitted at his camera's pupil would enable *Hubble* to regain much of the crisp, clean sight originally intended. It would be like putting contact lenses on Space Telescope, correcting its blurry vision of the Universe.

Other positive sentiments began to emerge. Although *Hubble*'s planned science program had been compromised, and the wide-field camera perhaps hardest hit, during the course of the afternoon a leading West Coast astronomer, Ivan King of Berkeley, stood unexpectedly, and in measured tones suggested that some aspects of his science program might be doable after all, and that he would like to use *Hubble* to take a peek into the cores of some dense star clusters. A Canadian planetary scientist, John Caldwell, chimed in, saying that despite *Hubble*'s crippled state he thought much good science could still be achieved regarding the study of surface features on some of the planets. Then another scientist stood without the chairman calling on him, and yet another, each declaring that a great deal of unique, world-class astronomy might still be accomplished with Space Telescope. A groundswell of optimism, spontaneous and feeding on itself, was sweeping the committee room. The shock had worn off, and many of those present began displaying renewed resourcefulness.

Unfortunately, Ed Weiler, NASA's program scientist for Space Telescope, heard virtually none of these encouraging sentiments, as he had left at midday to confer with headquarters officials and to participate in a dry run of the press conference. The conference itself would be at Goddard, but off-base in its Visitors' Center—again, the media would not be allowed to reach the scientists most heavily involved in the mission—and we halted our working-group meeting to watch it on close-circuit television. After some opening remarks from NASA managers who understood little of the science implications of *Hubble*'s purblindedness, or even the freshman-physics concept that had caused the problem—asked NASA's chief scientist on live television while turning to someone off camera, "What do you call the term? Oh, aberration"—we watched, somewhat appalled, as Weiler gave an overly alarming report. Obviously, he was emotionally drained; he had put a decade of his career into history's most heralded space-science mission, which now was at risk a mere two months after its launch. He emphasized that *Hubble*'s ultraviolet capabilities were partly intact, and he noted that the telescope's science program was not canceled but only deferred, pending a repair mission a few years away; but he also tersely stated for all the world to hear (and incorrectly) that, "No real science can be done with the telescope's main camera" and that *Hubble* "is running above [worse than] ground-based capability."

Apparently unable to come to grips with what he had just said, Weiler resorted to Orwellian doublespeak: "Although we are deferring science, 100 percent of the time available to Space Telescope will be spent doing highest-priority science." Then, as though he had confused even himself, he delivered in a quivering voice and with a pained face the evening's sound bite: "Our telescope now has bad myopia. It would be dishonest of me to say that the mood of the scientists is very happy right now. We're all frustrated." (Several newspaper and television reports packaged his comment in further negative tones, making him sound like he was pronouncing the death of *Hubble*.) None of the grim and somber NASA men on the press panel that afternoon made any mention of the more positive statements made at that day's Science Working Group meeting.

The next day's papers were full of sensational headlines—"Hubble Telescope Broken," "Pix Nixed as Hubble Sees Double," "Space Telescope Can't See Straight," "Hubble, Double, Toil and Trouble," and other lurid titles screamed across the tabloids—while the evening newscasters put their own spin on what they regard as objective journalism. The feeding frenzy was under way.

I came to recognize what some members of the fourth estate had warned me would be the norm, even before I joined the *Hubble* mission: The newspapers' bold headlines and the televisions' anchor lead-ins were exaggerated representations of articles and reports that followed, and that were otherwise well done—a common occurrence with the U.S. press, since the headlines are almost always chosen by editors seeking to sell the news, not by the more responsible journalists reporting on it. And, unfortunately, most of us tend to remember the headlines. Whatever, the news editors were apparently in their element now; there was human failure—particularly egregious human failure—and much hay was made of it.

To my mind—yet anathema among NASA officials—honest and open appraisals of *Hubble*'s commissioning problems might have resulted in a more balanced discussion of the aberration issue and of its possible solutions. The increasing suspicion beginning several weeks earlier that the telescope's optics were flawed should, in my view, have been shared with the public, thus diminishing the impact of the news. Why did NASA trot out its headquarters officials at this bombshell press conference, thereby making the announcement more dramatic than it might have been otherwise? Many of these government men were the same individuals with whom the media had had a tug-of-war during the prelaunch press conferences at the Kennedy Space Center, and some journalists were waiting to pounce when given the chance.

In their opening remarks, the emotionally drained officials immediately moved to raise the ante, even before *Hubble*'s ills were explained clearly, by announcing that NASA was impounding the relevant design records, data logs, and polishing machines of Hughes Danbury and was convening a formal Board of Inquiry to investigate the origin of the mistake—all of which had the effect of transforming a technical problem into a social crisis. And if NASA wished to hold back knowledge of *Hubble*'s mirror problem until it had unequivocal proof, then why didn't the space agency wait one more day in order to gain the Science Working Group's reasoned opinion of what could and could not be done with *Hubble* even in its crippled state? As noted, the mood of the working group changed considerably over the course of its deliberations, as most scientists began assessing what science could still be done rather than gloomily complaining about what could not. In the end, to my knowledge, NASA never did issue a carefully worded press release describing *Hubble*'s aberration and its implications for the successful prosecution of the mission. By this time, NASA's public-affairs machine had completely broken down. And so nearly had its management structure.

Hindsight is easy, and spherical aberration was far and away the most serious of *Hubble*'s defects to date. But the fact remains that NASA announced that *Hubble*'s cameras were close to useless without having tried to take a single science image, indeed without having even discussed with actual users of the telescope in more than a cursory way the prospects for doing science.

Still another factor combined with the bad timing to worsen the situation. Many journalists attending the fateful press conference had expected NASA to finally release some of the long-awaited science images taken so far by *Hubble*—some stunning findings, even if not outright discoveries—that would cause "our socks to be knocked off." Instead, they were greeted with the usual lineup of saddened, slumped individuals, who announced with unnecessary despondency that *Hubble* was gravely flawed.

The media knew well that the claimed-inoperative wide-field and planetary camera was the device likely to produce a steady stream of sensational pictures of the heavens. The prospect that there would, after all, be no pictures from *Hubble* hit the media hard; they had become wrapped up in this science project like perhaps no other in history. As things turned out, NASA's appraisal of the WF/PC's demise was decidedly premature—within a day it would be judged inaccurate—but vast public-relations damage had been done, for the subsequent reporting caused most of the American citizenry to sour on *Hubble* virtually overnight. The widespread perception was that the telescope was "broken," "blind as a bat," a huge "piece of junk uselessly orbiting the Earth."

Driving home later that evening, I worried about what I might say when the White House called. Throughout the spring and into summer, I had kept the president's people apprised of *Hubble's* shortcomings, all the while stressing its prospects for first-class science. But none of us was prepared for that day's obituary by NASA. Not until I reached the Annapolis city line and sited the Maryland State House, especially its resemblance to the U.S. Capitol building, did it dawn on me that the White House would not likely be calling me anymore. Instead, the executive branch of government would now move to distance itself from this project, all the while the 101st Congress would gear up—indeed, would demand an explanation.

"Perception is reality in this town," said a friend long ago about Washington, and the perception, at least inside the D.C. beltway and in the words of several key senators, was that *Hubble* "did not work," that it had "double vision," and that it was a "technoturkey . . . unacceptable to the scientists of the world and unacceptable to the taxpayers." Fortunately, Kevin Kelly, the senior aide to Senator Mikulski, who was then chairing the Senate Appropriations Subcommittee on Veterans, Housing and Independent Agencies (and the coiner of that very effective sound-bite, "technoturkey"), called the Science Institute to let us know that the powerful senator was not implicating us. But the space agency was about to undergo a "NASA-bashing" the likes of which it had not experienced in its thirty-year history.

Two days after the pivotal *Hubble* press conference, NASA's space shuttle fleet was indefinitely grounded, owing to the discovery of a leak in the fuel lines of shuttle *Atlantis,* then on Pad 39A awaiting liftoff and deployment of a secret military payload—the same kind of unresolved leak that had, a month earlier, forced shuttle *Columbia* off the launchpad and into the Vertical Assembly Building for repairs. What's more, the full Senate was beginning to debate NASA's appropriations bill for the next fiscal year; the timing could not have been worse. Clearly, it was going to be a long, hot summer for NASA on Capitol Hill.

While I was still at Goddard, on the second day of the Science Working Group meeting, the senior staffer of the Senate Commerce Committee, Steve Palmer, called my office at the Science Institute. Congressional inquiries were already in high gear and the Subcommittee on Science, Technology, and Space wanted institute representatives to testify the following morning on the impact of *Hubble*'s plight on the planned science mission. My secretary couldn't locate me—I refused to carry a beeper—and, with the institute's director safely abroad in Italy, referred the call to deputy director Pete Stockman, who agreed to attend if I would go with him. We didn't touch base with one another until 11 o'clock that evening, for, as things turned out, several of us had become ill after lunch at the Goddard cafeteria—"Is it psychological, or are they trying to kill us?" quipped a Hughes guy who also took sick. Thus, both Pete and I felt an urgent need to get away that evening from the *Hubble* project, its people, and its problems. (Pete went to the Baltimore Symphony and I took my four-year-old son to the Orioles game.) So, neither of us having testified in Congress before, we began drafting a statement shortly before midnight and polished it on the Metroliner from Baltimore to Washington around dawn. Throughout the train trip, the walk to the Capitol from Union Station, and a stop for much-needed coffee in the basement cafeteria of the Russell Senate Office Building, we peppered each other with nasty questions that might be asked of us. And, of course, we searched for answers.

Actually we were not overly concerned, for the Science Institute had had nothing to do with the mirror design or fabrication; all of *Hubble*'s optics had been built and tested before the institute was created a decade earlier. We agreed to confine ourselves to issues of science, and to defer all questions about *Hubble*'s hardware to NASA and its industrial contractors, and especially to the newly created Board of Inquiry (which would not be attending this hearing, given the reluctance of numerous people to chair this board, or even to serve when asked). Pete and I were confidently innocent—and innocently confident.

As we entered the Senate conference room, we were confronted by an array of bright lights, television cameras, and a long mahogony table surrounded by reporters impatiently drumming their fingers while staring at us. Our confidence instantaneously dissipated. These were the congressional-beat reporters, not the science journalists whom we had come to know and respect, and they would no doubt be looking for controversy. "How has it all come to this," I remember Pete and I said to each other in unison. Since there were only four chairs at the witness table, and three NASA officials were present to testify, I quickly deferred to Stockman and retreated to the gallery.

As it happened, things went well for the institute, if only because Senator Albert Gore, who chaired the subcommittee, smelled blood among the NASA witnesses—essentially the same headquarters group who had hyped the project early on, had botched the prelaunch press conferences at Kennedy, had dismissed their highly knowledgeable public-affairs chief, and had days earlier exaggerated the implications of *Hubble*'s aberration. Now, they testified, with a kind of bored demeanor, that the aberration problem was only a week old, that a 100-percent fix was possible, that the fine-guidance sensors were working perfectly, and that no other problems were interfering with the operation of *Hubble*—all these statements known (at least to me) to be incorrect at the time. Senator Gore didn't seem to be buying it either, for he impatiently interrupted a rambling discourse on what might have gone wrong by cutting bait: "In other words, [*Hubble*'s mirror] is perfectly wrong." To which Lennard Fisk replied matter-of-factly, "It's perfectly wrong, that is correct, a mistake ground in with great precision."

NASA's stiff upper lip was exactly what was not needed, for Gore then berated Fisk for allegedly refusing, the evening before, to appear at the same hearing with the space agency's two major *Hubble* contractors, Hughes and Lockheed, who did in fact fail to attend the hearing, citing short notice. Gore's attitude was unforgiving; consider a small but typical exchange that upset both parties—Gore: "I understand that you would like to design these hearings as you have designed the telescope—who appears, who comes to talk, and doesn't appear." Fisk, interrupting: "I trust the hearings will be more in focus." Gore: "Yeah, when we get both of the reflectors in line here in the same room."

We began to realize that conflict among NASA management and its industrial contractors was far more severe than the uneasy alliance between NASA and the Science Institute, and that the agency's relations with Congress were not too good either. Lecturing the now rattled NASA men at length in the tense, packed, hot hearing room, the obviously annoyed subcommittee chairman made clear in no uncertain terms that, when he reconvened the hearings later that summer, both the industrial contractors responsible for building *Hubble* and the NASA officials with authority for managing it would be present in the same room and at the same time—even if he had to subpoena all of them.

Referring to the *Challenger* tragedy, Gore angrily summarized his feelings: "This is the second time in five years that a major [NASA] project has encountered serious disruption by an inherent flaw that was apparently built into the project as much as ten years before launch and went undetected by NASA's quality-control procedures." He added, "The implications for the space station and the Moon-Mars effort, and for other large missions, are really quite significant." The drama, carefully orchestrated by Senate aides for the television cameras, was tempered only slightly by the not-so-hidden agenda of Senator Gore—he was launching his bid for the 1992 presidency.

As we left the Russell Building, Stockman exhaled, "Whew, what an introduction to the Senate!" Dale Bumpers, a southern senator who had been at the hearing—and a fierce space-station foe—overheard Stockman and kindly assured us that, "By comparison with the other guys who testified, you have little to worry about. As Al Gore said, NASA's eyes are bigger than its stomach," he announced while mixing metaphors, "and it needs to have its wings clipped." Added Senator Mikulski before jumping into a car, "I'm outraged at what has happened to *Hubble*. They've had ten years to put this thing together. They spent $2 billion to be able to get it right, and now we find that the telescope has a cataract."

The institute's testimony to the Senate subcommittee that morning was meant to be vanilla in tone and accurate in substance, an honest appraisal of the impact of *Hubble's* blurred vision on the science mission, without skewed commentary or false promises. It contrasted greatly with the emotional and misleading presentation made by NASA to the media during the fateful press conference hardly thirty-six hours before. And it filled a vacuum created by the fact that NASA, true to form, neglected to prepare any such statement for the Congress. Throughout that summer of inquiry, it remained the unofficial position paper of the Science Institute:

> First, I want to thank the chairman for giving me the opportunity to address the space subcommittee. I am pleased to provide a preliminary appraisal of the science mission of the Hubble Space Telescope in light of the optical measurements made over the last month and their analysis in which institute staff played major roles.
>
> At the outset, I would like to emphasize that many of the primary scientific goals of *Hubble* can be accomplished, even with the imperfect images which have been achieved to date. These are principally in the areas of ultraviolet spectroscopy, imaging, and photometry where the observatory's performance is only moderately affected. One of the key scientific projects identified by the Space Telescope Advisory Committee, the study of intervening gas and galaxies in the ultraviolet light of distant quasars, falls into this category.
>
> For a certain class of imaging involving high-contrast objects of moderate brightness, the sharp core of the images observed in the faint-object camera and the wide-field/planetary camera can be exploited to obtain nearly diffraction-limited images to address important morphological questions (e.g., the cores and spiral arms of galaxies, and features on planetary atmospheres).
>
> However, many critical scientific goals cannot be accomplished with *Hubble* in its current state. These are primarily in the area of faint visible imaging or deep imaging of crowded or complex fields. Two of the key projects fall into this category: obtaining the distance scale of the Universe by observing faint cluster stars in distant galaxies; and deep surveys of the sky to discover new objects and to obtain statistics of distant galaxies. Studying the faint environments around distant quasars and nearby stars has also been greatly compromised.
>
> We at the Space Telescope Science Institute have initiated a review of the general-observer science programs approved for the first year of routine scientific operation. Our preliminary analysis indicates that approximately 50 percent of the current program remains viable. Based upon the results of that review, we shall adjust the program accordingly. In the meantime, we shall communicate the relevant technical information we currently have to the scientific community.
>
> In considering the scientific objectives which cannot be achieved with the current imaging performance, we at the Space Telescope Science Institute are heartened by two facts:

1. The Hubble Observatory is healthy, with all five science instruments and the fine-guidance system passing their initial commissioning tests.
2. The nature of the current optical problem appears to allow a correction within the second-generation scientific instruments currently in development.

In particular, we at the Space Telescope Science Institute strongly recommend that following the report of the review board and further optical tests, NASA proceed with the necessary modifications to the wide-field/planetary camera replacement now in construction at the Jet Propulsion Laboratory and the two advanced instruments, and that these modified instruments be installed with shuttle servicing at the earliest possible opportunity.

We look forward to doing outstanding work with the Hubble Space Telescope in its present condition, during the next few years, and we also look forward to accomplishing all of the goals of the full *Hubble* mission, including the two deferred key science projects, mentioned earlier, during its planned fifteen-year lifetime.

In lieu of a formal statement, space agency officials elected to wing it, all the while resorting to their well-practiced aplomb. They didn't even flinch at the laughter in the Senate gallery when blurting out such gems as, "We [at NASA] know what we're doing. We're clever." What they desperately needed, of course, were pictures—intriguing cosmic images from Space Telescope that, despite shrill arguments to the contrary, they could well have had in hand, had there not been so much managerial confusion and project infighting, neither of which were much abating despite the unfolding tragedy.

5

Rocky Road
to the Imaging Campaign

Beyond the stars of the sixth magnitude you will behold through the telescope a host of other stars, which escape the unassisted sight, so numerous as
to be almost beyond belief, for you may see more than six other differences of
magnitude, and the largest of these, which I may call stars of the seventh
magnitude, or of the first magnitude of invisible stars, appear with the aid of
the telescope larger and brighter than stars of the second magnitude seen
with the unassisted sight. But in order that you may see one or two proofs of
the inconceivable manner in which they are crowded together, I have determined to make out a case against two star-clusters, that from them as a specimen you may decide about the rest.

Sidereus Nuncius
GALILEO GALILEI, Venice, 1610

For those of us closely associated with the *Hubble* project, the summer of
1990 amounted to a public-relations cruise through hell. The celebrated space
mission now constituted a national embarrassment and, quite possibly, a scientific catastrophe—a mission impossible, perhaps. Across the nation and around
the industrialized world, the Space Telescope became the butt of scathing editorials, derisive cartoons, and irreverent jokes on late-night television. Come-
dians on NBC's "Tonight Show" couldn't resist barking to the effect that,
"*Hubble* is working perfectly, but the Universe is all blurry." The project
became the subject of two "Herblock" cartoons in a single week, a dubious
distinction generally reserved for political boondoggles. And the word was
first heard around Washington as a verb—to hubble: to screw up in a mas-
sively unfocused manner. Edwin Hubble's good name had become mud.

Other more relevant parties were also maneuvering mere days following
NASA's shocking announcement that *Hubble* had nil picture-taking ability.

The chief executive officer of Hughes Aircraft wrote the *Los Angeles Times* that he was "obviously concerned about the Hughes name and our technology," and disavowed any invovement by his company in *Hubble*'s mirror fabrication. Lockheed moved immediately to delete mention of *Hubble* from its full-page advertisements in aerospace magazines; the project had gone from an optician's to a publicist's nightmare. A governor from a southern state, whose office had invited me to address the U.S. Governors' Conference a few weeks hence, called to politely disinvite me; when I reminded him that I had intended to speak more about education than about *Hubble,* an aide on the line interrupted not so politely, saying that it didn't matter, they wanted no one addressing the governors even remotely associated with the Space Telescope project. Minutes later, another call came in, this one from a producer at PBS's Maryland Public Television; contrasting greatly with the governor's don't-taint-me attitude, the educational arm of MPT made clear that, *Hubble*'s ailments notwithstanding, they still wanted to co-produce with us the *Starfinder* instructional TV series for precollege students. While the aerospace contractors were running for cover and the political operatives retreating, at least the educators, thank goodness, were willing to ride out the storm.

On Capitol Hill, the House of Representatives put a dent in President Bush's ambitious hundred-billion-dollar plan to send Americans back to the Moon and, by the year 2019, to Mars, deleting all such monies from its appropriations bill for fiscal 1991. Concerns over cuts in domestic programs, especially housing and veterans' affairs, were cited as the cause of the Moon-Mars termination, but certainly NASA's dwindling reputation for competency was near the heart of the issue. Congressional confidence in NASA and its large, multi-billion-dollar space-science projects, let alone its already controversial plans for a permanently manned space station, was clearly eroding. Complained a former NASA historian publicly, "They want Congress to give them extra money to put up a space station predicated on the assumption that they can make big expensive technology work the first time, to which *Hubble* puts the lie." And these were just the opening salvos.

Throughout much of the summer, numerous congressional committees grilled NASA officials in stormy Capitol Hill hearings that sometimes went on for hours. Nothing very factual emerged; the testimony was contradictory, the questioning ill-informed. The agency's managers knew few details about the *Hubble* project—neither its political history, nor its engineering R&D, and certainly not its science mission. One day, we heard the assertion that, "A full test of *Hubble*'s mirrors would have cost $200 million"; the next day, "Some simple, low-cost, low-risk tests could have been done." One day, *Hubble*'s cameras were said to be "inoperative, their science deferred for a few years"; the next day, "We can still do much good picture-taking." One day, the lawmakers were told that, "At no point in the minutes of the Science Working Group was there any concern expressed about mirror testing"; the next day, "Based on working group notes from 1979 and 1980, the issue was revisited and reaffirmed at various times during the development of the mirror." That NASA administrators were a bundle of contradictions was perhaps best typified by a single sentence of uninspired testimony by *Hubble*'s program manager: "The faint-object camera works well except that it can't see faint objects."

At least one Senate hearing included a closed-door, classified briefing on military-intelligence satellite technology, from which one junior senator emerged, admittedly reeling, "Frankly my head is spinning. Apparently not everything this country does in space is public knowledge." Sitting in the Senate hearing rooms or watching the proceedings on closed-circuit television, I had the sinking feeling that the Congress and the public at large must have been increasingly confused by the conflicting testimony and evident stonewalling. Skepticism regarding NASA's agenda was clearly on the rise. What really did NASA know and when did it know it? was asked repeatedly. The word *Hubblegate* was heard around town.

The executive branch of the U.S. government also signaled its dismay with the *Hubble* project and general unhappiness with NASA, suggesting a potential shift in space policy or at least space implementation. President Bush asked Vice-President Quayle and the newly reconvened National Space Council to appoint a blue-ribbon panel to review how NASA does business. (The Space Council, analogous to the National Security Council, which advises the president on military affairs, had been abolished following the age of *Apollo*.) The panel's charge, not stated in so few and undiplomatic words, was to examine if a wholesale shake-up of NASA management was needed—said a White House official, "to reinvigorate and restore public confidence in the U.S. space program, to get it back on track." The concern, another senior administration source told me privately, is that NASA is looking like "just one more big bureaucracy." This caused the considerable friction already extant between NASA and the Space Council to amplify, for the space agency considers civilian spaceflight to be its exclusive turf and it deeply resents outside advice.

The heat was intensifying from many angles and NASA Headquarters was not taking it well at all. NASA's top person, Administrator Richard Truly, threatened to resign in disgust, all of which he roundly denied on television's "MacNeil/Lehrer News Hour," but he did make his true feelings clear for all to hear: "This [*Hubble* investigation] is turning into a criminal witch hunt instead of a technical probe." A siege mentality set in at the beleaguered agency, with anyone talking about "space" in anything less than glowing terms being viewed as the enemy. To me, it was all reminiscent of a previous presidential administration that ineptly practiced damage control, maintained enemy lists, and embraced a reactionary posture—in this case toward everyone non-NASA. "I never want to live through a summer like that again," sighed one high NASA official more recently.

Not surprisingly, commentary inundated the op-ed pages, both pro and con, regarding the *Hubble* project. A national debate was under way—in and of itself a good thing—but the dialogue went well beyond NASA, questioning the very fabric of our technological being. Real doubt surfaced regarding the high-technology capabilities of the United States. *Hubble* became not just a metaphor for failure, but a flash point in the ongoing information-age race

with Japan and the European market. Were we a "sunset power," incapable of repeating its technological feats of the past? Maybe our workmanship *is* shoddy, our technical know-how rightly battered, suggested news media around the globe.

To others, *Hubble's* problems represented a setback to world science more than a blow to U.S. prestige. At the least, *Hubble's* adversity focused humankind's attention on the limits of high-tech gadgetry, and to some it meant that that gadgetry just plain didn't work. This feeling of high-tech helplessness, starting years ago with the mass production technology of mediocre cars and televisions and now culminating in the "hobbled *Hubble*" as the epitome of basic research technology, would not abate until the Persian Gulf War the following year, when the perception emerged that today's sophisticated technology really does work, albeit in the form of smart bombs and expert weapons systems. Even here NASA was the loser, for the average citizen heretofore had associated the best of America's technology with our nation's space program, yet owing to precision firepower and skilled leadership of the Gulf War, that expertise had now shifted to the military sector of society. In a decided about-face in public attitudes of the 1960s and 1970s, the military became the good guys while NASA was commonly viewed with fear and distrust.

The Science Institute, too, was the target of letters and calls from concerned citizens—though few like those before launch proclaiming that, "Your Hubble 'scope is well suited to confirm and photograph my unified field theory written across the firmament," and that *Hubble's* "vaunted eyesight on creation" would mean "the end of the world." Most letters we received in the summer of 1990 were from thoughtful, technically inclined people offering their own concoctions for what had gone wrong with *Hubble* and often proposing the most astounding Rube-Goldberg contraptions that might fix it. A 2-ton periscope lashed to the bow of Space Telescope was among the most ingenious, yet preposterous.

Generally, incoming letters fell into three categories. First, there were those commentaries that made a lot of sense to me, like this one from a Midwesterner:

> As for the flapping of the solar panels, this phenomenon is well known and understood amongst space scientists and it is difficult to see why it was not circumvented in the design of those solar panels and their mountings. The Hubble scope is not the first Earth satellite to flap its solar panels on crossing the terminator.

Then there were those who saw conspiracies, fraud, or sabotage abound. *Hubble* was being used for intelligence work in orbit, we were told quite often and not just in supermarket tabloids. Believe me, it wasn't; the "overhead elints" neither needed *Hubble* nor wanted it. Even so, who knows if we shall ever know the whole truth about the telescope's misshapen mirror, typical of this second class of letters we received:

> Every scientist who has ever ground a telescope mirror knows that mirror could not have been wrongly ground by accident. It would take a change in the laws of physics to produce the error as your people described, due to an "unnoticed" fault in the calibration instruments.

And finally, there were some slightly veiled threats such as this one from an irate Londoner who in the 1970s had proposed his own, and unfathomable to me, unified theory of the Universe:

> If it doesn't check out to be essentially correct, the Devil may have my soul, but if it does, but you won't use the Hubble Space Telescope to vindicate the general equation of physics, he may have *yours*.
>
> And that goes for all the rest of you obfuscating, bungling, conscienceless, proud and haughty men and women of organized science who have left the human family at risk from cosmic ignorance for at least twelve years longer than necessary.

With sarcastic print and broadcast editorials crisscrossing the country, indeed the world, the public's fascination with the *Hubble* project dissipated seemingly overnight. In the mind of the typical citizen, both the promise of stunning images of the heavens and the potential for great discovery instantaneously slipped away. NASA, so obsessed earlier with potential loss of control, had ironically lost complete control of events. My journal entry for July 13 summed up my feelings by appealing to black-hole physics: "We seem to be caught in a whirlpool, getting sucked deeper and deeper into the morass. With the media now fully on our case, there would seem to be few ways to escape without fatally crashing."

At the institute, those involved in the exhausting hours of analyzing the production line of variously unfocused images were looking increasingly deprived of sleep, some outright sick. Even I, who was not by any means under the most pressure, also felt my own health degrading. Surely that summer's air in the Baltimore-Washington area was uncommonly humid and the ozone levels bad—the first widespread awareness that the global greenhouse effect might be for real—but I noticed for the first time in my life that I was occasionally short of breath, especially while absorbing the sometimes pathetic media pronouncements and public reactions about the project on which I worked for a living. Perhaps it was entirely psychological, although more than once a senior colleague stopped me in an institute corridor and urged caution: "Don't let the robot kill you. It's not worth it."

Nor was I the only one internalizing the accumulating pressure, the loss of self-esteem. Chris Burrows, who had uncovered the mirror's flaw and who had worked night and day to convince the NASA/Hughes contingent, had virtually collapsed outside his office at the institute; later that autumn he needed several months off-site to recuperate. Several senior officials retreated to their analysts, trying to overcome their deepening depression with chemical uppers, downers, and lots of Prozac. And, among many others who experienced *Hubble*-induced illness, the astronomer Colin Norman, who had helped me greatly in dealing with the press of the media—and who had had months earlier visions of Agincourt—developed a heart irregularity and eventually needed hospitalization, unquestionably the result of on-the-job stress. Even in its idle, station-keeping mode throughout much of the fiendish summer of 1990, *Hubble* was exacting a toll among humans on the ground.

None of the lurid news accounts that summer—at least none that I saw—adequately addressed the twin issues of risk and complexity. The *Hubble* project, when stripped of all the propaganda, is an expression of human curiosity, an attempt to make a genuine technological leap forward by going well beyond what the civilian work force has previously been able to accomplish for space astronomy. This statement is true even though much of the equipment aboard *Hubble* is no longer state of the art, some of it plainly obsolete. No science experiment has ever before attempted to assemble such an intricate and many-faceted apparatus, put it out of reach, and then operate it remotely. And with this high degree of complexity inevitably comes risk. The more ambitious the mission, the more finite the risk of unanticipated riddles. Furthermore, risk was enhanced because *Hubble* in effect put all its eggs in one basket. In contrast to previous space extravaganzas, no backup Space Telescope waited to be wheeled to the launchpad if the first one ran into trouble. "We flew the prototype," was the cry most often heard around the project office.

The question is, What is an acceptable, informed risk? Reducing the risk of failure to zero would make the testing required for such a complex mission impossibly expensive. Abandoning risk altogether would be tantamount to abdicating tasks never before done, for risk taking is a hallmark of all great civilizations that become great because they occupy new ground, achieve victory, and accomplish notably new things—to go boldly where no one has gone before. The willingness to risk failure is an essential ingredient of most daring initiatives—the kind of grand ventures that leave legacies for our descendants.

The fact that risk is an intrinsic element of space operations is something the current NASA does not seem to understand either. Since *Apollo,* the space agency has tried hard to make both manned and unmanned spaceflight seem routine—an attempt that unfortunately has largely succeeded. It has sought to generate a perception that spaceflight is risk-free, that politicians can be launched safely, and journalists and teachers, too. In the *Mercury, Gemini,* and *Apollo* days, the public recognized full well that risk was an integral part of those missions, and responded to them with enthusiasm. Now, with NASA's hype and bravado reinforcing the perception that space missions are easy to accomplish and routine to undertake, the twin elements of adventure and exploration have been diminished and the public has responded accordingly by becoming bored with the space program. NASA, once more, has shot itself in the foot, for Americans would rather appreciate the truth about the risks of space science and engineering, and thus partake in that spirit of adventure.

So where were the science writers capable of a balanced and realistic appraisal of the complexity of the *Hubble* mission, the magnitude of the effort required, the element of inherent risk? The knowledgeable science journalists with whom we had previously worked closely—or at least their bylines—had nearly disappeared. Some were taken off the *Hubble* story completely, so that investigative reporters and congressional stringers could cover the unfolding events. These nonscience journalists were clearly out of their element in trying to explain the details and subtleties of space technology and *Hubble*-class science to the public. Eventually, at least one superb science writer, Luther Young of the *Baltimore Sun,* whose articles before launch we had widely admired at the Science Institute, quit the paper because he simply couldn't get his even, factual stories past obstinate editors. Apparently, his copy didn't accentuate the negative enough and thus wasn't printed.

Even among the science press, however, there were at the time instances of surprising unprofessionalism. For example, the astronomy trade magazine *Sky & Telescope,* from which we had anticipated seeing some of the very best science writing, quickly shifted into an investigative mode, with its journalists expecting the Science Institute to provide "deep-insider" information. Since NASA had put an absolute prohibition on the institute's speaking to the press about spacecraft problems, this caused some awkward encounters. After I spoke in July at an astronomical meeting in Boston about *Hubble*'s status—I refused to be anything less than truthful among scientific colleagues, NASA notwithstanding—one of the best technical writers for *Sky & Telescope* became furious when I wouldn't hand over my slides to him for reproduction in his magazine, accusing me and the institute, in language best not printed, of having been corrupted by NASA—an unpleasant irony, considering what we were going through with NASA at the time. Fortunately for all concerned, he way-

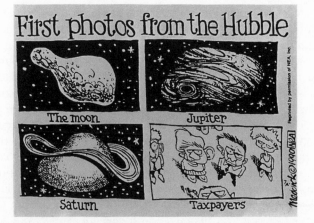

First photos from the Hubble

The moon | Jupiter | Saturn | Taxpayers

laid me privately, thereby avoiding an ugly scene, in contrast to when a NASA scientist attending an amateur astronomer convention that same month in Vermont was booed off the stage.

In fairness to *Sky & Telescope,* its editor-in-chief, Lief Robinson, had exceptionally clear vision when, in a scathing editorial written in July at the height of the fray, he blasted NASA for its handling of *Hubble's* problems. This excerpt also explains why its reporters were working the institute so hard for factual information:

> It has not been easy to get a good perspective on the telescope or its scientific potential. Journalists and the public alike were verbally flogged with NASA-speak by NASAspokesmen—jargon delivered in monotone by technical bureaucrats . . . They chose the same formula of obfuscation that served NASA so poorly after *Challenger* blew up.
>
> NASA's image suffers because it sends before the public scientists and engineers who are not skilled at communication. These people often appear either as arrogant, boring parodies of Dr. Strangelove or as talking potatoes. NASA needs a real communicator—someone who can express emotion, give perspective, *and* explain hardware. . . . I always come away from press briefings with the uneasy feeling that the speakers either knew at lot more than they were telling or didn't know much at all.
>
> . . . I suspect the press accounts were so very negative because HST had been sold to the public largely on its ability to deliver spectacular pictures revealing 'galaxies at the edge of the Universe' and other wonders. . . . Images sell astronomy; they can be appreciated by anyone. . . . Curiously, NASA seems to have a cavalier attitude about HST's images and how they affect the public's perception of the mission.

Those few science journalists still on the beat knew well that the only way out of our burial pit was to have *Hubble* take some pictures and to present them to the world at large. Witness a mid-July excerpt from a *Baltimore Sun* article, whose writer understood this subject's contentiousness: "NASA's 'pretty picture committee' . . . has recovered from the 'first-light' setback and is planning to strike again." Another piece in the *Washington Post* stated in part, "Republicans warned NASA officials that the agency needed dramatic pictures from the telescope to help rally support for its budget." And the *New York Times* argued in an editorial, "That, after all, was the main purpose of the mission, to use those laboriously crafted mirrors to concentrate light and bring the edge of creation nearer to the eye." This was the way some friendly reporters, who were genuinely eager to improve *Hubble*'s mien and for its mission to be successful, went about their business helping to rescue the project. Naturally, they had encouragement from some of us at the institute.

One of the saddest elements of this ongoing and tragic comedy was NASA's inability to jump-start the system—either to set in motion a robust repair of the telescope or to acquire some real and meaningful data with *Hubble*. So, the astronomers at the Science Institute began ratcheting up the pressure. Director Giacconi, on his own and independent of NASA, convened a panel of scientists and engineers (pointedly including none from NASA) to "cast its net widely [and to] identify and assess strategies for recovering *Hubble*'s capabilities degraded by spherical aberration." The group's charter, which severely torqued NASA, said specifically that, "The panel is appointed by and reports to the Science Institute director, who will take the panel's findings to NASA."

And we were also the ones, not anyone at NASA, who were spearheading the acquisition of science images before the project had dug the agency's grave deeper. "The institute's turning up the heat," wrote one worried NASA diehard in an electronic message forwarded to me by a sympathetic mole at NASA Headquarters. I recalled the first time that NASA had tried to muzzle me for saying, during a prelaunch press conference, that if *Hubble* ran into severe difficulties, the Science Institute would apply all due pressure on the space agency. The agency didn't like to hear that it needed any coaxing, or advice, from anyone. But that is precisely what we did apply and often, though throughout that summer NASA remained totally indecisive and unable to act.

Not least, the scientific community was in an uproar. The *Hubble* project had indeed become the center of the astronomers' Universe, but for the wrong reasons. E-mail messages were tossed back and forth at thousands of miles per second, with much heated commentary broadcast liberally throughout the network of electronically linked astronomers. Alternately commiserating and lambasting, and occasionally using nearly barbaric language, all the while wondering aloud how this, the greatest calamity of their professional lives, could have happened, the hundreds of scientists slated to be major users of *Hubble* were surely among the biggest losers.

Here, from my computer log of June 28, the day after the public declaration of aberration devastated the global community of astronomers, is a representative piece of e-mail that "beeped" scores of working astronomers' workstations while racing from the Midwest to the West Coast, into Canada, and back across the continent to the East Coast, where it entered my *Nuncius*. The message itself, written by the Illinois theorist Dimitri Mihalas, was largely incorrect gossip about what went wrong with the mirror polishing, yet ended by summing up the truly bitter feelings in the guts of astronomers mostly everywhere:

> i suppose it's "win a few, lose a few . . . " but i personally think that nasa, the government, and the people should stick it to PE [Perkin-Elmer] and TURN it hard until they agree to refund the cost of the mistake and of the repairs.
>
> i'm sick and tired of seeing defense and defense-related contractors get away with bloody murder and just get fatter and fatter on the profits.

Days later, a distinguished yet disgusted East Coast astronomer, Jim Gunn of Princeton, weighed in with a blistering diatribe—launched openly and internationally—that began a fiery behind-the-scenes debate that lit up the Internet circuit for weeks. Too long to reproduce in its entirety, his message was clearly one of self-flagellation owing to we, the astronomers, having so foolishly crawled into bed with NASA. It read in part:

> I think we are making a really dreadful mistake if we ignore the HST fiasco; it may or may not be too late already in more ways than one, but we have laid ourselves wide open to this disaster in the past and are asking for it over and over again in the future, and I personally doubt that the field [of astronomy] can absorb another, if indeed it can absorb this one.
>
> My view and point are very simple. We have lost all control of our destiny, having handed it to a bureaucratic agency which means well enough but is unable to handle large projects of its own (cf. the Shuttle) and certainly not ours. We were not "screwed over"—we have been exquisitely vulnerable to precisely this kind of thing happening for years; it is a part of our sorry heritage which began with gentlemen astronomers in their coats and ties at Mount Wilson, continued with the sorrier example of the national observatories, and has culminated with the first of the Great Observatories, probably the most expensive scientific failure in history.
>
> We are a discipline of technical incompetents, happy to let our or NASA's engineers build our tools to their desires, by and large, not ours. We are quite content to sit on the sidelines while the work is going on and bemoan the incompetence (often real, of course) of the people who are doing what should be our work, and crow about inflated expectations about what these tools will do, as if that would make it all real. The smaller NASA missions which have worked and worked well have had strong and close ties to the scientists, but even that has not prevented cost and schedule from climbing through the roof because of the ever-present aerospace industry, whose last gift to us we may not survive swallowing. And the bitter financial pill which I think is likely coming over this will affect everyone in the field, including all the people who have never had the stomach (or perhaps had too much integrity) to play the space game, so we who had anything to do with the project, including the not-so-few who worked very hard and the hordes who hardly worked, cannot claim that we have done in only ourselves.
>
> It was an ASTRONOMICAL failure; it was an ASTRONOMICAL satellite, and it does not matter a whit that it was probably some fool at PE [Perkin-

Elmer] that caused it and some entirely expected failure of NASA's criminally infantile QA [quality assurance] program that failed to catch it. It was two billion of astronomy's dollars that are flushed down the drain (let us not overreact—it may not entirely go down the drain, but the chances that it can be fixed PROPERLY are very small, and the ground [-based astronomy] is fast catching up). We must wait at least three years, making it, I think, an even ten years from the first projected launch date, no? and if I were a congressman, I would think twice about giving another nickel without some evidence that the discipline was at least aware of what had happened and had some notion of what to do the next time.

. . . NASA's style [of management] is killing/may have already killed us, and we will never get another chance to do anything about it, if indeed it is not already too late. Not that there would have been enough technically competent people in the field to have taken the two G\$ and done anything with it if we had been given it, but if the field had been up to snuff there would have been and it would have cost very much less.

All this is a plea for a plea to get our house in order, to somehow learn how to do these things before another disaster like this one befalls us—it may well be that we do not get another chance, but that does not, I think, absolve us from (or render it a waste of time to) spend some effort making ourselves slightly worthy of someone else's effort on our behalf.

If we choose to paper over this thing, as most of our colleagues seem to be doing as fast as they can instead of facing the questions and problems with some honestly and humility, I can only say that it is a sorry state of affairs.

To other scientists, the bitterness was not as evident, albeit their feelings strongly expressed, such as this excerpt from an e-mail missive intercepted from Rolf Kudritzki, one of Germany's leading astronomers:

Since the spherical aberration problem became known, I have had the opportunity to meet many colleagues from Europe and the USA . . . and all of them appear to be terribly shocked—so am I. The incredible lack of professionalism that has become apparent has almost broken our morale. All the statements to the effect that some amount of science can still be done, albeit under much

worse conditions with regard to exposure times, resolution and [sensitivity] than anticipated, do not help much (even if they are true). What the community needs is a clear statement that tells the truth, namely that in the most ambitious project in the history of astronomy NASA has failed in a way almost unbelievable for every astronomer. That is the one thing we really need if we want to uphold our work ethics. . . .

Radio astronomer Gerrit Verschuur, having no association with the *Hubble* project and thus neither personal loss nor axe to grind, well expressed my verdict of NASA's widespread lack of scientific understanding in a thoughtful *Washington Post* op-ed piece that said in part:

I am not a detractor of NASA science and am firmly behind their unmanned program. Their weakness lies in public relations, unconsciously structured to titillate and to perpetuate scientific illiteracy. Every new space venture is labeled in terms of superlatives; it is either the first, the greatest, the most important, the longest, or the most significant. There must always be a gimmick. There is seldom a word about complexity or subtlety. In pampering the media with sound bytes we are all done a disservice. NASA happens to do expensive science and not every mission has to be the first or the greatest. Above all, the goals are usually far more complex than can be effectively communicated in thirty seconds or in a few paragraphs. Most of the research can only be understood in a broader context, one that involves an overview of the work of thousands of ground-based planetary scientists, astronomers, atmospheric physicists, geophysicists, and biologists around the world.
    Let us raise the level of scientific literacy in our nation so that we will no longer have to resort to shoddy sales techniques in order to find support for that exciting adventure of the collective human spirit, the scientific search for knowledge.

A harsh statement, to be sure, yet I, too, concur that NASA's public posture and political agenda not only fail to ameliorate our nation's growing scientific illiteracy; they also often contribute to it.
    In retrospect, Jack Brandt, perhaps the most taciturn of *Hubble*'s principal investigators, probably had the right idea. He told me later that summer that he just stopped admitting to anyone what he did for a living. "Given that almost every negative statement you could make about the management of HST is true, it's just not worth explaining," he said cooly.
    Adding to the wheeling and dealing within the *Hubble* project at the time, a handful of astronomers were exercising damage control in their own self-interest—maneuvering to bolster those parts of their observing programs that could still be done, while protecting the rest of their intended observations from rival astronomers, whom they regarded as poachers. The guaranteed time observers (GTOs) were especially active astropolitically, pressing NASA for additional telescope time. Before launch, each of seven teams of GTOs had been granted 350 hours of observing time, but science with the impaired *Hubble* now required, on average, exposures three times as long as

previously estimated, in order to partly compensate for its aberration; so each team would need about a thousand hours of guaranteed telescope time to achieve those original scientific objectives still viable. If each team got such an increase, the wider community of astronomers in the wings—the general or guest observers (GOs)—would not gain access to Space Telescope for many years.

Those GTO observations impossible to undertake with the aberrant *Hubble* presented an even thornier issue. Several GTOs wanted to wait for replacement instruments, yet didn't want anyone else using *Hubble* to observe "their" targets in the meantime. Should such prime targets be locked up for several years, with everyone else forbidden to take a look? What if *Hubble* was never repaired? Would those astronomical regions be off-limits indefinitely? The permanently contentious issue of proprietary data and protected celestial sources became even more politicized than feared before launch, a genuine nightmare for science managers. At the crux of the problem, once again, was perception: Some scientists viewed their research program and its associated funding as a right, rather than an opportunity that comes with a responsibility. The federal grants system, of which the *Hubble* project is part, had become a form of social welfare for some scientists, an assumed entitlement program. To be sure, it was then as now the grant money—legal bribes, of sorts—that has kept many astronomers from speaking their minds about NASA.

Worked up agonizingly over the clearly political issue of proprietary data restrictions, NASA program scientist Ed Weiler blurted out his true feelings one afternoon at a marathon meeting at NASA Headquarters: "I've given up trying to please everyone. My intent is to make everyone equally unhappy." In the end, NASA capitulated by revising its policies, and the GTOs secured about 30 percent more time for themselves. They also won permission to extend their proprietary rights to cosmic sources for a year or two after *Hubble* might be fixed—which meant that they would remain the controlling influence on the science agenda until at least the mid-1990s. This kind of telescope mortgaging became a "hot button" in everyday conversation at the Science Institute, for it outright prohibited Space Telescope observations of numerous celestial targets certain to yield much good science despite the telescope's myopia. NASA, then in desperate need of real scientific findings, had become a victim of its own restrictive policies.

Two things of note emerged from the Science Working Group meeting at Goddard at the end of June: The Early-Release Observations (EROs) Committee was resurrected, and a Science Assessment Team (SAT) appointed. The first group would do what had been formally proposed nearly two years earlier—take several snapshot images of spectacular cosmic objects and quickly share the fruits of Space Telescope with the general public. This group was composed of the six principal investigators and myself, and was chaired again by the leader of the seventh team, Bob O'Dell. The second group would oversee the acquisition of more subtle Space Telescope data, in order to assess how well *Hubble* could observe faint objects and crowded fields, and especially the former *in* the latter. Such an appraisal was critically needed, and urgently too, if we were to make recommendations about the telescope's surviving capabilities to the hundreds of astronomers who had won the right to use *Hubble* once commissioning ended.

These early science observations, to be made as quickly as possible during the following week or two, would be used to understand specifically which of the approved Hubble science programs were still feasible and to refine the software algorithms needed to computer enhance the observed images. Of equal import, all these early data would rapidly enter the public domain, the hope being to regain some support among taxpaying citizens.

Still, and somewhat astonishingly, an objection was lodged by the astronomer who all along had been the greatest critic of securing anything scientific mainly for the public. While clear to just about everyone at the meeting that this scientist was originally loathe to let others even peek at some of his favorite objects, everyone was both puzzled and flabergasted why this man would persist in these elitist ways, especially in light of the technical and political troubles then facing the project. Most of us thought that everyone should now pull together as a team, if not magnanimously then at least grudgingly, in order to save the Space Telescope project from the very real possibility of cancellation.

Given at the time the deep sense of depression among scientists, the onslaught of extremely negative press, and the genuine anger within the Congress, it was unclear if the U.S. government would elect to fix *Hubble* or turn it off. A ship at sea is abandoned when it takes on enough water to flood a critical number of compartments, and with the pointing and aberration problems among others plaguing the telescope, it was not unreasonable to conclude that *Hubble* was beginning to "list to starboard." As the chairman of the Science Working Group diplomatically struggled to prevent the meeting from entering its infamous nonlinear mode by pleading that, "These are no longer normal times," Roger Lynds of the National Observatories in Tucson, one of the world's most skilled astronomers, stood and summarily ended the argument by bluntly shouting, "Now is the time to suppress personal possessiveness in the interest of the mission."

But the days grew into weeks and no imaging was scheduled with Space Telescope. Growing contentiousness and lack of cooperation on the part of a few astronomers had both committees stymied. Corridor talk and boardroom discussion proliferated; indecisiveness ran rampant. Said even a normally staid flight status report issued by the Goddard project office, "Right now we are in the phase where committees, groups, tiger teams, and the such are springing into being at a phenomenal rate." And although many attempts were made by various scientists to write formal proposals to observe this or that astronomical object, ensuing arguments made it clear that the issue of proprietary rights was still alive. We astronomers just couldn't seem to get our act together.

Eventually, during the second half of July (and largely by ignoring all the complaints about protected sources being imaged "prematurely"), the EROs and SAT groups managed to agree on some realistic, nonstellar targets, and the institute's User Support Branch accomplished in record time the extensive preparations needed for *Hubble*'s computers to execute our objectives. The observations were now ready to be uplinked to the telescope—but more days became more weeks and still no science imaging was accomplished.

It seemed that no one was willing or able to give the command to start. NASA Headquarters was rudderless, tossed every which way by attacks from various influential senators, the Vice-President's Space Council, and an irate public and press over the ongoing *Hubble* saga—let alone the grounding of the shuttle fleet and the vacillating design status of the space station. And on the front lines, the *Hubble* project lacked a strong technical champion, an operationally minded czar willing or able to take charge. Some three months after launch, we were still in the first part of orbital verification—"OV-1"—an initial testing period originally expected to last a couple of weeks.

Nor was there any evidence that the conduct of the mission was improving—although the project office at Goddard had been clearly humbled enough to file internal flight-status reports that were more technically truthful than before. A mid-July report was typical, and did not exactly instill confidence among those of us waiting to use *Hubble* for initial science observations:

> Things have been unexpectedly "exciting" this week at HST. The biggest event since the last report was a spacecraft safing event. . . . All scientific instruments and spacecraft subsystems are fine, recovery is underway. The cause of the event was human error: a series of six fixed-head star tracker updates failed resulting in a small, 100-arc-second attitude error, a real-time slew was requested to correct the pointing, but somehow in PASS/OPS [part of the operations infrastructure at Goddard] the final attitude of an earlier slew entered the system as the final position of the short real-time slew, and checks of the manuever failed to uncover the error. . . . As a result a request for a 90-DEGREE slew was sent to the spacecraft, the slew was executed and upon completion of the slew the spacecraft onboard computers noted that the sun was not normal [perpendicular] to the solar power arrays, and concluded that the ground controllers did not know what they were doing and safed itself. . . . A day before this safing event there was a science instrument safing: the faint-object spectrograph safed because of a microprocessor reset. . . . Efforts to get faint-object camera images in support of the focus analysis have failed completely due to a combination of human error and fine-guidance sensor acquisition failures. . . . Because of the safing event a number of scheduled science instrument tests never executed.

Day after day in July, the institute schedulers scrambled to respond to engineering requests to revise *Hubble*'s observing calendar; new focus tests were uplinked to the telescope and new data downlinked—all of which showed nothing more than that the telescope suffered from spherical aberration. It was as if the engineers still refused to believe that *Hubble* was purblind. Hughes, who felt most responsible for the debacle, had reinforced its efforts to unravel the precise optical status of the telescope by putting yet dozens more engineers on the job, partly to inject fresh blood into the fray and partly to relieve those recent reserves who, in turn, had become exhausted while struggling to decipher *Hubble's* signals nearly round the clock for the past weeks.

As we watched from a distance the daily activities in the Goddard operations areas, we became increasingly disillusioned, and soon flat-out critical, of what was going on. Did all these engineers, seeming at times to operate the telescope covertly, have any kind of rational plan of attack? Did one day's tests follow logically on the preceding day's results? Just who was in charge of this project?

Tests were also under way to characterize further the jitter problem that plagued *Hubble's* pointing when the spacecraft entered orbital day or night. The solution to this problem—a piece of software that would essentially drive the spacecraft's gyroscopes ever so slightly in the opposite sense to its pesky seesawing—was scheduled to be uplinked to the spacecraft by mid-July. But mid-July came and went with only a brief (private) admission from NASA that its installation would be delayed a month or more, and perhaps indefinitely. We also learned that the extent to which the jitter would be damped would surely be less than earlier promised, and that the spacecraft would not be able to meet its design spcifications for pointing—another Level-1 failure.

These delays and lowered expectations of SAGA—the solar array gain augmentation—combined with the continuing loss of lock on guide stars during terminator crossings made us worry that we might have to avoid scheduling during terminator crossings those science programs requiring the finest spatial resolution—another hit to science efficiency. I thought back to my earlier intelligence briefing when it became clear to me that military space vehicles "of this size and shape" were long known to experience oscillations of the type now plaguing the *Hubble* telescope. Again I fumed, why hadn't they informed us? And how could I make contact with that part of the government which I had been told would be able to help us? I made some phone calls to some friends in the defense establishment, and got nowhere—or so I thought.

Less than a week later, I was contacted discretely by the Central Intelligence Agency. The morning-long report on the status of Space Telescope to eight very tight-lipped, clean-cut individuals would not be their last expression of interest in what they called "your little project." Although they said little during essentially a monologue, it became obvious during the course of the meeting that they were myopically focused on three issues: instrument calibration, data handling, and image compression—the latter a computer technique designed to compact vast quantities of digital bits and move them rapidly along the orbit—ground—analysis pipeline, but they quickly became unimpressed with what we knew about it, at least compared to what they knew. At any rate, the CIA-types were clearly not interested in those secrets of the Universe we were hoping to unlock with *Hubble*. This was an operational briefing, stressing the ups and downs of driving bus-sized vehicles in low-Earth orbit.

Only once during the session did I learn anything that I can publicly report. Interestingly, one of the attendees blurted out that their vehicle—presumably the most sophisticated of the *Keyhole* spacecraft, known to some as *KH-12* but which is really an advanced *KH-11* (or *KH-11*$^+$), alias "son of Big

Bird"—spends most of its time intentionally in safemode. We civilians find it worrisome when *Hubble* enters safemode because such an idle vehicle cannot perform useful science observations, but in retrospect it is reasonable for military space telescopes to station-keep much of the time. Their observing targets coincide mostly with potential trouble spots on the globe, such as the Middle East or Eastern Europe, and thus the *Keyhole* handlers apparently put their spacecraft into safemode when not in use during many consecutive orbits.

The meeting ended on a sour note when at a working lunch I asked the unamused attendees half-jokingly if any of their photos of missile garages and submarine pens have halos around them. From the looks on their faces, either these guys lack a sense of humor or they, too, suffer from aberration, having been victimized by the same team of industrial contractors that gave us a defective *Hubble*. (Actually, the phenomenon of aberration would not appreciably degrade orbital images of Earth since such data comprise reflected radiation passing up through Earth's atmosphere, which somewhat blurs their images anyway.) As regards spacecraft jitter, I learned nothing useful—other than an offhand and cryptic comment to the effect that we were "using the wrong solar panels."

Toward the end of July, the Space Telescope project had seemingly run aground. We were now three months into the mission, completely wrapped up in an endless cycle of engineering tests, and still not a single science observation had even been attempted. The project office at Marshall was purportedly controlling *Hubble*, thus keeping Goddard at bay even on Goddard's turf, but for a good part of the time since launch the vehicle had in fact been idle, doing nothing much at all except mutely orbiting Earth. The project office at Goddard was mostly alienated and insecure, just as frustrated with the lack of progress as we were at the Science Institute. The program office at NASA Headquarters, which had circled the wagons and virtually ceased communicating with anyone in the project, had become diplomatically desperate in its reports to Administrator Truly: "The program continues to evaluate all possible alternatives offering any rational possibility for improving the current situation." And the institute was plotting to declare its independence, threat-

Ground features are hopelessly blurred when imaged by Space Telescope, and this has nothing to do with its aberration problem; rather *Hubble*'s pointing-control system is not designed to track and focus on terrestrial objects. These four images of Earth, each exposed for the shortest possible time of 0.1 second, were taken with *Hubble*'s wide-field camera at about 5,000 angstroms. Even in one-tenth of a single second the spacecraft, with an orbital velocity of nearly 18,000 miles per hour, overflies a half-mile of Earth's surface, thus grossly smearing anything in its field of view, which here shown equals about one-tenth of a square mile. The dark streaks across the otherwise blank images must be prominent ground or atmospheric features or even man-made objects— small islands, clouds, or perhaps a truck or building, for example. At the time, *Hubble* was skirting the Pacific Ocean and was aimed toward Hawaii. Such exposures, called "flat fields," are taken occasionally to help calibrate the telescope's cameras and have no military or intelligence value whatever. *[NASA]*

ening to break out of the pack publicly to demonstrate the urgently needed leadership.

Then, spontaneously and without warning, management of *Hubble,* which had slalomed from crisis to crisis, reached a climax. During a hastily convened "shoot-out" at the Science Institute among pivotal *Hubble* players from Goddard, headquarters, and the institute, discussion boiled over and became even more heated than usual. Institute personnel went beyond saying quietly in the hallways that no one was in charge; several of our speakers wrote it out on their Vu-Graphs. One speaker cast the words in large block letters, and, much like the president mouthing during the 1988 election campaign, "No—New—Taxes," ended his presentation by stating loudly and clearly, "No—One—In—Charge." The Hubble wars had broken out into the open.

Days later and without explanation, on July 23, we were told that the Marshall managers and engineers had left Goddard. Suddenly, those who were supposed to be in charge of orbital verification had packed up and gone back to Huntsville. Had the Goddard managers, in a burst of enthusiasm, mutinied and taken over the spacecraft, unceremoniously dumping the Marshall contingent overboard? Or had the Marshall team just thrown in the towel, effectively abandoning ship, as was rumored? Or had NASA Headquarters finally acted like a headquarters should by directing a change in management?

The resulting shift in bureaucracy—a kind of beltway musical chairs—was initially confused when the head of the *Hubble* project at Goddard, among others there, was removed from his duties ("exiled to work on the space station," said a colleague), while several new people were detailed to Goddard from NASA Headquarters to help manage the remainder of the commissioning period, including the Lockheed and Hughes people who would now begin taking orders from Goddard. To this day, I do not know the full details of this episode; NASA's next flight-status report merely noted that, "Transition to a totally Goddard-managed operation has been proceeding with some difficulty."

Thereafter, we heard nothing from Marshall until about a month later, when its Public Affairs Office issued a "NASA Fact Sheet" that was so upbeat, positive, and wrong as to be technically dishonest. Resuming the bold hype and bravado for *Hubble* that Marshall had so staunchly doled out before launch, their "facts," sent to educators nationwide to inform America's schoolchildren at the start of the fall term, read in part:

> The activation and fine-tuning of the telescope's onboard instruments, called "orbital verification," proceeded well, with only a few minor problems arising. . . . What was supposed to be a simple engineering test turned out to be a pleasant surprise for scientists . . . a "pure" spherical aberration, a problem relatively easy to correct much like the way an eye doctor corrects poor vision with spectacles.

Verification proceeded *well? Few minor* problems? *Pleasant* surprise? *Easy* to correct? Where does NASA get the people who write this stuff? It was as though the Marshall folks had, after a three-month hiatus, resolutely resumed their campaign of disinformation. All the while, I couldn't help but wonder about the technically knowledgeable personnel and especially the engineers from Marshall who had lived through the nerve-racking days and nights of initial orbital verification. Granted they were now back in their comfort zone in Huntsville, but were those people also part of the overt hyperbole that so regularly emanates from that space center? Not pausing to find out, we at the institute and I think Goddard, too, essentially elected to break diplomatic relations with the Marshall contingent.

Like most of my colleagues, I took the change of command to be potentially positive. The Goddard personnel were at a disadvantage running the spacecraft since their Marshall counterparts had kept to themselves much of the insider knowledge needed to conduct gross vehicle verification. Thus, there might be a window of opportunity for science observations before the new team began its own series of engineering experiments, which might go on for who knows how long. So, I began prodding, via phone and electronic mail, central elements of the Space Telescope system, urging anyone who would listen to join me in pushing to have some EROs and SAT science images acquired. During a single day at the end of July, I took three strong swings to activate the inaugural imaging campaign—and summarily struck out.

I first approached the project's leading NASA scientist, Al Boggess, a low-keyed yet effective person on the Goddard team who had been helpful to me during several earlier end-runs on the management. I pleaded with him to order up some science imaging, but he replied that he was unsure how that could be done until the spacecraft jitter was fixed. I protested that once the jitter had damped during each orbit, telescope pointing was acceptably stable for nearly a half-hour, especially during orbital night, and furthermore that the jitter problem might never be fixed until a shuttle servicing mission years later. Boggess then argued that the aberration would blur images of interesting cosmic objects, but again I protested that the EROs and SAT groups had been charged to address just that—to seek, by trial and error, how well we could image real, nonstellar celestial objects in spite of the aberration. He then expressed concern about slewing Hubble toward different positions in the sky, at which point it became clear that project engineers were afraid to command the vehicle away from the two or three sparse star fields where most of orbital verification had been conducted to date. Our discussion ended when Boggess confessed himself unsure of who had the authority to order that science imaging be done.

By contrast, Ed Weiler, the *Hubble* program scientist at NASA Headquarters, was eager to try some science imaging. But he complained to me that the SAT group was again bogged down in controversy about protected cosmic sources. "Proprietary data rights are killing us," he said. Putting aside feuds among members of the SAT group and now referring to the EROs group, he blurted out, "Where is your pretty-picture committee now that I need it?" I told him we were ready to go—just give us the word and the telescope schedulers at the institute would arrange the relevant observations on the calendars for uplinking in the next few days. He then retreated, saying that he couldn't give the order to do so. Despite Goddard's takeover of the project, he claimed that no one was in charge—no leader had been appointed by upper management or had emerged naturally. Weiler suggested that perhaps I could conspire with the spacecraft calendar builders at the institute to sneak some imaging onto *Hubble*'s schedule. Things were getting ridiculous.

Third, I called Bob O'Dell, the chairman of the EROs committee, reminding him of the charge we had been given at the previous Science Working Group meeting, more than a month earlier. He acknowledged the need to acquire some science data, but said that, "Things are dragging on the planning of the EROs observations [because of the need] to involve other proprietary scientists." He also now thought we needed to stress the nonimaging instruments aboard *Hubble*, arguing that spherical aberration would hopelessly distort pictures taken with either of the cameras. He thought it made more sense to do spectroscopy and photometry, though he agreed with me that explaining these exotic kinds of data (plots and graphs of squiggly lines) to the public would be something of a challenge. O'Dell seemed uninterested in computer enhancement and urged me to forget about taking worthwhile pictures with *Hubble*. Instead, he told me to stand by for a new list of cosmic objects toward which mostly spectra might be acquired sometime in the fall or winter, after a few more months of testing. His lingering concern about the EROs program surfaced when he asked rhetorically—and, to me, unbelievingly—"Who will own these data?"

My day was finished when I received a phone call from the senior staffer of the Senate Appropriations Subcommittee. Kevin Kelly wanted to give me a heads-up that by Labor Day's end of the congressional summer recess, there would likely be a motion to abandon *Hubble*—"to shut it down," he said. "The project is jinxed. You're burning up dollars furiously, yet have nothing to show for it. We've not seen a single pretty picture." As I hung up, I experienced a terrible sinking feeling, the same kind of total deflation that I felt years ago when that Red Sox first baseman let the ball go straight through his legs in the ninth inning of what should have been the winning game of the '86 World Series.

Disgusted with the disorganized state of NASA management, I was, for the first time, genuinely worried about the fate of the *Hubble* project. Disgust, of course, differs from frustration, and it was indeed a kind of disgust I then sensed—a nauseating repugnance that made me wonder how and why I had become associated with such a technically thin and managerially clumsy organization. I again and aloud asked while looking at a huge wall map of the Universe prepared to chart the progress of the EROs images, "How the hell has it all come to this?" and then let out a primal scream that caused a staff member on the floor above to race down and inquire if I were all right. But what to do? Does one just ride along with such a huge project, all the while realizing, despite a whopping $250-million annual operating budget, that it is floundering for lack of leadership and technical expertise? "It puts bread on the table," was the typically jaded response I often heard from government employees who were mostly marking time. Even more frequently something to the effect, "Hey, I've only got another decade until retirement. Get off my case." Or does one blow the whistle, knowing full well that those who did so during the *Challenger* investigation essentially got their heads chopped off?

Judging that being fired might actually be preferable to adapting to this system, I decided to write a memorandum. But to whom? If there was no one in charge, to whom should I write? No matter, like Chris Burrows had found earlier, it would be mentally therapeutic just to write an "open memo to the system"—even if it never went beyond the hard disk of my *Nuncius*. I entitled it, "A Specific Plan to Recover from Space Telescope's Poor Image in the Public Domain," and throughout the ensuing week or so, I occasionally revisited and polished it, getting out of my gut what I wanted to say. As things turned out, it was a pivotal memo, but since it was rather lengthy I'll reproduce only a part of it here:

> The Hubble Space Telescope may well be safely orbiting overhead, but its image before the public is in a tailspin. Despite repeated assurances by project scientists that *Hubble* can perform better than any ground-based telescope, the general public thinks the telescope simply doesn't work, precollege educators are beginning to growl that *Hubble* is damaging, rather than enthusing, our nation's youth to study for careers in science and engineering, and even many of our colleague astronomers are skeptical that *Hubble* can do much good science. As long as these misperceptions persist, we shall continue to see an erosion of public support for NASA's ambitious space programs—an erosion likely to accelerate shortly after Labor Day, when congressional recess ends and educators return to their classrooms.
>
> In the absence of any bona fide demonstrations that *Hubble* is working, a news vacuum exists. The media are filling this vacuum almost daily with negative "investigative" reports about "*Hubble's* troubles and NASA's problems," many of them unfounded. No matter, in Washington perception is reality, and unless a dramatic campaign is mounted to share with the American public the technical truth with a positive spin, then we should prepare ourselves for an endless stream of newspaper cartoons, late-night TV talk-show jokes, and blistering postmortems by columnists the world over. All of this will be exacerbated when, in a couple of months, a visually stunning "first-light" image is released from the new Keck telescope, an event that will likely trigger another round of reports that emphasize the lower cost and demonstrated science returns from new active-optics telescopes at ground-based sites.
>
> There is a critical need, *now,* to demonstrate decisively and unambiguously that Space Telescope is a functioning, viable space observatory—that it can do exciting, unique *science*. ... The Educational and Public Affairs Office at the Science Institute is prepared to mount, *immediately,* an attractively packaged program of early science releases into the public domain. These would take the form of print and video media giveaways of *Hubble* images as soon as such images can conceivably be acquired. We stress that, despite *Hubble's* superb spectroscopic and photometric capabilities, *only pictures will impress the taxpaying public* at this time.
>
> A dozen targets, approved before launch by the Science Working Group for early imaging of astronomical sources, were selected because they are colorful,

contain structure, and are otherwise aesthetically intriguing for the nonscientist and scientist alike. They include, for example, the Pluto/Charon and Saturn/Titan systems, the gaseous nebulae 30 Doradus and Eta Carinae, a colorful star-forming region, the structured supernova 1987A, and several peculiar galaxies. All these are reasonably bright objects, requiring relatively short exposures.

At the 27–28 June 90 meeting of the Science Working Group, two other committees were established to acquire science data, both for publicly oriented demonstration purposes and for use by deconvolution experts who are eager to explore various techniques of image enhancement. These two groups are moving slowly, are not necessarily emphasizing imaging (which, we repeat, is the only way to reverse the public's negative perception of Space Telescope), will not likely have much data to release until the fall, and are partly plagued by a panoply of proprietary data rights and NASA policy concerns.

Our specific plan, if someone could so direct it, is to acquire before Labor Day seven colorful images of spectacular cosmic objects, and to release them in a carefully packaged and specifically choreographed manner that will cause a dramatic turnaround in the public domain. This specific plan calls for, during one week in early September, a pair of *science* press conferences and several media releases timed in the following manner: a colorful image released on a Monday, a press conference and two images on a Wednesday, escalating to a press conference and four images on a Friday—all within one week. Spectrographic, photographic, and astrometric data, very carefully and attractively packaged to be understood by the general public, as well as more imaging, must continue to enter the public domain at the rate of about one release per week throughout the fall.

We stress that these press conferences would emphasize *science,* would feature the principal investigators, and would be conducted in a manner that is both positive and understandable to the general public. In particular, these conferences would not address spacecraft jitter, optical aberration, pointing problems, or a host of other negative issues that seem to have center stage in the congressional hearings and weekly media telecons.

The Space Telescope project need not wait for better telescope focus, for there will not likely be appreciably better focus given *Hubble's* aberration, nor for a fix to the jitter problem since there are nearly thirty minutes of stable pointing after the jitter has damped during each orbit's night-side pass.

The Science Institute's Educational and Public Affairs Office is prepared to orchestrate high-resolution glossy releases of the early images, as well as carefully crafted explanations of their science implications. The glossies released to the print media should be of the highest resolution, processed at the Science Institute through our Data Reduction and Analysis System and Photo Lab areas. We don't care whose logo is on them, just that they be of the highest resolution and quality, accompanied by a clear, understandable description of the scientific significance of each image.

The memo went on to describe specific plans to make clear to precollege educators that *Hubble* can, as promised, be the intellectual vehicle to enthuse children to consider careers in science and engineering, as well as to share with our colleague astronomers the results of the *Hubble* science mission and thus win back the support of the university scientific community. It especially addressed the need to recover the public's trust in the project—and in NASA—by disseminating Space Telescope science widely and accurately. The memo ended with what I intended to be a clarion call for "a magnanimous effort to rise above individual concerns" about proprietary rights and to work together as a team for the good of the project.

I read the memo two or three times a day, wondering what to do with it, if anything. I more or less resolved to send it to NASA, keeping my criticisms within the system. But, again, to whom should I address it?

In early August, the project began testing *Hubble*'s high-resolution spectrograph. Colorado's Jack Brandt, the instrument's principal investigator, wisely approved a request from Dennis Ebbets, a scientist on the Ball Aerospace contractor team that had built the spectrograph, that *Hubble* be aimed toward a distinctly more interesting cosmic region than any the telescope had examined to date. This was possible not only because a strong ultraviolet source was needed for the inaugural test, but also because their team as a whole was not overly obsessed with proprietary paranoia; Ebbets, in particular, had been a strong supporter of our EROs program early on.

As the first target, a bright and cluttered area in the Southern Hemisphere known as the Large Magellanic Cloud was chosen. Named after Ferdinand Magellan, whose voyage brought back word of the glowing southern skies, the Large Magellanic Cloud is considered by most astronomers to be a small companion galaxy to our Milky Way; however, since its distance is "only" 170,000 light-years away, I have always preferred to think of it as an exceptionally rich region of star formation in the extended halo of our own Galaxy, and not a dwarf galaxy at all. At any rate, although the specific target to be examined was a rather isolated star in the outskirts of the cloud, attention quickly focused on a nearby complex of gas, dust, and bright young stars called 30 Doradus—the thirtieth member of the constellation Dorado, the Swordfish.

To many astronomers, the 30 Doradus complex is one of the most remarkable star-forming regions in the sky—a stellar nursery as vibrant as any thus far found either in our Galaxy or any other. Colloquially called the Tarantula Nebula, it is the largest and most luminous gaseous nebula known, and is readily visible with the naked eye despite its great distance. Harlow Shapley once computed that if 30 Doradus were at the distance of the "nearby" Orion Nebula (about 1,500 light-years away), it would be bright enough to cast shadows on the nighttime landscape of Earth. Its core, a source called "R136" (the 136th object in a catalog compiled by South Africa's Radcliffe Observatory), was long thought to be a single, supermassive star, containing as much as 3,000 times the mass of our Sun. Such a gargantuan star is difficult to reconcile with current ideas about stellar evolution, however.

Theory suggests, and virtually all observations to date concur, that stars are not likely to maintain their structural integrity with more than about a

This is a ground-based image of the 30 Doradus complex in the Large Magellanic Cloud, some 170,000 light-years distant. The star cluster known as R136 is at the core of the nebula, made luminescent by vast quantities of stellar ultraviolet radiation sufficient to ionize the surrounding gas. The three viewing domains of the fine-guidance sensors are superposed, inside of which guide stars were used to facilitate *Hubble's* first science observation. *[Space Telescope Science Institute]*

hundred solar masses and will probably collapse and shatter under their own weight. Within the last few years, a special active-optics observing technique known as speckle interferometry had implied that R136 was not just one star but several—perhaps as many as a couple of dozen stars—but this technique works only for a small pencil-beam field of view of a few arc seconds across; thus, the results were inconclusive. Even so, some members of the cluster would still seem to be very massive—as much as hundreds of solar masses and still possibly at odds with theory. Uncertainty prevailed, since the stars appeared too close together to be individually resolved from the ground; if they were really more numerous, their mass values would fall. Would a telescope with sharper acuity show many more stars? In principle, only *Hubble* could address this problem. But in practice, with its flawed optics, would *Hubble* be up to the task?

Since the high-resolution spectrograph had never before been used, Brandt and his colleagues were uncertain precisely where it would be pointing. Its exceptionally small apertures, designed to take advantage of *Hubble's* exquisite resolving power, had not yet been calibrated relative to the fine-guidance sensors. So the first step would be the acquisition of a target image with the wide-field camera, in order to see just where *Hubble* was aimed. Then the spectrograph would be brought into play.

Not surprisingly perhaps, as things turned out, largely because of the pointing difficulties, this first attempt to use the spectrograph was unsuccessful. Space Telescope was still suffering from pointing errors of between 5 and 10 arc seconds, and since this was larger than the 0.25-arc-second and 2-arc-second apertures of the high-resolution spectrograph, we were effectively lost in space, at least as regards spectroscopy. It would be many more months before we learned how to use the onboard spectrographs. However, the target-acquisition image—a 40-second exposure—was indeed successful, owing to the wide-field camera's much larger field of view. The specific target was somewhere in the camera's sight. And what the camera saw surprised us all.

My heart skipped a beat when I first spotted the glowing image on a computer monitor in the data-analysis area upstairs at the Science Institute. Any astronomer could see new features and novel patterns in the celestial region in and around R136. The raw image was blurred by aberration and corrupted by cosmic rays that had streaked across the camera's CCDs during the exposure. And the camera had not yet been well calibrated nor had its instrumental effects been removed, both of which would make the image a lot cleaner. Computer enhancement would doubtless provide even sharper contrast and greater acuity. Even so, and for the first time, one could see virtually into the core of the nebula. Despite spherical aberration, many more stars could be seen than ever before. What's more, their cores measured only 0.1 arc second across, and this image quality was sustained over most of the full 160 x 160 arc-second field of view. Science! This was science!

Events began transpiring rapidly. The first real science image, obtained serendipitously to pave the way for a spectroscopic test, amounted to a breakthrough. On the third floor, several scientists were feverishly manipulating and analyzing the image in different ways. Colleagues were trotting up and down the corridors, poking their heads into different offices, eagerly consulting with each other and comparing the various computer manipulations then under way. The institute was finally starting to hum, its inhabitants bounding with enthusiasm. This was undoubtedly the most exciting time since deployment.

Unfortunately, a squabble broke out that would last for days. Institute scientists and those from Goddard were at odds regarding the precise star field displayed in the image. We were apparently *still* lost in space, even as regards

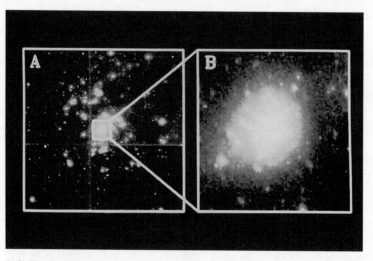

Left: This is a portion of an image made with *Hubble*'s wide-field camera, exposed for forty seconds to blue light (approximately 3,700 angstroms) and centered on 30 Doradus's core. The camera captured four adjoining sky regions simultaneously, which are assembled here as a (clipped) 90 x 90 arc-second mosaic.

Right: This is an enlargement (9 x 9 arc seconds) of the central part of the *Hubble* image at left. Shown here is the compact star cluster, R136, consisting of numerous twelfth- to sixteenth-magnitude stars spread across nearly 10 light-years. The cores of the individual stars are only about 0.1 arc second wide; the fuzziness is largely due to the accumulated aberration from many stars.

Neither of these images has been deconvolved—that is, computer enhanced, to remove some of the aberration. They are "raw" images, meaning approximately those seen directly with *Hubble*'s camera. [NASA]

imagery. With a telescope pointing error of uncertain amount and a rendering of a small piece of the sky showing many new stars never before seen, it was surprisingly difficult to decipher exactly where *Hubble* was aimed. No matter, I kept urging our staff at the institute to forget about the pointing problems for now, and to get on with image enhancement and analysis of the stars and nebulosity seen. It was the high quality of the image—the first real view of deep space with terrific clarity—that was so promising, regardless of where *Hubble* was looking.

Further complicating the investigative process at work in the hallways and offices at the Science Institute was the BBC. Weeks before, the British Broadcasting Corporation had arranged to visit the institute to film the second in a series of documentaries about the Space Telescope project. As if he had telepathy, their savvy film director, Alec Nesbitt, was almost perfectly positioned on the third floor when the excitement began. I and many others thought it was wonderful that his camera crews were capturing the essence of science—researchers at work at an opportune time, showing how we go about employing the scientific method, subjective baggage and all. Better yet, the film makers were recording for posterity a triumph against all odds—a minor discovery of sorts on *Hubble*'s first science image!

Soon enough, though, some scientists were complaining to me that the media should not be allowed even a peek at the data. "These are proprietary data," one member of the wide-field camera team told me sternly. Another astronomer, an affiliate of the high-resolution spectrograph team that had taken the image preliminary to their (failed) spectroscopic observation, argued that the R136 data should not be the property of the camera team—at which point the camera team moved to deny everyone further access to the data, a dispute that lasted for several more days. Yet another scientist was waving his arms frantically, literally yelling at the camera crew, "No, no, no, you can't film the image on my computer screen." I couldn't believe this spontaneous outburst of selfishness. Given the deep yogurt in which the project was then floundering, why would anyone not want to share immediately this thrilling development?—to hell with proprietary rights! Besides, as is worth repeating, the BBC and other media organizations were not as interested in the newly acquired 30 Doradus data per se, as much as the human beings who were using their heads to decipher the cosmos.

The event then under way was an example of what all of us had anticipated from Space Telescope—acquisition and analysis of new and spectacular views of the heavens that simply couldn't be obtained from the ground. We had expected that, day after day, we would be treated to celestial splendors that had not been seen before, or even imagined. This is what had drawn many of us to the Science Institute, to the Space Telescope project. Now, for the first time since launch, it seemed that perhaps *Hubble*, after all, was going to live up to some of its prelaunch billing and fulfill its potential as a world-class scientific instrument.

The evening of the breakthrough observation coincided with our monthly open night for the citizens of the Baltimore-Washington metropolitan area. Since the announcement of aberration, the crowds attending these public lectures had decreased greatly in size, and those who did come were disappointed and skeptical of the project, though not really hostile. Understandably, no one else on the staff wanted to face the public, so I gave the talk myself, and although I knew the new image of 30 Doradus would become an antidote to the many criticisms of the *Hubble* mission, I dared not say anything publicly about what was happening elsewhere in the building. NASA would become convulsive if I prematurely displayed the first science image and it wouldn't be fair to those scientists who had made the observation—even if they had acquired the image only as part of a test prior to their main non-imaging objective. Even from outside the building, several people noticed the unusual number of lights burning in the Institute's restricted areas. I could hardly wait for the public session to end so that I could get back upstairs to join the ongoing analyses.

In the early evening hours, at least three groups began to computer-clean the observed image of unwanted noise. Each group agreed to use a different numerical method of deconvolution in an attempt to rid the image of much of the corrupting aberration. It almost became a contest: Whose technique would yield the cleanest image—and the most science? Which technique would render the image most faithfully, without introducing artificial features that weren't really present in the astronomical object? How much computing power, and at what expense, would be needed to undertake this sort of image reconstruction on complex, crowded fields of stars?

At about ten o'clock that evening, I saw one of the deconvolved images. It was stunning! Whereas earlier in the day there was but a handful of new stars, now one could see scores, perhaps a hundred. Through *Hubble*'s eyes, the purported supermassive R136 blob had become clearly resolved into numerous stars of ordinary size—stars so tightly packed that all of them would fit into the space between the Sun and Proxima Centauri, our closest neighbor, some 4 light-years from Earth. This image was not just intriguing to look at; it definitely contained new science. And damn it all, it *was* pretty! Said *Hubble* astronomer Sally Heap, proudly, a few days later, "We now have the finest family portrait of stars beyond our Milky Way. As the hottest and heaviest stars known . . . the stars in this cluster define the limits."

Knowing the mixtures of stellar masses in a cluster, astrophysicists can deduce basic information about how stars form and how they create the chemical elements present in space. The most massive stars probably explode as supernovae, spewing the elements from which other stars, planets, and even life itself are made. By determining how many stars of each mass are present in 30 Doradus and similar star clusters in more distant galaxies, astronomers can infer more accurate information about how the chemical elements have become enriched over the course of time throughout the Universe.

After listening to a spontaneous tutorial in the corridor about the significance of the day's findings—one of many impromptu scholarly discussions under way that evening—I returned to my office shortly before midnight. I was due to leave early the next morning on a speaking trip to the Midwest. I didn't want to go, given the excitement at the institute, but I thought it

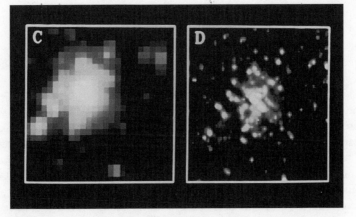

Right: This computer-enhanced ("deconvolved") image of the R136 object, taken with the Hubble Space Telescope, has sharpened the star cluster's appearance considerably. The undesirable fuzzy halos around each of the stars as seen in a previous figure are now substantially reduced. The stellar cores remain about 0.1 arc second wide throughout the 9 x 9 arc-second image. [NASA]

Left: This ground-based photo of the same region and on the same scale shown at right is not as clear as the *Hubble* image. Not a typical ground-based view of R136, it was obtained under rare atmospheric conditions and is among the best obtained to date with any telescope on the ground. The star images measure about 0.6 arc second across and were recorded with the 2.2-meter Max Planck telescope at the European Southern Observatory in Chile. [ESA and G. Meylan]

important to honor commitments to talk on Space Telescope. To cancel would send precisely the wrong message, especially given that day's very positive development. Besides, one of the groups I was booked to address was the National Association of Legislators, which was meeting that summer in Tennessee, where Senator Gore was expected to be in attendance.

While in my office, *Nuncius* beeped at me, indicating that someone had just sent me an electronic message. It was from the co-principal investigator for the high-resolution spectrograph, Sally Heap at Goddard, where evidently the "midnight oil" was also burning brightly. Her message said something to the effect that, "We are sitting on a pot of gold," and went on to interrogate me about my knowledge of the 30 Doradus star complex. She also inquired how we might jump-start the system in order to do more of this kind of science imaging.

I looked at the *Hubble* image of R136. And I looked at the memorandum that I had written earlier that week. Now was the time to send it. By the next day, word about *Hubble's* success would have spread into the bowels of NASA—and then there was the advantage of being out of town when my memo arrived. I addressed it to the Goddard project office for Space Telescope, as was the appropriate next step in the formal chain of command, but I also broadcast it liberally throughout the NASA system, including the highest levels of the food chain at headquarters in Washington. Never before had I written such an open letter, yet I feared that the memo might be circularly filed at Goddard, much as had previous suggestions to that space center's management. It was about time for me to put up or shut up. *Carpe diem,* and all that. And since I was concerned that the memo would not go anywhere at headquarters either, I expressed a copy directly to my contact at the White House. Then I left town.

On the second day of my Midwest trip, I received an urgent call from the institute director. At the time, I was at NASA's Lewis Research Center, in Cleveland, where I had just finished giving a colloquium about Space Telescope to an audience of several hundred NASA employees who—astonishingly, but like the public at large—thought that *Hubble* was broken. Giacconi, half out of breath and half screaming as had become his stressful telephone style, told me to return to the institute immediately: "Your memo has somehow reached the highest levels of the government," he said. "NASA has pulled the plug on the spacecraft [meaning that more engineering tests scheduled for uplinking had been canceled]. The telescope will be ours in a matter of days. Get the hell back here to coordinate the imaging campaign!"

I hastily left for the airport, and after setting up shop in a telephone booth in the departure lounge, I consulted with various groups at the institute who were preparing the EROs and SAT observations. I still chuckle at the scene I must have created in the airport—sitting for more than an hour in the cramped phone booth, outside of which the floor was strewn with computer printout, reams of calculations, and mistake-filled balls of crumpled paper.

My memorandum had, it seems, caused a revolution in the Space Telescope project—"jerked the system" as I was later unfortunately though correctly quoted in an Associated Press story. The memo had quickly traveled up and down the hierarchy at NASA Headquarters, where at least two associate administrators had embraced it summarily. The order reportedly came roaring back from on high: "Do it. Do it. Do it!"

When I returned to the institute that evening, the bottom floors of the building—the operations areas in and around OSS—were abuzz with activity, indeed much like an engine room. While the rest of Baltimore slept, scientists, programmers, and technicians toiled in a controlled frenzy, compiling new flight SMSs (telescope schedules) that contained the EROs and SAT astronomical targets. This was an enormous amount of additional work for the planning and operations crews at the institute, and it meant that they would have to double their efforts around the clock, but everyone agreed that it was worth it; within a few days, we would be imaging a wide variety of cosmic sources—we would be doing astronomy, or trying to.

Some will say that my memo was unnecessary, that *Hubble* science data eventually would have been acquired without it. And perhaps this is true— eventually. But I was convinced then, and remain so now, that the early-release observations meant to be shared with the public never would have been scheduled, and that it would have been months, and maybe longer, before the (mostly) nonimaging observations of the Science Assessment Team were uplinked.

I was further worried that whenever normal imaging operations did begin, certain members of the camera teams would promptly activate their proprietary rights, thus bottling up *Hubble*'s early findings for another year. The sole point of my memo was that unless the bureaucratic-engineering log jam was broken by Labor Day, after which Congress reconvened, there might not be an operating Space Telescope capable of yielding *any* science data. As fate would have it, however, the replanned SMSs would not be ready to be uplinked for a few days, and in the meantime there was one more hurdle to clear—another Science Working Group meeting.

The Science Working Group was back in true form. The same room at Goddard where, a few months before launch, the program to take a few early Space Telescope pictures had been torpedoed, would now, August 14, be the scene of more discussion of the very same issue. And what a wild meeting it was. First came the scientists' guests—representatives of the engineering community who had built the spacecraft wrong and who were now scrambling to fix it right. As we expected, the presentation on telescope jitter which the engineers had originally claimed would be easy to repair—"a piece of cake" being the phrase used most often—was not encouraging. The software patch needed to drive *Hubble* in directions opposite to its oscillations would not fit into the spacecraft's small 48-kilobyte flight computer. If we were willing to accept a partial solution, then a new piece of software that did fit might conceivably be concocted—but the NASA engineer giving the briefing wasn't sure, which sounded to us like another hollow promise.

The briefer then meekly noted mounting evidence that the jitter might be worsening, and, in any case, that the Marshall team, now gone home, had essentially given up working this problem. A new committee of Goddard engineers would attempt to discover a better solution, although they had much less experience with *Hubble*'s intricate pointing-control system, whose design had been closely managed by Marshall—and working relations with Marshall were, to put it mildly, strained. One scientist present was moved to groan aloud and urged impatiently that we move on to discuss the deficiency of *Hubble*'s main mirror.

Here, we were addressed by a Goddard engineer previously unknown to us and representing yet another new committee, the Optical Review Panel, which had been formed to oversee various tests in order to characterize the telescope's optical performance. His group had been directing repeated repositioning of the secondary mirror in small increments (called the "minisweep") and large increments (called the "big sweep")—apparently trying by sheer horsepower to establish the best focus setting for each of the onboard instruments. We suddenly realized why no science imaging had been scheduled in the nearly two months since the previous working group meeting. The engineers—now a whole new group of them—were continuing to monopolize the bird, just as they had since launch.

Next into the U-shaped well was John Mangus, head of Goddard's Optics Branch and member of the six-man Board of Inquiry—now called the Allen Board, after Lew Allen, its recently appointed chairman. (I thought it amusing that the space agency had stacked half the board with its own employees. "Is NASA capable of evaluating itself on this issue?" complained a skeptical congressman.) Mangus told us that the main pieces of equipment—tools, optical benches, measuring devices, virtually the entire laboratory setup—used to manufacture and test the *Hubble* mirrors had been found intact at Hughes Danbury, just as they were left nine years before, when the firm was part of Perkin-Elmer. This confirmed a story related to me by an insider at Hughes, that when company personnel reentered, during the previous week, the cyphered laboratory where *Hubble*'s main mirror had been polished in 1980–1982, it was as though they had negotiated a time warp; nothing had been touched, including the coffee cups, whose contents had been left to evaporate.

Apparently, the Perkin-Elmer engineers had evacuated the room rapidly and had never gone back, leaving behind technological fossils of their handiwork. More likely, given the management wranglings, deadline pressures, and mounting cost overruns in the early years of the *Hubble* project, Perkin-Elmer officials simply wanted to get this package out the door as quickly as possible. Their celebrated involvement with the Space Telescope project had become an administrative and technical nightmare. ("Why be bothered with NASA, when it's much more pleasant, lucrative, and technically interesting working with DoD people, on the other side of the wall," a well-placed Perkin-Elmer manager later told me.) That the fabrication and testing gear were found in precisely the condition used during the period of mirror manufacture, Mangus said, suggested that it would be a straightforward job for the Allen Board to determine not only the source of the error but also the actual shape of the mirror—a good thing, since Perkin-Elmer's log books ("ground truth") and NASA's quality-control documents were either in shoddy condition or missing altogether.

Incredibly, Mangus also reported that at least two tests of *Hubble*'s main mirror had shown evidence of spherical aberration, and that the results had been ignored. One was a simple "back-of-the-envelope" test—called a "refrac-

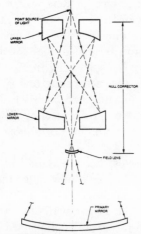

NOTE: THIS DRAWING NOT TO SCALE

This schematic diagram shows key parts of the "reflective technique" or interferometer template used in the early 1980s both to guide the polishing and to test the huge slab of glass that would become *Hubble*'s main mirror. The procedure begins when a laser beam atop a test tower is split into two, half of which is sent to a perfectly flat mirror, the other half to *Hubble*'s mirror blank (at bottom). The first beam of light goes directly to a sensor, which can be a piece of photographic film or a computer. The second beam passes through a "null corrector" (at center), which contains two small mirrors and a tiny "field" lens precisely positioned in an aluminum cylinder about the size of a beer keg, causing the beam to bounce uniformly off all parts of the main mirror's curved surface. The light reflected from the main mirror then returns to the top of the apparatus and intentionally interferes with the first beam—hence, the name, interferometer. The sensor compares the two interacting beams, creating a superprecise contour map—called an interference pattern—of the big mirror's surface. In the actual device used at Perkin-Elmer, the small lens in the null corrector was positioned less than a tenth of an inch too low. Thus, although the expected pattern was seen by the detector, because this testing template was incorrectly assembled, the final mirror was also incorrectly shaped. [Hubble *Board of Inquiry*]

Three tests of *Hubble*'s main mirror were made in 1981–1982 at Perkin-Elmer. None of these tests used conventional tools such as micrometers, calipers, or other mechanical gauges. Instead, sophisticated optical measurements were made using precise devices known as laser-driven interferometers that can measure extremely small distances equaling fractions of a wavelength of light, or about a millionth of an inch. Such optical metrology is so sensitive to vibrations that the techniques could be used only in the middle of the night when nearby traffic was minimized and only with the plant's air conditioners turned off. The result in each case is known as an "interferogram"—an interference pattern of dark lines (or "fringes" looking a little like close-ups of a zebra's hide) that amounts to a map of a mirror's bumps and valleys, in effect showing how well or badly its surface has been shaped. Two quick-and-dirty tests, one of them repeated more than two dozen times, showed a problem, as illustrated by the two curved and unevenly spaced patterns superposed on *Hubble*'s mirror disk at right and center, but both were discounted as somehow inherently flawed or inaccurate. The third test produced an interferogram (left) with a series of nearly straight and evenly spaced lines—the ideal pattern of a perfect mirror—but *this* was the test setup that was flawed owing to a procedural error that caused a testing lens to be mispositioned. Apparently, none of these warning signs was known to anyone outside of a small group of opticians at Perkin-Elmer—at least not until *Hubble*'s blurred images were telemetered to the ground for all the world to see. *[Adapted from data released by Hughes Danbury]*

tive-null" technique—that any amateur astronomer could and often does render when making small telescopes for backyard viewing of the heavens. This test was performed in 1981 during the rough-grinding phase, and records clearly prove that it showed the mirror's faulty signature. A second test, performed in the same year, clearly indicated a problem in the manufacturing routine used at the time, but this test, too, was discounted as itself being flawed. Either of these incidents should have run up a red flag indicating that something was wrong with the mirror.

Instead, a more sophisticated (and therefore presumably more accurate) measurement called a "reflective-null" technique was apparently relied upon exclusively and showed no evidence for aberration. Much like a dress maker who uses a dummy and a pattern to help refine delicate cloth cuttings, the main part of this optical technique—ironically called a "corrector lens"—was used to see a pattern that, in turn, determined how to fine-cut the glass. This test was done during the final shaping and polishing of the primary mirror, and it formed the centerpiece of the contractor's plans for developing the "perfect mirror."

But, as fate would have it, the testing equipment—part of an elegant new technology borrowed from the esoteric art of optical "metrology" (or ultra-precise measurement of light), and practiced handily by the military-intelligence community—was assembled incorrectly. Either while the widget was being used earlier to refine spy-satellite mirrors, or while it was being transported from the "military side of the house" and reassembled on civilian territory—the device was almost surely "used goods" since by then the production line of mirrors for the "black" *Keyhole* project was nearing an end—some paint was knocked off the end of a 2-foot-long (Invar metal) measuring rod that the Perkin-Elmer engineers apparently used upside-down while building the apparatus and, in any case, attempted to correct for the twin mistakes by wedging a few small, 20-cent household washers, uncalled for in its design, into the otherwise exquisitely devised million-dollar apparatus—all of which caused a 0.05-inch (1.3-millimeter) spacing error in the test setup. That's about half the thickness of a dime, or the size of the tip of a ballpoint pen.

Moreover, the device used to guide the polishing of the mirror's surface into its final shape was part of the same flawed test instrument; thus, there would have been no independent check on the grinding of the mirror's shape. "That's a huge error, just unbelievable. In an environment this precise, we're usually talking about micro-inches, or a millionth of an inch. The lack of independent check looks more and more like an unforgivable sin," said an academic optician, the same NASA advisor who in 1982 pronounced Perkin-Elmer's accomplishment "easily the finest mirror ever made."

Don't be fooled by the small numbers. The opticians didn't merely miss by a fifth decimal place somewhere. In today's world of precision optical engineering, they had in the words of a distraught scientist, "missed by a country mile." Actually, scaling things to our everyday world, the machining error would be roughly equivalent to a whole inch. An astronomer colleague put it in context: "Can you imagine trying to hang a door in its frame if it's an inch too wide? Can you imagine making that mistake after planning for years to make that door?" Knowing full well the oft-paranoid working conditions as well as time and money pressures of near-crisis proportions gripping the *Hubble* project inside the Perkin-Elmer firm at the time, board member Mangus flatly refused to comment on a pointed question from Giacconi: "Were the P-E metrologists simply sloppy in the early eighties?" Instead, he read us the essence of a summary paragraph excerpted from the board's "Optical Systems Failure Report," which would be released several months thereafter:

> During 1981 and continuing through early 1982, the HST program was beset by many difficulties. The estimated cost of the Perkin-Elmer contract had increased several-fold and the schedule had slipped substantially. The fine-guidance sensors were having serious technical problems, and the severity of the challenge to keep the mirrors sufficiently free from contamination to meet the specifications in ultraviolet light was just being recognized. The program was threatened with cancellation and management ability was questioned. All these factors appear to have contributed to a situation where NASA and Perkin-Elmer management were likely to be distracted from supervision of mirror fabrication.

In other words, human error. Needless to say, former employees of Perkin-Elmer claimed to the press that NASA was informed of certain test discrepancies, yet the space agency roundly denied, also to the press, having had any such knowledge. Later in the commissioning period, potential litigation was abetted between Hughes and the U.S. Justice Department when a timely mirror test result—called an interferogram—was uncovered, apparently showing evidence for aberration in an internally published 1982 Perkin-Elmer document, but which seems to have been cropped around its edges or otherwise doctored to minimize the evidence of that aberration. This suggested to many colleagues the possibility of a cover-up, fraud, or even worse—harking back to a Marshall engineer's remark, "as though someone had done it on purpose"—sabotage! Given all that has transpired in the *Hubble* project, such seemingly farfetched notions are not inconceivable; indeed, they have led to ongoing investigations that will doubtless keep more lawyers solvent for years to come. Owing to a peculiar combination of national security concerns and corporate obstinacies, I seriously doubt if we shall ever know the full story.

The Science Working Group also learned that the Allen Board had further determined that both the amount and character of the observed image blurring can be explained by the 0.05-inch positioning error of a lens—namely, the corrector lens noted above—within the testing device used a decade ago. Such, again, is big science: a measurement error of a fraction of an inch can lead to a dimensional flaw nearly fifty times thinner than a typical piece of writing paper—admittedly 200 times larger than the design tolerance—which, in turn, can cause nearsightedness in a multi-billion-dollar telescope.

Mangus also confirmed that aside from extensive testing done on the innermost few percent of the primary mirror's total area (mostly to check for smoothness, which did exceed design tolerance; the error hugs the mirror's outer edge), no "end-to-end"—that is, overall—systems test of the fully mounted optical telescope assembly containing both primary and secondary mirrors was ever performed by Perkin-Elmer, or required by NASA. Various factors were cited: there wasn't enough time, despite long delays in the *Hubble* project during the early years; there wasn't enough money, despite massive cost overruns; there was an overly confident mindset among engineers, despite a history of spherical aberration afflicting mirrors of the Ritchey-Chretien type; there was an attitude that *Hubble* could be serviced by astronauts whenever needed, despite the infrequent flights of the backlogged space shuttle.

By contrast (to my knowledge), no such U.S. *Hubble*-class spy satellite has ever been sent into orbit without a full systems test of its many mirrors in tandem. Such secret test equipment (built partly for the *Keyhole* project) is not necessarily designed to grant the subtle precision expected of the *Hubble* mirror—spy satellites have less rigorous optical specs than *Hubble*, for they are meant to look down through Earth's murky atmosphere—but the *Hubble* error is so large that such military test equipment would readily have detected it. Such comprehensive testing, maintained NASA, would have cost the agency "several hundred million dollars"—a number hotly disputed by many experts as far too high, but in any case could have been done for little if any cost by the Air Force which, according to reliable sources, did offer its testing services to NASA. In fact, the mirror imperfection is so large that it probably could have been detected with a simple "knife-edge" test, comprising equipment not much more elaborate than a flashlight, a few razor blades, and some masking tape.

The bottom line, starkly stated but no less true, was that the correct tests were not done, and those tests that were done, were done incorrectly. Although NASA rather foolishly defended its decision not to verify the viability of *Hubble*'s two-mirror system, even Senator Gore made clear during the previous month's congressional hearings why a shakedown cruise is useful: "When you build a ship, you test it in the water. There's something about a final assembly test that gives you a reality check that can catch problems that get through individual component tests." It bothered some of my colleagues that a politician would be offering advice on how to conduct a relatively simple sanity check—even if he was known as the technically minded "senator from planet Earth." What bothered me was that he was right.

A discussion followed in the working group about the "other main mirror," which no astronomer to my knowledge has ever seen. A second *Hubble*-class primary mirror had been built and tested by the Eastman Kodak Company, even though the latter firm had lost the original bid to manufacture *Hubble*'s optics. Given that Perkin-Elmer had proposed to use a new, sensitive polishing technology brought "over the wall" from the military-intelligence community, NASA thought it wise to build a backup using more conventional fabrication and test methods. To do so, Kodak used a traditional grinding instrument—a mechanical "lap" or polishing disk nearly as large as the intended mirror—whereas Perkin-Elmer used a small, spinning abrasive pad, whose grinding speed, pressure, and direction were controlled by computers obeying a curvature formula having numbers, or "conic constants," with as many as nine significant digits—the most advanced optical manufacturing technique in the world. But in an odd management decision, NASA directed Eastman Kodak to build its mirror as a subcontractor to Perkin-Elmer and not as part of an independent competition. Kodak completed the grinding and polishing of its mirror on time and within budget in 1980, but Perkin-Elmer had no incentive whatever to use a competitor's product, however better that product might have been. And since Perkin-Elmer was claiming that its mirror exceeded specifications and was also building *Hubble*'s fine-guidance sensors, NASA management apparently went along.

Exclusion of uncleared personnel—and that included most people at NASA—from the Perkin-Elmer laboratories, where defense-related projects were also housed, doubtless contributed to the lack of close management supervision, technical review, and quality control. Criticism of Perkin-Elmer's operations division often focused on its turf-minded hubris and "closed-shop" mentality, wherein the polishing team disdained paperwork and outside interference. At the same time, the Defense Department severely restricted the number of NASA employees allowed to "penetrate" the Perkin-Elmer complex in Danbury, given the obsessive nature of the highly classified work then under way at the facility. "When armed guards escort you to the rest room," a colleague sent to Danbury to eyeball the mirror in the early 1980s told me, "you know they must be building more than astronomical telescopes."

In fact, Perkin-Elmer had been a prime source of optics for the *Keyhole* project, and by 1981 when work on the *Hubble* mirror began, had provided large mirrors—from 5 to 8 feet in diameter—for more than a dozen space telescopes designed to reconnoiter Earth. This already tooled production line in effect determined the size of *Hubble*'s mirror. Contrary to popular belief, it is not the largest mirror that can fit into the shuttle's cargo bay; rather, NASA, strapped for cash in the mid-1970s yet eager to boost its floundering shuttle development program with a science package, descoped an LST—a large space telescope of 10-foot or 3-meter diameter—by piggybacking onto the Pentagon's impressive array of neat optical products.

In any event, few questions were asked by NASA, fewer reviews demanded, and even less test data seen. After all, the Perkin-Elmer opticians had come "certified" by the Air Force, and the firm had successfully done this kind of work before. The upshot was that the Kodak mirror was never seriously considered for use, although there is now every reason to believe that it far surpassed the quality of the Perkin-Elmer mirror eventually used onboard *Hubble*.

Where is the Kodak mirror today? As of this writing, the wooden shipping crate in which it was delivered a decade ago to Perkin-Elmer is still at Hughes Danbury. Unconfirmed reports from some intelligencers maintain that it was used years ago for a *Keyhole* mission—perhaps, ironically, one of those looking glasses rapidly enlisted to help the Western coalition fight the Persian Gulf war in real time. Other sources, equally as reliable, have told me that some sort of mirror definitely sits in that crate.

In accord with congressional testimony by the board's chairman, Lew Allen, that the inquiry will be "very embarrassing to someone or some group," the board's official report eventually stressed that both Perkin-Elmer and NASA should share the blame for the mirror's flaw ("100 percent each," said Allen)—the former because its engineers ignored several warnings of the problem, and the latter because its managers understood and accepted the fabrication and testing methodologies used by those engineers. Management was called into question at both institutions—especially their organizational structures, which were said to have "inhibited communications" and provided "inadequate cross-checks."

Not long thereafter, the "current" NASA again began circling the wagons while blaming the "old" NASA, indeed trying to deny any culpability at all. In an op-ed piece published in the trade journal, *Space News,* NASA's Lennard Fisk accused, ironically, the Perkin-Elmer opticians of being arrogant. This same chief scientist, in a formal Washington press conference, also described *Hubble*'s optical flaw as "a minor inconvenience," whereupon astronomer Roger Angel, an optical expert of some renown and an Allen Board member, responded by calling it "the single largest mistake that's ever been made in optics."

The meeting of the Science Working Group turned distinctly sour when someone drew attention to money matters noted in that summer's *Congressional Quarterly Report.* Despite NASA having paid Perkin-Elmer $440 million for work done on an original $69.4-million "cost-plus-fee" contract, and despite the cost overruns, the missed deadlines, and the failure to meet specifications just described, the space agency had also paid the optics firm an additional $11 million in bonus award fees! Later confirmed by astonishing NASA statements released by the agency's Public Affairs Office, this bonus was "a reward based on meeting deadlines, technical performance, and cost containment."

Moreover, both Lockheed (which had also already received a bonus fee of $14 million) and Hughes are eligible for further bonuses based on *Hubble's* "on-orbit performance." At which point the already outraged scientists became absolutely livid. "The accountability stinks," said aloud a distinguished astronomer. "Someone ought to be arrested," yelled another. More than one stood up in disgust and left the room. NASA's credibility had bottomed out, its public communiques to the taxpayer fully eroded. And its procurement system, once a model of purchasing efficiency outshining all other federal agencies, had truly run amuck.

None of this hindsighted finger-pointing did much to neutralize our, by that time, strongly held conviction that much of the *Hubble* project had been victimized by shoddy aerospace engineering and botched systems management. But by then the engineers and managers had become easy targets. Instead, some of the working group members turned on their fellow scientists, and began interrogating those whom they thought might have been partly responsible for the fiasco.

Two members of the group had been assigned a decade earlier to oversee the optical aspects of the *Hubble* project, and a third member had been at the time of mirror construction the lead project scientist at NASA; these three were subjected to rapid-fire cross-examination: Did you ever see the discarded test data? Were you ever aware of any such tests that even hinted at spherical aberration? To what extent were you involved in any tests at all at Perkin-Elmer in the early 1980s? Not surprisingly, considering the hostile atmosphere in the room, which by then rather awkwardly resembled a kangaroo courtroom, the answers all around were negative. None of the principal players had known of any such mirror tests, the compartmentalization at Perkin-Elmer had been that effective. Or so we have been led to believe.

To the annoyance of many members of the Science Working Group who felt slighted by the Allen Board, the final report of this federal commission later carried, among a dozen judgmental statements, the following:

> The NASA Scientific Advisory [Working] Group did not have the depth of experience and skill to critically monitor the fabrication and test results of a large aspheric mirror. However, this group should have recognized the criticality of the figure of the primary mirror and the fragility of the metrology approach, and these concerns should have impelled them to penetrate the process and ask for validation.

In the end, the Board of Inquiry published its findings in a document that was a model of technical investigation and science writing. The "Optical Systems Failure Report" clearly and systematically outlined the "cause of the telescope's flaw, how it occurred, and why it was not detected before launch." But it also did something else: it gave advice to NASA.

To his credit, Lew Allen, the retired Air Force general and spy-satellite expert who chaired the board, ignored that part of his committee's narrow charge, given by NASA's Fisk, which stated, "This group is not established to render advice or make recommendations." Instead, responding to pressure from a key congressional committee, the board penned an entire chapter on "Lessons Learned" which, among other things, admonished NASA and its industrial contractor for poor judgment, documentation and communications, and especially for NASA condoning an apparent philosophy that "considers problems that surface at reviews to be indications of bad management," to wit, "A culture must be developed in any project which encourages concerns to be expressed and which ensures that those concerns which deal with a potential risk to the mission cannot be disposed without appropriate review, a review which includes NASA project management."

The report read like the *Challenger* investigation all over again. And although its findings were widely admired in industrial laboratories and technical companies worldwide—the director of the multi-billion-dollar superconducting supercollider accelerator project made it mandatory reading for all his managers—NASA resented the board's conclusion that the space agency must share fault for *Hubble*'s poor sight and especially didn't like being given unwanted advice. NASA does not like uncontrolled, outside advice; never has.

The working group members' patience now having been spent, this long and revealing discussion was, alas, followed by a report of yet another committee newly established since the public announcement of *Hubble*'s myopia nearly two months earlier. NASA, finally realizing despite the long-held belief at the Science Institute that computer-enhancement techniques could gain us back much of the lost resolution, had formed a group of its own to explore the very same issue. Twenty minutes into a pompous lecture on deconvolution methodologies that yielded absolutely nothing more than what we had months earlier tried to share with NASA, Giacconi interrupted loudly: "Stop! Enough! There is an absolute proliferation of paper-generating committees, most of them smoke screens created by NASA management. And they are not helping. There is a time for talk and a time for action. We simply *must* acquire some real science data with the telescope."

Most of the scientists present were by then saturated with talk of technical foul-ups and management caveats, and were eager to begin the imaging campaign. They realized the seriousness of our predicament—especially with the congressional hearings and appropriations committees due to reconvene shortly after Labor Day. And they, too, were tired of the parade of excuses for faulty spacecraft devices and ongoing tests without rationale. But a few of them were irked that the Science Institute had taken the initiative, and the lead, to trigger the start of science observations.

One scientist—the same influential astronomer who all along had denounced early images for the public—was incensed that a program he had vehemently opposed was about to go forward. I suspect he was angered partly because he felt insufficiently consulted about which cosmic targets would be examined and mostly because he was not positioned to derive maximum credit from the early observations. This time, he didn't target his wrath toward me as much as the chairman of the Science Assessment Team, who was resolutely ignoring him.

Admittedly, this was heady stuff; just a handful of us scientists using all our knowledge of astronomy chose some two dozen previously approved science targets that Space Telescope would look at first—all of them derived from scientific programs sanctioned by the six principal investigators, who in turn would present *Hubble*'s initial findings to the world and publish the results in scientific journals. And besides, as Giacconi had intoned, this was a time for action, not for committee consultation that had, especially among members of this Science Working Group, usually degenerated into confrontational miasma.

Even so, at the beginning of the two-day working group meeting it was widely expected that the original opponent of the imaging campaign would express his outrage in another of his famous "3-sigma reactions." But, given the reality of a disabled *Hubble* on orbit and a host of other real-world politics at NASA, the alliances had shifted. Before launch, this distinguished scientist had, for various political and personal reasons, enjoyed the support of NASA Headquarters. Now, desperate for a positive development, headquarters had uncharacteristically entered into a rapprochement with the institute, which was championing the imaging campaign.

The strong-willed objector, finding himself isolated and deeply frustrated, finally snapped and became hysterical; he shoved some papers to the floor, shouted aloud that he was resigning from the group, and stormed out of the room, slamming the door just about as hard as one could without breaking it. It was pretty dramatic, apparently just as he had planned. But the telling thing was that no one went running after him. One of the guest engineers, who had taken an earlier beating at the hands of the impulsive scientists, blurted out with glee, "I'll be damned if some of you guys don't need a child psychologist." A few fed-up members concurred in low but clearly audible voices that his resignation was opportune. "We astronomers really can be spherical bastards," said one scientist. Not having heard the term before, I leaned over and inquired what he meant. "Spherical bastard," he repeated, "a term left over from Edwin Hubble's day to describe a malcontent from any angle."

Naturally, lots of anxious phone calls ensued later that evening, among principal investigators, NASA officials, and the institute directorate. The Science Working Group had lost one of its founding members and the rest of the group, out of sheer mental exhaustion as much as deep frustration with the status of the project, seemed about to disintegrate. Would NASA Headquarters "do a 180" and reverse itself once more, capitulating virtually on the eve of the imaging campaign? Among the many e-mail messages copied to the Institute that evening was a concern of one Hubble scientist to another about "the Institute short-circuiting the system"; the second scientist responded that whether the Institute was a maverick or a leader was a matter of perception. When I received a brief phone call at my home from Giacconi around midnight to the effect, "Stay out in front of the curve," I took that to mean, to keep the pressure on. Still, given the heat throbbing across the telephone and e-mail circuits that night, I was concerned that the extensive efforts we had made to schedule ten days of exciting scientific observations might be tabled or even canceled. As things turned out, they were not, largely because of two developments on the second day of the meeting.

First, the scientist who had resigned on the previous day showed up at mid-morning and asked the chairman for permission to address the group. I thought he was going to apologize for his uncivilized behavior on the previous day. But to my surprise, to use a NASA phrase, he went ballistic. He viciously attacked NASA's project management and resigned a second time in an identical manner—by storming out of the room and slamming the same poor door again, this time in fact breaking one of its hinges. Any sympathy for the man's arguments that his proprietary rights had been infringed upon or that he had been consulted insufficiently about the imaging campaign now completely evaporated. If only the budding-scientist kids of the world—those who think and are taught that the conduct of modern science is objective and unemotional—could have witnessed this meeting among some of the world's most eminent astronomers. What a revelation—more so, what a shock to their psyche—it would have been.

The tension in the room was now palpable, like a powder keg about to explode. Most of us had been up much of the night and were both dazed and spent, physically and mentally, unable to absorb an iota more of anything. A recess materialized without anyone having called for it. I was quickly intercepted by a seventh-team member resembling an elastic band near the breaking point personified, and lecturing me sternly that the institute's insensitivities were "causing the situation to go unstable." A principal investigator stepped between us, interjecting to me harshly, "Ignore him. Let's get on with it." Some pushing ensued, but no punches were thrown this time. Giacconi came to the rescue by motioning me to follow him. When I reached the lobby, he was pacing furiously, smoking like a chimney, and merely said to me, "It's going to be all right." I returned to the meeting room, this time my director following me. He sat next to me and whispered in a loud voice the same message, "It's going to be okay." Then he rose and began pacing again. The word *uptight* had been invented for times like these.

The recess ended suddenly when Lennard Fisk arrived unexpectedly from Washington to speak to the group. Never, to my knowledge, had this chief scientist attended a Science Working Group meeting for this or any other NASA mission. Entering the meeting room quickly, he greeted no one. Rather he strode directly to the well, then paused impatiently while people gathered in their seats and the room stilled. Moments later, the newly appointed director of the Goddard Space Center rushed into the conference room, a small entourage in tow. As he sat down heavily, I overheard him ask an aide, "What's he doing in our domain?" Clearly, the presence of both of these powerful government men was indicative of the perilous state of the project, and much of NASA too, for the Hubble Space Telescope was the centerpiece of NASA's space science program. It also typified the adversarial relationship between a headquarters organization unable to lead and a NASA field center unable to manage.

Then Fisk began speaking deliberately, telling us some facts about "political reality inside the beltway." With the meeting room exceptionally quiet, he slowly and carefully reviewed the precarious position in which NASA found itself. The entire shuttle fleet was grounded for leaky fuel lines; the *Magellan* mission to Venus was unpredictably losing contact with Earth; the *Galileo* mission to Jupiter was plagued by a balky antenna; the design and utility of the space station was under steady attack; and *Hubble* had caused the agency a black eye. He suggested that NASA was experiencing the deepest crisis of its thirty-year history.

"*Hubble* must appear to the public to be more of a winner than a loser during the next thirty days," Fisk pleaded, referring to upcoming tribulations with the deficit-conscious Congress. Perhaps because he was saying what I wanted to hear, though perhaps also because he was among his scientific peers, I found his Rockne-like speech most eloquent. This chief scientist who had, in my view, performed poorly at crucial press conferences and congressional hearings, was now making crystal clear the need to "obtain pictures to demonstrate to the public that *Hubble* works," adding dramatically, "and if you're concerned about proprietary data rights, then you're worrying about who is stealing deck chairs on the *Titanic*." He committed NASA to a repair mission to fix *Hubble* at the earliest opportunity, and concluded with the bold declaration that, "The future of the space program—not just of the *Hubble* project—is in your hands."

6

Inaugural
Science Observations

The next object which I have observed is the essence or substance of the Milky Way. By the aid of a telescope any one may behold this in a manner which so distinctly appeals to the senses that all the disputes which have tormented philosophers through so many ages are exploded at once by the irrefragable evidence of our eyes, and we are freed from wordy disputes upon this subject, for the Galaxy is nothing else but a mass of innumerable stars planted together in clusters. Upon whatever part of it you direct the telescope straightaway a vast crowd of stars presents itself to view; many of them are tolerably large and extremely bright, but the number of small ones is quite beyond determination.

... Further—and you will be more surprised by this,—the stars which have been called by everyone of the astronomers up to this day nebulous, are groups of small stars set thick together in a wonderful way, and although each one of them on account of its smallness, or its immense distance from us, escapes our sight, from the commingling of their rays there arises that brightness which has hitherto been believed to be the denser part of the heavens, able to reflect the rays of the stars or the Sun.

Sidereus Nuncius
GALILEO GALILEI, Venice, 1610

By mid-August of 1990, the Space Telescope project had reached its high-water mark. The engineers' insatiable desire to play with the technically sweet gadget had been suspended. The scientific personalities who had an aversion to sharing results publicly were damped. The original technomanagers had given up and gone home. The astronomical community had virtually abandoned us, the government seemingly disowned us, and the taxpaying public thought *Hubble* was broken. Those who were still standing had finally warmed to the idea that it was now paramount to secure images, preferably ones that could be understood by almost anyone. Promises were unacceptable, more hype out of the question. What we needed were data, scientific data.

I packed my family onto the Amtrak for Acadia National Park in Maine and told them to stay there for two weeks. Numerous others at the Science Institute also canceled prearranged plans at the height of the vacation season. *Hubble's* initial exploration of the cosmos that should have, and could have, commenced weeks after launch—spacecraft jitter and optical aberration notwithstanding—was finally about to begin. We were entering one of the most exhilarating periods in the history of astronomy.

On Thursday, August 16, at 11:12 A.M. Eastern Daylight Time, the imaging campaign began. At the time, *Hubble* was overflying the Indian Ocean on very nearly its seventeen-hundredth orbit. The plan was to take about a hundred and fifty-five images of some twenty-five targets, using many different filter settings and exposure times. The cosmic sources would span the spectrum of known objects in the Universe: planets, clusters, nebulae, galaxies, and an assortment of active and peculiar objects such as quasars, supernovae, and a comet. We had no idea that the campaign would be so tri-umphant, such a milestone in the history of *Hubble* on orbit.

The first celestial object we chose to image was a classic spiral galaxy called NGC 925. Like our Milky Way, this galaxy spans about 100,000 light-years and houses about 100 billion stars. It is estimated to be roughly 20 mil-lion light-years away, almost half the distance to the great Virgo Cluster of galaxies. This initial observation would use the wide-field camera to acquire data toward a crowded and already well-calibrated star field, and thus deter-mine how well we could measure the brightnesses of stars regardless of the telescope's aberration. A routine science exposure, with no unusual require-ments, and yet it would test the extent to which we might be able to pursue one of the three key projects planned for Space Telescope—nothing less than a measurement of the size of the Universe. If we could establish better the dis-tance scale of the Universe, we could derive, by implication, the rate of cosmic expansion and the age of the Universe. What's more, the results of this project might eventually tell us the ultimate fate of the Universe. But could it be done in the light of *Hubble's* poor vision?

Currently, astronomers reason that the Universe is about 15 billion years old—that is, it has been expanding from a point (the "Big Bang") for the past 15 billion years. This age is based on the "Hubble constant," the rate at which galaxies recede from one another in the Universe. First measured by Edwin Hubble and his colleagues in the 1920s, the expansion rate of the Universe has been refined by astronomers over the last several decades. Most of us now take that rate to be 22 ± 7 kilometers/second/million light-years.

Now, if you think about it for a moment and in your mind reverse the recession of the galaxies, the inverse of Hubble's constant is an estimate of the time during which the Universe has been expanding. Taking the inverse and canceling the units properly, we find that the Universe has an age of 15 ± 5 billion years, meaning that it could be as young as 10 billion years or as old as 20 billion years—an uncertainty of some 30 percent, and perhaps more in the minds of some astronomers.

This uncertainty results largely from our inability to measure well the dis-tances to celestial objects, including some galaxies within relatively nearby realms of several tens of millions of light-years. In particular, if we could better determine the distance to the Virgo Cluster—a huge swarm of more than a thousand galaxies about 50 million light-years away—then we could refine the Hubble constant and gauge better the age of the Universe. The expectation before launch was that Space Telescope should, during the first few years on orbit, acquire enough data to determine the distance to Virgo and thus the age of the Universe to an accuracy of about 10 percent. This would be a remarkable achievement for astronomy, a subject that often involves order-of-magnitude estimates.

Our current knowledge of the distance scale of the Universe depends on measurements of stellar parallax—the angle subtended by the apparent to-and-fro motion of nearby stars as Earth travels in its orbit about the Sun. Stellar parallax is, oddly enough, based on radar measurements of Venus, which can be used to find the relative orbits of Earth and Venus, and then the distance between the Sun and the Earth. To date, astronomers can measure parallax well enough to determine the distances to those stars that are within about a few hundred light-years.

During the next several years, both the *Hubble* telescope and especially the European *Hipparcos* spacecraft, which is a small satellite launched in 1989 and dedicated to the measurement of stellar distances, should permit us to extend the useful range of parallax to as much as a thousand light-years. But these distances are still well within the confines of our own Galaxy. To measure the truly cosmic distances of millions and even billions of light-years to other galaxies, and thus to establish better the rate of universal expansion, another technique is needed—photometry, which is concerned not with the apparent motions of stars but with the apparent brightnesses of stars. Photometry is a bread-and-butter aspect of astronomy, one for which many observations and much patience are required. Many issues in astronomy depend critically on accurate measurements of brightness, and if this could not be done well owing to telescope aberration, then a significant fraction of *Hubble*'s science mission would be compromised.

The stars known as Cepheid variables expand and contract—thus appearing to vary their brightness—in a regular, almost rhythmic manner. Those that are intrinsically dim change their brightness rapidly, often over hours or a few days, whereas those that are intrinsically more luminous pulsate more slowly, with a period on the order of weeks and sometimes even years. By knowing the rate at which a Cepheid star is pulsing—which can be easily observed with ground-based telescopes out to a distance of several million light-years—we can then determine such a star's intrinsic (or absolute) brightness. And comparison with its apparent brightness yields a distance, in much the same way that we often infer the distance to a traffic light whose true brightness we know on the basis of how bright that light appears to be. The Cepheids have thus become "standard candles," or reliable yardsticks, for measuring intergalactic distances.

The Space Telescope was originally expected to be able to image Cepheid stars in the galaxies of the Virgo Cluster, a region about ten times farther than those stars can be examined with ground-based telescopes. It was important to test to what extent this could still be done—or whether the key project

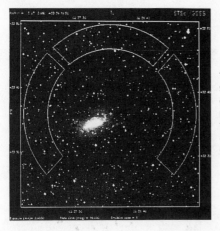

This is the ground-based photograph of the NGC 925 region used by targeters to plan *Hubble's* inaugural observation of the August 1990 imaging campaign. The horseshoe-shaped areas delineate the fields of view where pairs of guide stars were selected to keep the spacecraft pointing steadily at specific locations within the spiral galaxy. *[Space Telescope Science Institute]*

of gauging the size and age of the cosmos would have to be deferred until *Hubble* was repaired. Doubting that Virgo could now be probed effectively, the Science Assessment Team chose NGC 925, a galaxy roughly half as far away.

The objective was to image the northwest part of NGC 925's outermost spiral arm. In a second observation, the telescope would be stepped farther to the north (by about 30 arc seconds), thus capturing more of the arm. From studies of neighboring galaxies, we estimated that a healthy *Hubble* should be able to identify a couple of dozen Cepheid stars in the galactic coronal region beyond the arm, much as Cepheids are found above and below the plane of our own Milky Way. (By "neighboring," I mean those twenty or so galaxies considered members of the "Local Group," which inhabits a relatively nearby space a few million light-years in extent.) The questions were: How many Cepheids could we find? Could we measure their brightnesses? How well could we monitor their subtle variations in brightness? In particular, could we do these things given the crowded field of stars and the overlapping halos from each of the individual stars within the spiral arm?

This initial observation was a gutsy one. Previously, *Hubble* had imaged only isolated stars in and around our own Galaxy. Now, on the first shot of the imaging campaign, we were trying to capture light from a specific part of a tangled region more than a hundred times farther away than 30 Doradus—a full-fledged galaxy far beyond the Local Group. The exposure times would be tricky to estimate, since we did not yet know how the aberration would affect crowded and distant star fields. The camera's filters were still uncalibrated, its myriad observing modes yet untried. Fine-lock on guide stars had more often than not been lost, owing not just to the jitter of the spacecraft but also to the aberration of the guide stars themselves in the fine-guidance sensors. So, to hedge our bets against faulty long exposures, we took several fifteen-minute exposures, which could later be accumulated into a longer exposure if needed.

Even fifteen-minute exposures had not yet been attempted with *Hubble*, but long exposure time was itself an integral part of this science assessment. If we could expose for no more than, say, seconds at a time, as was then the practice among the *Hubble* engineers, the efficiency of science observations would be dealt a serious blow. We opted to undertake the first series of observations on coarse track; this operational mode would keep us pointed well enough to address the science objective without running the risk of losing fine-lock during the exposure.

As the observation was about to begin, *Hubble* entered orbital day, which meant that solar heating would shortly induce spacecraft oscillations. The observation commenced, guide stars were successfully acquired, *Hubble* found NGC 925 in its sights, and the wide-field camera shutter opened. As the jitter began, we watched the glowing amber computerspeak on the telemetry monitors follow the oscillation of all three gyros—positive, negative, back and forth—again and again as the gyros tried to stabilize the spacecraft by compensating for the flapping of the solar panels. Since the jitter typically lasted some ten minutes, the majority of the fifteen-minute exposure would be taken with the gyros fighting the jitter. If the gyros moved beyond a certain limit, the guide stars would be lost, the camera shutter would automatically close, and the observation would abort. If any of the gyroscope movements exceeded yet another limit, Space Telescope would enter safemode—certainly an inauspicious start for the imaging campaign.

Fortunately, since we hadn't requested fine-lock on the guide stars, the telescope was able to coarsely track them throughout the jitter period. Much to our delight, the observation executed as planned, if a little shakily. Another fifteen-minute exposure was taken, after a small slew to the north, with *Hubble* aimed at a slightly different part of the same spiral arm; that, too, went well. Since *Hubble* was at the time in direct contact with TDRS-East, both exposures were transmitted to the ground almost immediately. The images were breathtaking—yet surreal, because of the fog inundating the myriad, blurred stars.

Staring at NGC 925 intently, I felt as though I was out there in space, millions of light-years distant, floating among the realm of the galaxies. Other scientists in the institute's OSS area crowded around the monitors in rapt silence, marveling at the sheer beauty of the object's outermost spiral. Console technicians pulled up chairs to get a close look at it. And, this time, the spacecraft schedulers and computer gurus came into OSS, too. The latter had

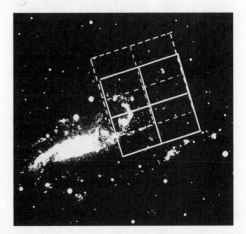

Those portions of NGC 925's outermost spiral arm scheduled to be examined with Space Telescope are shown in the boxed areas, some 2 arc minutes northwest of the center of the galaxy. (North is at top, west at right.) This is essentially a finding chart, the photograph having been taken with the 200-inch Hale Telescope on Palomar Mountain—a significant barrier crossed in and of itself: we had entered an age of exploration where photographs acquired with the best ground-based telescope had become guiding charts. The four superposed solid-line squares show the area imaged during *Hubble*'s first observation of the imaging campaign, whereas the four dashed-line squares denote a 30-arc-second offset area imaged during a subsequent observation. (The wide-field camera contains four silicon chips within its CCD array, which yield four electronic images simultaneously.) *[Adapted from* Atlas of Galaxies, *compiled by J. Bedke and A. Sandage]*

been working very long hours since launch, repeatedly rescheduling *Hubble* to suit the unpredictable whims of the project engineers. Now their efforts seemed to mean something, for we were at last doing science. It was a good beginning to the imaging campaign.

Initial comparison of *Hubble*'s view with the ground-based views of NGC 925's outer spiral arm was encouraging. Details and structure that cannot be seen from the ground are readily apparent in the *Hubble* image. But this observation called for more than a pretty picture. The intent was to assess how well astronomers might be able to measure the brightness of Cepheid variables in a galaxy of intermediate distance. And here, upon closer inspection, the results were disappointing.

As noted, *Hubble*'s distorted vision causes about 85 percent of the light from a given star to be spread over an area as extensive as several arc seconds in diameter. Although superb resolution is achieved with the remaining 15 percent, which is focused, as designed, within an area some 0.1 arc second across, the trade-off is that a great majority of the acquired light must be discarded. This causes a decreased sensitivity, or, in astronomers' parlance, a "loss in limiting magnitude," all of which means an increased difficulty in detecting and measuring the brightnesses of stars. Such measurements are especially tricky in crowded fields and star clusters—often just those places where Cepheid variables seem to congregate.

Additionally, the halos around each of the stars make detection of faint sources problematic, both because such sources tend to blend with the background noise and because they can be confused with the false structures (tendrils, rings, etc.) of the halos themselves. As if that were not troublesome enough, the halos differ in different parts of the cameras' field of view, and there is some evidence that their structures also vary with time (even during the course of a single orbit). The optician's nightmare that had transformed into the publicist's nightmare had, in turn, ultimately become the astronomer's nightmare.

All these factors combined to make measurements of subtle brightness

This is a copy of a raw image of part of NGC 925's spiral arm, precisely as telemetered in real time to the OSS "mission control" at the Science Institute. A fifteen-minute exposure with *Hubble*'s wide-field camera at the central visible wavelength of 5,550 angstroms, it was the first image seen at the start of the imaging campaign. These data therefore are not calibrated or corrected for instrumental response. Nor are they computer enhanced in any way to minimize aberration. Even with this unprocessed data, one can already see details and structure not visible in the ground-based photograph. One can also see a great deal of whitish haze, much like fog, engulfing the arm; this is due to the accumulated aberrations of innumerable stars. [NASA]

changes in Cepheid variable stars in distant galaxies doubtful, or at least controversial. In fact, although the faintest stars detected in the imaged parts of
NGC 925 are about the twenty-fourth magnitude (with an uncertainty of
about 15 percent), no Cepheids could be found unambiguously. Based on
aforementioned studies of nearby galaxies (especially M33, a well-known spiral member of the Local Group), Cepheids should be evident in NGC 925 at
about the twenty-sixth and fainter magnitude. Since we would therefore need
to reach a sensitivity about two magnitudes deeper, useful observations of
Cepheids in the even more distant Virgo Cluster were judged virtually impossible—at least without inordinate increases in exposure time.

Our conclusion was that one of *Hubble*'s most important tasks—to determine the size, scale, age, and fate of the Universe by refining the Hubble constant—could not be done reliably and unambiguously. Said Sandra Faber of
the wide-field camera team while contemplating the glowing images of NGC
925, "These data reveal both the glory and the tragedy of *Hubble*."

On a positive note, if the telescope could be repaired, then *Hubble* will
yet achieve its originally designed goals for both resolution and sensitivity, and
one of the greatest of its expectations might still be realized. Given that the
Universe is approximately 15 billion years old, a few more years of waiting
shouldn't matter much. However, such an on-orbit repair mission would not
be trivial and might never be accomplished well enough to improve measur-
ably the science capabilities of the orbiting observatory.

In a related observation with a similar objective, our next science target
was the open star cluster NGC 1850. Like 30 Doradus, this moderately
bunched group of stars is part of the Large Magellanic Cloud, and is thus estimated to be some 170,000 light-years away. Its stars are scattered over hundreds of light-years and it is thought to be only a billion years old—relatively
young in astronomical terms. The purpose of this test was to provide further
realistic data for stellar photometry in a relatively crowded field of stars.

Several observations were made with the wide-field camera, varying from
as short as ten seconds to as long as two serial eighteen-minute exposures
(roughly the maximum possible in a half-orbit before Earth occults the target). Between each of these observations, the *Hubble* spacecraft was deliber-

This is a raw *Hubble* image of a small part of
NGC 1850, precisely as displayed on the OSS
monitors on the first day of the imaging campaign. Taken with the wide-field camera, the
image represents an 1,100-second exposure at
5,550 angstroms and measures 75 arc seconds
on each side of the four CCD chip fields shown
here. The confluence of the chips approximates
the area selected for in-depth study; disappointingly, the faintest stars are only of the twenty-
third to twenty-fourth magnitude—not even
close to the twenty-eighth magnitude expected
before launch. *[NASA]*

ately rolled in order to assess the effect of aberration on different parts of the camera's field of view. The two wavelengths used—5,550 angstroms close to the center of the visible band and 7,850 angstroms in the near-infrared—gave us some sense of "color," yielding not only the relative brightnesses of individual stars but also their differing brightnesses at widely spaced wavelengths, a useful photometric indicator.

The results of this test were only moderately encouraging. Although twenty-third-magnitude stars could be seen to an unprecedented 0.1 arc-second level, the loss in sensitivity makes problematic many programs requiring accurate measurements of photometric magnitudes. Again, this is especially troublesome in crowded fields, where the halo of one star can confuse and interfere with another star's core, thereby making deconvolution techniques tricky. Work in such crowded fields was intended to grant *Hubble* an unchallenged advantage relative to similar work done with ground-based telescopes. Better computer algorithms might grant us some improvement, but the sensitivity loss of two magnitudes (for isolated stars) or three magnitudes (for crowded stellar fields) is absolute and could not be recovered unless and until *Hubble*'s spherical aberration was eliminated.

The next observation also went well but did not yield useful science results. What's more, it clearly implied that certain types of important astronomical work would not be possible with *Hubble* in its impaired condition. Although the observations were executing well, the prognosis for doing exciting science was looking bleak.

Our objective was to use the faint-object camera to image AP Librae, an active galaxy that had been previously well studied from the ground. (Active galaxies, as opposed to normal galaxies, emit much of their radiation by means other than those of ordinary stars—such as from the disks around black holes or as electrons spiral through magnetic fields; in fact, it is not known for sure if active galaxies have many stars.) Estimated to be some hundred-million light-years away, AP Librae is one of the closest active galaxies, and is known as a BL Lac object. Such objects, so named since the first in its class discovered in the 1920s is the ninetieth variable source in the constellation Lacerta, are now thought by astronomers to be intermediate in energy output between normal galaxies like our own Milky Way and the more active galaxies, such as quasars. Some colleagues in fact regard them as Rosetta stones, capable of providing us with at least one key to the secrets of the powerhouse quasars.

What are quasars? Answer: the denizens of the Universe. The word itself is a shortened version of "quasi-stellar objects," so named for their appearance on the sky and in photographs resembling ordinary stars. But stars they are not. The precise nature of quasars remains controversial, although most astronomers now take them to be the violent, active cores of very distant galaxies. Their light is extremely red-shifted, indicating via the Doppler effect that they are traveling away from us at very high speeds—which in turn implies that they are more distant than any other known type of object.

Some quasars recede, along with the expanding fabric of the Universe, at speeds of as much as 80 or 90 percent of the velocity of light; accordingly, they must be near the limits of the observable Universe. Yet despite great distances of many billions of light-years, quasars appear reasonably bright in the sky; therefore, they must be prodigious sources of energy. Some quasars emit a trillion times more radiation than our Sun. Modern astronomy has reached a reluctant consensus that the only conceivable energy sources capable of powering quasars are supermassive black holes—one of the Holy Grails of modern astrophysics, to which we shall later return.

Hubble's observations of AP Librae were designed to explore the regions surrounding its core, seeking evidence for any structure—such as spiral arms—in its glowing corona, also known as "fuzz" to the ground-based observers who can dimly perceive it. Astronomers the world over had anticipated that Space Telescope's heralded vision would be able to address quickly and decisively this most vexing problem in modern astrophysics: What is the relation of galaxies to quasars? A more complete picture of a core-halo object, such as AP Librae, might reveal common denominators between normal and active galaxies—such as a massive black-hole candidate, which is the hallmark of an active galaxy, sitting within a bunch of ordinary stars, which typify normal galaxies.

To seek out that better picture, two filters were commanded to rotate into the optical axis of *Hubble's* faint-object camera. A blue filter (~4,000 angstroms) captured images designed in part to reveal evidence of outlying galactic structure, such as spirals or jets emanating from its core. An ultraviolet filter (2,740 angstroms) in principle should have yielded data that effectively omitted the outer part of the galaxy, thereby allowing us to examine its inner nucleus for structure.

The results were disappointing. The scale of AP Librae's "fuzz" matches closely the extent of image blurring caused by the telescope's laxity. The two effects—one in nature, the other in the instrument observing nature—simply cannot be distinguished clearly, and thus no claim would be made by any reputable astronomer for observable structure in the underlying galaxy. Even after diligent subtraction of multiple aberrated halos, one cannot be sure whether the minute wisps of light that remain are real or not, for the process of deconvolution itself can introduce artificial structures on a very faint scale. The most optimistic of the expert researchers examining the data—Dave Baxter, a British astronomer loaned to the institute staff by the European Space Agency—later confided that with much longer exposures, precise calibration, and extraordinarily careful computer enhancement, one might be able to "solve the quasar fuzz issue" for a few such enigmatic objects, but that he doubted his colleagues would believe a result so hopelessly entwined in "*Hubble's* own fuzziness."

The attempts to search for dim features of light surrounding a bright central source were doubly depressing to many of us present on that first day of the imaging campaign, for at that moment we realized that another of astronomy's Grails could not be addressed by *Hubble*—the search for planets around other stars. The halos engulfing any star imaged by the orbiting beast would scatter light into precisely those locations where we had all hoped we might be able to spot another world beyond our own solar system—something that has never been unambigously accomplished by any means. The prelaunch chances that *Hubble* might identify a point of reflected light perhaps a billion times dimmer than the planet's parent star were small; now it became obvious that the chances were nonexistent.

These findings were a considerable setback for the *Hubble* science mission. All programs calling for a search and inspection of dim light sources nearby a strong light source seemed compromised. Rather than waste valuable telescope time, all such planned observations were abandoned, a taxing yet necessary judgment made almost immediately by the Science Assessment Team, and later approved by the Time Allocation Committee. Fortunately, we soon learned that these kinds of programs are among only a few types of science objectives that the impaired *Hubble* could not address.

The next object targeted for *Hubble*, NGC 7457, was a more normal galaxy, although one at a considerable distance—some 40 million light-years away. NGC 7457 is a so-called S0-type galaxy, meaning that it has a smooth central bulge and a relatively flattened elliptical shape wherein a few hundred billion stars are distributed about a plane extending across the usual hundred thousand or so light-years common to most normal galaxies. In contrast to our Galaxy, however, egg-shaped S0-type galaxies display no spiral structure, nor any trace of interstellar dust. Astronomers judge them to be quiescent, as they are not known to have any of the violent events found in the hearts of active galaxies, such as quasars. Before *Hubble* took a look, NGC 7457 was considered a dull, uninteresting, poorly known galaxy more suited for calibration work than for scientific investigation.

As is the custom in astronomy, I prepared for this observation by gathering finding charts of the cosmic real estate around NGC 7457. I used several sources of information, including my own personal computer on which I have a number of simple sky maps, as well as an elegant computer program that calls up digital scans of the guide-star plates stored in the vaults at the Science Institute. I also consulted the *Hubble Atlas of Galaxies* that I found in the institute library. This is a large set of photographs taken by Edwin Hubble, compiled decades ago into a convenient volume often used by galaxy researchers around the world.

Because I was having trouble reconciling conflicting pieces of positional information between my finding charts and the published photograph of NGC 7457, I trekked to the dim-lit basement of the Science Institute to consult with John Bedke, the manager of the photo laboratory. There, amid numerous dark rooms, photographic machinery, and huge color printing presses, a small staff of high-tech photographers works closely with the institute's astronomers to produce prints and slides for journal articles and scientific talks.

Studying my problem with the published *Hubble Atlas*, Bedke, who formerly operated photographic services for the Mt. Wilson Observatory, said, "Oh, 7457, I think I can do one better for you." He happened to have in his possession the original negative of the NGC 7457 region taken in July 1951,

This twenty-five-minute exposure of the galaxy, NGC 7457, was taken by Edwin Hubble in 1951 with the 200-inch telescope on Mount Palomar in California. The small square box drawn around the galaxy's center outlines the area imaged with the Hubble Space Telescope. [*J. Bedke and Carnegie Institution.*]

at the 200-inch Palomar telescope by Edwin Hubble himself. Carefully pulling the negative from its envelope (which had Hubble's handwriting all over it), we laid it on a large light table and saw the image of the galaxy more clearly than in the atlas. We quickly solved my scaling dilemma and simultaneously recognized the larger significance of what we were about to do: we would be able to compare the best ground-based image of NGC 7457 taken by Hubble, the man, with a similar image of the same galaxy taken by *Hubble,* the telescope.

The objective was to image the central region, or nucleus, of NGC 7457 by collecting light in the middle of the visible band—that is, at about 5,500 angstroms—in several exposures from 40 to 800 seconds long. The image would be extensively examined for structure near its nucleus, and also searched for bright stars, globular clusters, and previously undetected interstellar dust beyond that nucleus. Because we wanted to achieve the highest possible (visible-light) resolution in the galaxy's core, the planetary mode of the wide-field and planetary camera was used, giving us twice as good seeing as with the wide-field mode. (The planetary camera is not just for planetary observations, although its design is ideally optimized for images of some of the giant gas planets.)

We also commanded *Hubble* to acquire guide stars on fine-lock, which was a gamble, in that fine-lock could easily be lost, causing an exposure to end prematurely. But, again, throughout the imaging campaign we wanted to conduct some of the observations in much the same way as might normally be done in *Hubble's* operational era. We wanted to be conservative enough with some of the observations to be sure to acquire good data, but it was also important to conduct occasional risky observations consistent with some science objective.

In an ironic twist given the earlier NASA stipulation that the early-release observations should address no new science, real-world circumstances had now driven us to the conclusion that the imaging campaign had better not just acquire pretty pictures; rather, if at all possible, the results should indeed demonstrate novel science. And science we got! The results of this observation were spectacular, implying the presence of a supermassive "something," possibly a black hole, at the very core of NGC 7457.

The guide-star fine-lock mode held nicely and all the exposures went well. When the data were telemetered to us, all we could see in the field of view was an elliptical blob of starlight, much as expected. The inner parts of NGC 7457 seemed much like its outer parts—lots and lots of stars in a well-defined ellipsoid practically devoid of structure. But when the gray scale on the computer screen was adjusted (essentially like turning the contrast knob on your television set), we could see, astonishingly, that a significant fraction of the galaxy's light was coming from an extremely small, pointlike source at the very center of the object. Apparently an exceptionally bright and compact nucleus is embedded in the diffuse background of the rest of the galaxy.

Enormous quantitites of energy are pouring forth from a region so small that even *Hubble* is unable to resolve it. The energetic central region must be less than 0.05 arc second—roughly, the pixel size (or "diffraction limit") of the planetary camera—which at the distance of NGC 7457 amounts to about 10 light-years across. Never before had optical astronomers been able to study at this level of spatial resolution the core of any galaxy outside the Local Group. We were making good on the honest prelaunch claims of taking a

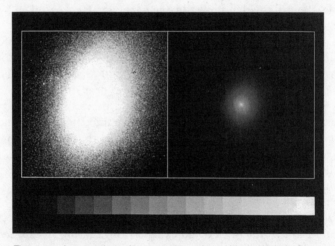

These two images show the same data taken with the Hubble Space Telescope as it peered into the core of NGC 7457, but each displays a different representation of those data. Both show the same field of view on the same scale; the square frame of each image displays what was seen in the area (17 arc seconds on a side) outlined in the square superposed on the previous ground-based image of NGC 7457. The difference between these two frames is only one of contrast (or gray scale); no computer enhancement has been done on either image.

Left: The image with low contrast shows the central part of the full galaxy, where, predictably, the density of the starry population increases smoothly toward the galaxy's core. The fuzzy appearance results from the combined light of billions of stars.

Right: The contrast has been adjusted to reveal a surprisingly high concentration of stars, spread over an area no larger than 10 light-years across and pinpointed exactly at the galaxy's core. Such a sharp spike of light is called by astronomers a "luminosity cusp" and in this case was quite unexpected. [NASA and T. Lauer]

great leap forward in clarity of view: we were moving into a new realm of cosmic exploration.

The young postdoctoral fellow, Tod Lauer, who had responsibility for analyzing the data from NGC 7457, experienced the time of his life while trying to decipher the newly captured light. As the data were being downlinked from *Hubble,* he had adjusted his computer monitor for high contrast. And when the data appeared on his screen, he merely saw a point of light—which he thought might be the image of a star. Lauer momentarily worried that *Hubble* had suffered a severe pointing error and had missed the intended target entirely. But when he adjusted the contrast on his monitor, he was astonished to see the galaxy itself "grow" around the bright point at its center—at which time he was reportedly seen floating, unsuspended, about 3 feet above the floor. This is known in science as a Eureka experience.

Here, then, was a surprise. A vast amount of energy emanates from an extremely small region—so compact that even a Space Telescope with perfect optics could not have resolved the central structure. Evidently, stars are at least hundreds of times more tightly concentrated at the center of NGC 7457 than was previously expected— except in the idiosyncratic work of a young Caltech astrophysicist who had earlier taken his own life rather than endure the pent-up frustrations of the job. What's more, the density of stars in the central pixel is estimated to be at least 30,000 times greater than that in our own stellar neighborhood. To imagine what that means, we would have to visualize about 30,000 stars between our Sun and the Alpha-Centauri system 4 light-years away—a proposition so wild as to make our nighttime sky unimaginably different. A key question here is how such a tight, compact core might be gravitationally assembled. One way to grant structural integrity to such a swarm is to invoke a black hole, but it would probably need to be one containing a mass of about a million suns. If such a huge black hole does inhabit the nucleus, then its indirect signature, paradoxically, would be a sharp spike of light—not coming from the hole per se but from the outlying accretion disk of matter spiraling inward toward it.

So is the heart of NGC 7457 a black hole, whose surrounding environment radiates huge amounts of energy before the hole gulps perhaps entire stars? Or could it be something less exotic, yet still surprising, such as an exceptionally rich star cluster? No one knows for sure, for the *Hubble* images alone cannot provide the answer. Only *Hubble's* spectrographs, potentially able to measure the velocity of the gas and stars swirling within the core and thus to infer the total amount of mass concentrated within the galaxy's nucleus, could do it; but we later learned that aberration badly affects the use of those instruments for this purpose, making the issue moot unless and until *Hubble* might be fixed.

These findings are significant because they suggest that compact cores might more commonly populate normal galaxies than previously thought. Despite *Hubble's* degraded optics, astronomers could surely use Space Telescope to probe the mysterious centers of many other "dull and uninteresting" galaxies in our continuing search for black holes. Lauer summed up these pioneering efforts in a smart and balanced way: "NGC 7457's striking nucleus is a surprising finding—not a real discovery, but surely more than just pushing back a decimal point."

We were beginning to feel more hopeful that the imaging campaign was going to be a success. This was good for just about everyone concerned, not least for me personally since an awful lot of senior people were watching my moves closely. Impatient for news, Giacconi asked me constantly about the status of the campaign—"It's going to be okay," he kept saying while pacing in his office, my office, or our imaging lab. And, untrue to form, officials at NASA Headquarters were calling the institute regularly—something they had never before done—asking for progress reports of completed observations and appraisals of upcoming science targets.

By contrast, NASA personnel at Goddard seemed essentially "out of the loop," showing little interest in these inaugural science observations; said the project's flight-status report for this week (the same week that Goddard informed the institute that we would henceforth need to fill out time cards for each hour worked): "We do not have a whole lot of information on these observations. . . . We will not be issuing a report the week of 20 August—vacation."

Pressing on, we next extended our probing to a much greater distance—out to about a billion light-years. There had been much confusion in the popular press, and even among some scientists, too, regarding how far the deficient *Hubble* might be able to see. Lots of people took the inane statements made by NASA that *Hubble* would now be limited to studying bright objects to mean that it wouldn't be able to see very far. They equated brightness with nearness, dimness with remoteness, largely I suspect because NASA's slick, million-dollar PR brochure on *Hubble* had made this same mistake, despite our prelaunch written objections. (Relatively nearby objects can be sometimes dim, such as potential planets around neighboring stars, whereas distant objects can be sometimes bright, such as faraway but powerful galaxies.) The confusion was best exemplified when one of the nation's most highly respected science journalists called me to ask, "Will *Hubble* be able to image any objects outside our solar system?" He was surprised and pleased to hear that even the nearsighted *Hubble* would be able to do significant science out to very great distances, perhaps nearly as far as originally planned, close to the limits of the observable Universe.

The target was called PKS 0521-36, an active galaxy that had been well studied by radio astronomers, who years ago detected a prominent jet of plasma (ionized gas) streaming from its core, much like water from a garden hose. (The designation "PKS" refers to the Parkes Radio Observatory, in Australia, and in this case its numbers describe the location of the object in the sky.) It is within the cores of such active galaxies that we suspect abnormal, even violent events, unfold, the most dramatic of which are manifest in the mysterious quasars. Jets, in some objects a single, fingerlike beam and in others two such beams on opposite sides of the core, are peculiar by-products of extraordinarily energetic affairs occurring within the "central engine" of some active galaxies. Whatever their true nature, astronomers reason that such jets provide an escape route for vast quantities of energy generated by the galaxy's powerful nucleus—in effect, cooling it lest it blow up.

This test would determine to what extent *Hubble* could unravel a contorted and relatively faint astronomical structure. The bulk of the starlight in PKS 0521-36 is about the sixteenth magnitude—a reasonably bright target, considering its billion-light-year distance. But the jet, detected (perhaps) optically a year before by the New Technology Telescope in Chile, is estimated to be of a dimmer twenty-first magnitude. The challenge was to distinguish the active nucleus from the host galaxy, and to map the jet over its several-arc-seconds extent. To maximize the chance of unraveling the target's subtle structure, the faint-object camera was exposed to blue light for two 25-minute intervals—the longest exposure yet taken with *Hubble*.

Since the high-resolution operating mode of this camera has such a small field of view—normally 11 arc seconds across—*Hubble* was commanded to take a target acquisition image before the first long exposure began. The vehicle was then experiencing significant pointing difficulties, and this quick look was an essential prerequisite to confirm that PKS 0521-36 was really in the camera's sights. Such aiming checks required *Hubble* to be in real-time contact with the Tracking and Data Relay Satellite System—another observing "first," and a necessary part of many normal science operations. All was copacetic, as we watched the telemetry signals indicate that the camera shutter had opened on schedule, that the exposure had occurred, and that the data were being downlinked to Earth. About a minute later, the target-acquisition image appeared on a high-resolution monitor in the institute's OSS area, centered neatly in the middle of the field. The computer software needed to communicate with *Hubble* in real time via the TDRSS network—something we had fretted about at length before launch—had been exercised and had worked well, and the longer science exposure began as scheduled.

The results, when analyzed by an unassuming British scientist, Bill Sparks, in residence at the institute, provide astronomers with the best view yet of the luminescent jet of plasma hurtling into space from the bright core of PKS 0521-36. The processed image clearly shows the jet extending to a surprisingly large distance of about 30,000 light-years from the galaxy's core. (For scale, this is the distance between the Milky Way's center and the galactic suburbs where we live.) Unlike ground-based images, the faint-object camera picture shows fine detail in the jet, including two bright knots of gas. The bluish light in the jet is known from radio studies to be highly polarized—that is, the detected waves of radiation have a preferred orientation. Astrophysicists take this as evidence of "synchrotron radiation," produced by fast-moving electrons spiraling through a magnetic field. The name derives from the high-energy particle accelerators known as synchrotrons, which speed subnuclear particles to nearly the velocity of light.

The loss of energy as the electrons slow down is manifest as radiation of a nonthermal kind. (As if "synchrotron radiation" were not enough of a fifty-dollar word, working scientists often use the tongue-twister "magnetobremsstrahlung" to denote braking radiation caused by particle deceleration in a magnetic field—another of those overly technical terms designed to keep the beginners out.) Acceleration of electrons to nearly the speed of light must require violent events deep within the cores of galaxies such as PKS 0521-36. And because the decay times for synchrotron radiation at visible wavelengths are much shorter than those at radio wavelengths, optical measurements provide information about more recent acceleration within such jets.

The energy needed to create and maintain the gargantuan jet in PKS 0521-36 is generated deep within the core of the galaxy. The energy source is thought to be smaller than the volume of our solar system, and thus is far too small to resolve even with *Hubble*. The favored mechanism behind these cosmic fireworks is a supermassive black hole purportedly fueled by a steady infall of nearby gas and occasionally whole stars. This strictly gravitational process, whereby infalling and accelerating matter collects in an intensely hot, turbulent accretion disk around a black hole hiding upward of billions of solar masses, is much more efficient at converting mass to energy than the thermonuclear-fusion events that power individual stars.

Theory suggests that much of the accreted matter spirals into the black hole and is lost from the observable Universe, never to be seen again. But the extraordinary high pressure and temperature generated near such a black hole would cause some of the infalling gas (not yet inside the hole per se) to be

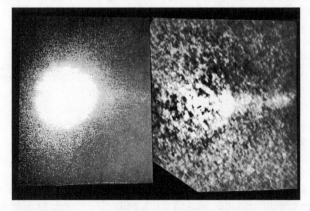

Left: The raw data acquired with *Hubble*'s faint-object camera shows only a hint of a jet protruding from the core of PKS 0521-36, an active galaxy about a billion light-years distant. The wavelength sensed was 4,300 angstroms, the exposure time twenty-five minutes, and the field of view a clipped 10 arc seconds.
　　　Right: In this image, on roughly the same scale as at left, the disk and nucleus of the host galaxy have been subtracted through computer processing to reveal more clearly the jet's structure, which has in turn been computer enhanced to diminish the effects of *Hubble*'s aberration. For the first time, the 6-arc-second-long jet can be followed almost all the way into the nucleus of the galaxy. Furthermore, the optical emission in the jet displays some structures—"hot spots" as small as 0.2 arc second across—that brighten probably where electrons are now being accelerated, perhaps due to instabilities at the edges of the gas flow along the jet. *[NASA, ESA, and W. Sparks]*

This is a single frame grabbed from an animated video produced by a graphics artist in the astronomy visualization lab at the Science Institute. It depicts one possible scenario for the "central engine" of an active galaxy: Two jets of matter are shown moving outward perpendicular to a flattened accretion disk surrounding a supermassive black hole. The rear jet is mostly blocked from view by the accretion disk, which possibly explains why PKS 0521-36 displays only a single jet protruding from its core. *[D. Berry]*

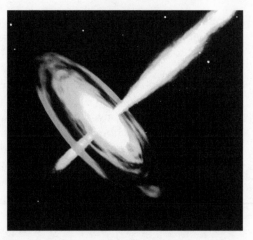

reversed and ejected parallel to the hole's spinning axis, thus creating huge plumes, or jets, much as observed. Exactly how such beaming occurs is not yet known, though powerful electrical currents must surely be involved, generating twisted magnetic fields along the hole's spin axis, which provides an escape route for the high-speed electrons. Nor has it been proved that black holes do in fact lurk in the hearts of active galaxies. Detailed studies of the optical morphology of PKS 0521-36's jet will surely aid our understanding of the galaxy's core, the acceleration mechanism along the jet, and why the jet remains so well defined and narrow across the span of tens of thousands of light-years.

This science-assessment observation was most encouraging, and suggests that astronomers have much to gain by using *Hubble* to observe faint jetlike features as diagnostic probes of the central engines of active galaxies. Studies of such galaxy outbursts will give us more information about a variety of energetic objects in the Universe, including the remote quasars, which are widely regarded as the most enigmatic objects of all, anywhere.

We were now three days into the imaging campaign, on a roller coaster of highs and lows. I should have known that some law of averages would offset the confidence we were beginning to build. As *Hubble* slewed across the sky to find its next target—another active galaxy known as 3C66B—the spacecraft balked for some unknown reason and entered a moderately deep safemode. It was the eighth or ninth time that *Hubble* had "safed" since launch—including one worrisome episode when the telescope somehow received a spurious command from someplace other than a NASA facility—but it was the first time it did so for no apparent reason. Contrary to the dire implications regularly implied by the press, safemode is a perfectly practical status for any spacecraft, especially, I suppose, when it feels that human controllers on the ground are getting a little cocky. But in this case, we were uneasy because we didn't understand why *Hubble* had sensed the need to "safe" itself.

The operations crews at Goddard, who had by now become quite practiced at recovering *Hubble* from safemode, began the lengthy and demanding process—a little like bringing an intensive-care patient out of unconsciousness and eventually back to full, productive health. It would be a couple of days before we could resume the campaign.

I came up for air, so to speak. For the last several days, I had gotten little sleep while trying to stay on top of all aspects of the imaging campaign. Having the family happily vacationing in Maine was working out fine, as I was able to come and go at odd hours, much as I remember while undertaking round-the-clock, radio astronomical observations for my doctoral thesis nearly twenty years ago. Heavily preoccupied with Space Telescope issues since launch, I had gone months without much awareness of the real world. Suddenly, I was surprised to find that the United States was sending troops and supplies to the Middle East in an attempt to convince the Iraqis to relinquish Kuwait. Even more jarring, I spied during a quick trip to the District black-suited commandos lurking astride several of the great oak trees on the White House lawn, blending nicely but ominously with the dark bark of their environment.

My greatest concern was that fighting would break out in Arabia and the media wouldn't even notice the encouraging results we were now obtaining with *Hubble*. Earlier in the summer, nothing especially interesting was happening in the world and the media were starved for news—especially, of course, sensational news that involves human shortcoming. Now that our mission was beginning to yield some excellent science, I feared that the opinionated fourth estate would ignore us.

Another concern was that the imaging campaign would run aground because of heavy use of the TDRSS network by military satellites then doubtless scrutinizing Iraq. We later learned from Air Force intelligence that those who inhabit the "Blue Cube"—that huge, windowless edifice in Sunnyvale having more upward-looking antennas than any place west of the National Security Agency—were toiling feverishly overtime while commanding and controlling an unprecedented fleet of five and possibly six *Keyhole (KH9-11$^+$)* space telescopes, amidst a virtual armada of other spaceborne "elint" assets.

I wondered whether *Hubble*'s mysterious safing event could have been triggered by the plug having been pulled on us because of heavy customer demand on TDRSS. We had at the time been using TDRS-East, which in its geosynchronous position hovering above the Atlantic was also the prime relay satellite of those lowly orbiting vehicles inspecting the Persian Gulf. With the buildup of Allied troops in Saudi Arabia and Iraqi forces in Kuwait, the need for overhead reconaissance must have rated the highest degree of TDRSS priority. Thereafter, whenever possible, our *Hubble* schedulers at the Science Institute requested use of TDRS-West, which was in lesser demand even when the fighting broke out the following January.

Fearing distractions from the concentration needed to monitor every detail of our imaging campaign, I nonetheless scanned my backlogged electronic and paper mail. And distractions I got. On e-mail, a string of "nastygrams" from Goddard's technical managers directed us to make an exception to their own off-limits restrictions in order to arrange, in the institute's operations areas, a public-relations stunt by a photographer for a corporate contractor—all the while NASA had steadfastly refused entry to responsible journalists at the height of the imaging campaign when their presence could have been most effective. We side-stepped the directive, though even this was not trivial; said Ethan Schreier, the institute's associate director and one highly accomplished at this sort of thing, "Ignoring middle management at Goddard takes a helluva lot of work."

Regarding the old-fashioned kind of mail—on paper—a large brown envelope with the return address, "Public Affairs Office, NASA Headquarters," made me sit up and take note. My few previous interactions with the agency's public-affairs officials had been largely in the form of complaints about the Science Institute's public outreach activities, especially our educational programs. Now, in the midst of the intense imaging campaign and other activities designed to save the *Hubble* project from possible cancellation, perhaps they were about to make peace with our efforts to prove that the *Hubble* mission was not lost altogether—and maybe even help us disseminate the scientific fruits of our labors. Alas, inside was a twelve-page, professionally printed, highly glossy booklet outlining the guidelines and requirements for use of the NASA logo. The covering letter made clear that, "The success of the NASA Hubble Space Telescope identity program depends on the correct and consistent use of the logo." By now, absolutely disgusted with NASA's posture of self-aggrandizement, I tore up the document in a rage and threw it away.

If there is a common denominator among middle managers at NASA, indeed evidently even at the top of the agency, it is surely the use of the NASA logo—the old NASA "meatball" symbol of a spacecraft orbiting around Earth, which was replaced years ago by the new "worm" signature spelling out in sleek, high-tech form the four letters, NASA, only to be more recently replaced with the old meatball by another NASA administrator, Daniel Goldin, one of whose first acts was to lead the agency into another meaningless debate about NASA's logo. It's as if the new NASA hierarchy thinks that by adopting the previous logo of the old NASA that the latter's glory days will once again emerge magically. Reinforcing their parochial attitudes expressed on numerous occasions before *Hubble*'s launch, NASA public-affairs officials had become nearly religiously obsessed with the "proper" use of the NASA logo rather than any attempt to explain clearly the ups and downs of the *Hubble* project.

Be that as it may, on August 21, the day before the imaging campaign was to resume, a meeting was quickly convened at NASA Headquarters in Washington to assess the science results garnered thus far. Upon entering the headquarters building on Independence Avenue, I was astonished at the fortress mentality that reeked throughout. Clearly, the space agency was under heavy media and congressional attack; many people I met that day seemed overwhelmed by the pressure and stress of their jobs. My concerns shot far beyond the *Hubble* project when I noticed several NASA officials physically shaking, their eyes bloodied, their insecurity maximized. The sight of a fax machine belting out a steady stream of incoming public-affairs traffic directly into a wastepaper basket didn't help either. My mind repeatedly flashed back to the astronauts; why would those brave men and women want to place their lives in the slippery hands and misguided judgments of such a reactionary system of management?

About a dozen people attended the meeting, among three disparate groups. The scientists among them were happy with the images obtained to date, but only in view of the afflictions of the vehicle; the results thus far were not what we had originally expected to achieve with the spacecraft. Secondly, a few of the government's public-affairs officers had virtually nothing relevant to say save for some hollow plans and promises for the Hubble project a few years hence; they were quickly cut off by other NASA personnel who rightfully wanted to address the present lack of public confidence. And there were the *Hubble* managers, who only wanted to know what discoveries we had made.

The issue of "discovery" was of paramount concern to all these NASA types, for they apparently now felt that, with scientists effectively in control of the telescope for ten days, even the hobbled *Hubble* should yield scientific discoveries upon demand. We cautioned repeatedly that an interesting, even unique, scientific "finding" or "result" can differ greatly from an outright discovery. But NASA officials were having none of what they took to be pedantic academic subtleties. Instead they insisted that we agree to announce the discovery of something or other from *Hubble*. Desperate for a positive morsel of news, their browbeating was especially intense regarding the ongoing analysis of the core of 30 Doradus, where *Hubble* had found dozens of new stars in a region where the best ground-based scopes had only hinted at the presence of a small star cluster. "How many stars are there here—40, 60, 100?" one of the bureaucrats asked impatiently, adding, "How many do you need before you can agree to call this a major discovery?" Referring to what was mostly noise amid the salt-and-pepper pixels in the new *Hubble* image, a particularly obnoxious manager impressed his fingerprints all over a glossy print while he urged, loudly, "Looks to me like there are lots of small white specks in this image. Aren't they all stars? Hell, there must be hundreds if not thousands of stars in this cluster. Isn't that enough to announce a discovery?" At times, the dialogue resembled a cross-examination, at others a deranged diatribe. In reality, it was a renewed attempt at hype. In the end, this and similar attempts by NASA to assign more importance to the data than they warranted alienated many astronomers—including several leading NASA scientists, among many others who were not even present that day—all of which greatly hurt our subsequent efforts to work closely with Hubble astronomers who, quite naturally now, worried that their scientific findings were going to be falsely sensationalized.

Fallout from this melodramatic meeting at NASA Headquarters occurred on many fronts, not least the need to harness those zealously in search of media-feeding discoveries. Several colleagues realized that I had overlooked an important feature of any early-release observations program. An editorial review board would now be necessary to guarantee the validity of new science results emanating from *Hubble,* lest we potentially fall prey to a cold-fusion scenario and begin doing science at press conferences rather than among peers. The upper end of the NASA hierarchy reluctantly agreed, still clearly apprehensive about losing control to a bunch of scientists. Fortunately for those seeking a balanced appraisal of the Hubble mission, Colin Norman, the knowledgeable head of the Institute's Academic Affairs Division, was appointed chair of the review board, which functioned well throughout much of the remainder of the commissioning period.

The attitudes and reactions I saw that day were typical of management at NASA Headquarters. Seldom did its personnel understand the technical issues or the science objectives, and they thought that yelling and posturing was the way to ensure bureaucratic discipline. Maybe intimidation works with secretaries or airport limo drivers, but the computers don't care if they are screamed at or threatened with firings. The science data are completely indifferent to all this explosive and misdirected human emotion. NASA's management had clearly not learned the lesson taught by the great physicist Richard Feynman who, as a member of the *Challenger* Investigative Board, had so aptly put it: "Reality must take precedence over public relations. Nature cannot be fooled."

Upon returning to the Science Institute, I found a small crowd surrounding my workstation, *Nuncius,* including Riccardo Giacconi and one of the STS-31 astronauts, Bruce McCandless, who had kindly offered to lend a hand in planning the *Hubble* repair mission. All were peering into the screen at the sweeping spiral arm of the galaxy, NGC 925, where efforts were still under way to determine the extent to which we might be able to measure Hubble's constant. Aside from McCandless chiding me about the extracurricular reading he saw on my desk—I had temporarily switched from Emerson to Machiavelli, whose *Il Principe* Riccardo had earlier jokingly urged me to reread "in the original"—it was a breath of fresh air to be back in the midst of rational people attempting to honestly unlock secrets of the Universe, the bitter arguments with flip-flopping NASA officials, who one day had to be dragged kicking and screaming to the imaging campaign yet now were demanding scientific discoveries at will, confined safely inside the beltway.

Much worry abated when *Hubble* exited gracefully from safemode without incident, on August 22. The astronomical source named Arp 220 was the new target—another attempt on 3C66B was postponed—and would be examined with one of *Hubble's* highest-resolution modes. Ground-based photographs of Arp 220 suggest two galaxies in the act of collision, one of which might be a quasar. So the faint-object camera would be used to capture radiation in the blue and ultraviolet portions of the spectrum, thereby diminishing much of the source's confusing infrared emission from dust spread across the object's complex central region.

The observation failed, however, because the wrong position was sent up to *Hubble*. More than two years before, a European astronomer on the staff at the Science Institute had given me slightly incorrect ("unprecessed") coordinates for the object, which we used in preparing the proposal to observe Arp 220. Those same incorrect coordinates were duly recorded throughout many steps of the Proposal Entry and Processing System at the institute. None of us thought it necessary to double-check them, so when the observation was made and the data transmitted to the OSS station, the field of view was blank. I immediately interrogated the institute's database of astronomical photographs, feeding in the same coordinates that we had used with *Hubble;* the field was indeed blank. Arp 220 was about 40 arc seconds away—not a very large error if we had been using a ground-based telescope with a larger field of view, or even *Hubble's* wide-field camera. But the faint-object camera—like a pencil beam probing the sky—was unforgiving.

Strutting into my office at a pretty good clip, the institute's chief of operations, Jim Crocker, the most capable engineer I ever met in the space program, was furious, and rightly so, for this was a case where the scientists messed up and the beleaguered engineers could point fingers. (On balance, it was a small, recoverable error that forfeited but a single orbit—compared to the hundreds of orbits in which *Hubble* had been idle during the preceding months, while managers deliberated what to do next with the balky spacecraft.) We'd been burned by the trade-off made decades ago by *Hubble's* designers: This observatory in the sky would have superb angular resolution—in fact, better than any telescope on the ground—but its field of view would be much smaller than most other telescopes. This, I believe, was a proper sacrifice, for resolution is the advance that more than any other factor has driven astronomical discovery. I also believe that we should have gotten our coordinates correct.

Surprisingly, wrong coordinates chosen by astronomers became a frequent source of human error throughout the commissioning period and beyond. This attention to detail was so poor and caused so many observations to fail that the institute eventually declared a new policy: everyone's coordinates would be checked against our *Guide Star Catalog* before uplinking any science instructions to *Hubble.* Astronomers can be so meticulous, even picky in many respects, but when it comes to something simple like choosing correctly the position of a celestial target, they seem to have a common shortcoming. Ironically, some of the same people who were so concerned earlier about "wasting" minutes of valuable telescope time to secure a few images suitable for public release would aim *Hubble* at wrong positions for upward of a half-orbit, and obtain no useful data.

We quickly recovered by acquiring another picture—this one of unprecedented clarity—of R136, the core of the 30 Doradus complex in the Large Magellanic Cloud. Earlier, R136 had been observed with the wide-field camera; this time we were attempting to resolve further the luminous midsection of R136 by probing it with *Hubble's* finest seeing—that provided by the faint-object camera. It was also an opportunity to assess how well *Hubble's* two cameras could complement each other—one providing a bigger picture (160 arc seconds across) while the other honed in on only a small part of that picture (typically 11 arc seconds across).

The observations went well, as a series of short (ten- to fifteen-minute) exposures was taken at both optical and ultraviolet wavelengths. No pointing difficulties were encountered this time, and the resulting images showed

Although Space Telescope has no zoom feature on its cameras, by observing in different modes one can approximate an increasingly finer view of a given region. Each of the images illustrated here has been computer enhanced in order to clean the image somewhat of its corrupting aberration. Keep in mind the rule of thumb that at the distance to the Magellanic Clouds, 1 arc second equals approximately 1 light-year.

Top: This is part of the image of R136 shown earlier and taken with the wide-field camera. The image covers only about 1 percent of this camera's full field of view, displaying a resolution of approximately 0.1 arc second across some 10 arc seconds.

Middle: This image of the central part of the same region was taken in blue light at 3,460 angstroms with the faint-object camera. The field of view shown here is only 6 arc seconds across (as seen by the camera in its f/96 mode), but the increased resolution (approximately 0.07 arc second) permits us to see the central stars more clearly. Note in particular the V-shape (on its side) string of stars, barely delineated in the top image and now fully resolved here.

Bottom: This image, an even higher magnification of the central part of the middle frame, was taken at the ultraviolet wavelength of 2,530 angstroms with a special f/288 mode of the faint-object camera. Here, the very finest resolution designed for the Hubble Space Telescope has been nearly achieved—0.04 arc second across some 3 arc seconds field of view. (Note that the bottom two images were taken with different filters and exposure times so that the intensities of the starlight in the two images are not the same.) [NASA and ESA]

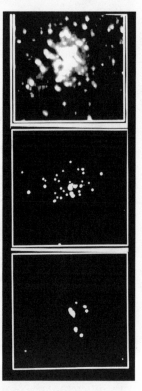

pretty much what most of us expected—the blobs of unresolved light previously seen by the WF/PC were now divided into yet more stars, demonstrating that the heart of R136 is indeed a very dense star cluster. In particular, the brightest object in R136 actually consists of two stars separated by about 0.06 arc second—a minute spatial domain close to the theoretical best seeing of the telescope.

The sharpest resolution designed for the Hubble Space Telescope had thus been nearly achieved, and the result was that an old and vexing problem in stellar astronomy had been solved—a tremendous accomplishment for a telescope that had been written off only two months before as a technoturkey. To be sure, an accomplishment that could have been realized within little more than a month after launch, had the originally planned early-release observations program not been axed by scheming scientists and bungling bureaucrats.

Understanding the details of young clusters like R136 is important, not only because such clusters are the nurseries where stars are born but also because the kinds of stars contained in such clusters play a key role in the evolution of galaxies. R136 is especially interesting, since it appears, even through *Hubble*'s high-resolution eye, to contain some of the most massive stars known; some might conceivably be a hundred times as massive as our Sun and millions of times brighter. According to modern theories of star formation, stars of this immensity are thought to be near the maximum possible. Our ability to study individual stars in such cluttered regions will greatly improve our understanding of stellar families and the process whereby vast interstellar clouds become stars. The high-resolution capability of the Hubble Space Telescope's two cameras will likely contribute handsomely to this area of active research in modern astronomy.

A small objective, the next observation required less than two orbits but yielded results that did knock our socks off. The plan was to image a peculiar star system using three wavelength filters, which together covered the full range of sensitivity of the faint-object camera—from 5,000 angstroms in the middle of the optical band to 1,200 angstroms deep in the ultraviolet. Since the star system was known to be embedded within nebulous plasma, this set of observations would also test the ability of the faint-object camera to detect dim and diffuse features in the presence of bright compact objects. (Plasma— the "fourth state of matter" after solids, liquids, and gases—is usually a hot, thin ensemble of atoms so highly excited as to be ionized, or stripped, of their electrons.) The science agenda of this observation was clear and ambitious: to look closely into the core of a belching star system for the first time in history.

The cosmic object targeted was the unstable and occasionally erupting star system, R Aquarii, which is the eighteenth member (hence "R") of the constellation Aquarius and one of the most puzzling regions in the sky. At the relatively nearby distance of 800 light-years, R Aquarii is the closest member of a unique class of double stars known as "symbiotic" systems. Symbiotic stars are characterized by a hybrid spectrum suggesting the presence of two very different types of stars—one hot, the other relatively cool.

Until *Hubble* imaged the region, the double-star nature of R Aquarii was inferred only from its unusual spectral features, since the two stars cannot be individually resolved in ground-based telescopes. Astronomers reason that symbiotic systems consist of a cool red giant star surrounded by an envelope of extremely hot, ionized gas that is energized by radiation from a much smaller and hotter companion orbiting about it. Making any interpretation even more challenging, the larger member of R Aquarii is an M-type variable star, meaning that pulsations of its outer envelope occur every 387 days.

Interestingly enough, R Aquarii was studied in the 1930s and 1940s by Edwin Hubble himself in an effort to unravel the mechanism that powers such peculiar star systems. The instrumentation then available yielded fuzzy images of nebulosity in the region that led him and his colleagues to conjecture that R Aquarii had erupted about six centuries ago. It has since become clear that R Aquarii has actually undergone a series of violent eruptions, in the process spewing out huge quantities of forged nuclear matter into the surrounding interstellar space.

Today, we know that R Aquarii, like other symbiotic star systems, undergoes semiperiodic outbursts that resemble the sudden surge of energy commonly seen in novae. One possible explanation for its peculiar behavior is that the red giant is spilling some of its matter onto its companion star. If the companion is a compact stellar remnant—perhaps a white dwarf, or neutron star whose atmosphere has been shed long ago—it would have a tremendously deep gravitational "well" into which matter can fall. Fortified with fresh fuel, the companion experiences an extremely rapid burst of nuclear burning, akin to the fusion reaction in a hydrogen bomb. In the process, not only would a great deal of kinetic energy be released in the form of heat and light but the energy would also power the ejection of a good part of the outer layers of the star at speeds of up to several hundred thousand miles per hour. (Classic novae are caused by the interaction of a red *dwarf* and a white dwarf in a binary dance, but symbiotic systems usually have a giant star as one of their members.)

That the outbursts occur periodically (or nearly so) suggests that the companion might follow an elliptical path that sometimes brings it very near to the red giant. At close encounter, the companion's gravitational field would pull matter from the giant's tenuous outer atmosphere. Since R Aquarii is also known to be a pulsating object, an alternative explanation is that the giant red star periodically swells up and engulfs the orbit of its suspected companion. In either scenario, gas would presumably be pulled rapidly onto, or at least near, the surface of the companion star. Rather than taking a straight path down onto the companion, infalling matter would spiral, forming a flattened vortex of hot gases called an accretion disk. A sudden surge of infalling matter could make the disk so hot that the outward radiation pressure (from the heat alone) would overcome the gravitational pull of the companion. This "over-load" on the disk would allow matter to escape in an explosive outburst.

Yet another idea is that, due to gas friction, the accretion disk might become so hot that it emits ionizing radiation in short spasms. A combination of these events and possibly other unknown interactions might account for the truly bewildering array of activity observed over the past fifty years toward R Aquarii.

Further complicating this object is a large expanding outer nebula that surrounds R Aquarii, as well as a smaller inner shell of gas most likely ejected in successive outbursts on several occasions—both distinct from the gaseous enve-lope just mentioned. Recent analyses of the spewed nebular matter, based on the speed at which the loose gas is observed to be expanding, place one of the major explosions about 640 years ago (confirming Hubble's earlier conjec-

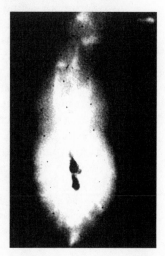

This is the first Space Telescope picture of a nova. It shows the "reactor core" of the symbiotic system R Aquarii as well as numer-ous filamentary features of hot gas (plasma) ejected at high speeds from the binary pair. The glowing plasma emerges like the flame of a stellar blowtorch as a 200-billion-mile-long luminescent geyser that is twisted by the force of the explosion and channeled upward and out-ward by strong magnetic fields. The observed pattern of flowing mat-ter suggests structure in the jet, including a fascinating helically shaped gas feature never before seen. This image, made mainly in the light of doubly ionized oxygen (at 5,007 angstroms) and not deconvolved for fear of introducing artifacts difficult to identify in such a complex region, shows 0.2 arc-second seeing across the entire field of view of 11 arc seconds (vertically). Separated by 0.8 arc second, the two dark knots at the center of the image possibly outline the binary star system itself; they are dark because the faint-object camera saturates (i.e., severely overexposes) when observ-ing very bright objects. Reseau marks painted on the camera optics (photocathode) have not been removed and are seen in this image separated by 1.2 arc seconds. *[NASA, ESA, and F. Paresce]*

ture), another only some 185 years ago. The latest such eruption might have occurred as recently as the late 1970s. Both nebulae have a characteristic hour-glass shape—structures consistent with the notion that a sudden explosion has blasted matter outward from the environment of an accretion disk. The hour-glass would have been produced because matter preferentially escaped fastest above and below the disk and slowest along the plane of the disk, where the higher gas density would be expected to restrict the flow of gas.

Hubble's faint-object camera was aimed at R Aquarii during the graveyard shift, early in the morning of August 23. Hardly a single orbit was needed to acquire the best of the R Aquarii exploratory images, a twenty-minute expo-sure emphasizing the strong 5,007-angstrom emission feature arising from doubly ionized oxygen. When the observations were telemetered to the insti-tute's OSS station, an image of bright arcs, knots, and filaments flashed across the computer screens, astonishing everyone present. One did not need sophis-ticated computer massaging to realize that this object is special. R Aquarii's distorted disk fully painted the monitors in intricate detail, unlike anything we had yet seen with *Hubble*. The image was electrifying—cosmic fireworks glow-ing before our eyes.

Two distinct, very bright knots of emission (which might be the binary pair itself) are seen separated by less than 1 arc second and surrounded by faint circular structures of "fog" created by the spherical aberration of *Hubble's* primary mirror. Bright, geyserlike filaments can be discerned emerg-ing from the inner regions of the symbiotic system. Prominent arcs, streaks, and knots are superposed on fainter, diffuse nebulosity resembling smoke from a chimney and extending throughout much of the field of view. This is clearly plasma that has been ejected at high speeds from the "fusion reactor" on or near the binary pair. The entire nebulosity glows because it is bathed in high-energy radiation released at the time of the eruption. The plasma emerges in the "polar regions" as geyserlike streams twisted by the force of the explosion are channeled upward and outward by strong magnetic fields.

In the northeastern part of the image is an obstruction in the path of the flowing matter, impeding the expulsion in the equatorial plane, the plasma here apparently acting like a stream curling around a rock, even bending back on itself in a somewhat spiral pattern. This subtle helical feature, not repro-duced here as well as in the original data, fascinated us, if only because it extends for more than 100 billion miles—more than a thousand times the dis-tance between Earth and the Sun, or roughly ten times the diameter of our solar system! The total extent of the emitted matter traceable beyond the core is at least twice that—extraordinary dimensions even in astronomical terms.

The scientist in charge of these observations, Francesco Paresce, the ami-able son of a Roman diplomat, thought it miraculous that this early-release observation had gone so well. Greatly moved by the image of R Aquarii, he felt impelled right then and there in the OSS control room to write an emo-tional essay attacking the then public mindset that assumed everything about the *Hubble* project was dead. He sent it to several leading newspapers world-wide, for consideration on their op-ed pages, but apparently only one ran a synopsis of it. Given that it was a balanced statement lacking sensationalism or stories of human failure, the media were apparently uninterested. His essay was a good and honest appraisal of the way we astronomers mostly felt halfway through the imaging campaign:

Okay fine, you say, no use crying over spilled milk; we can't bring *Hubble* back any time soon to fix it so what can we do in the meantime? Is is even worth the time, effort, and cost to continue operating it or should we just close it down and wait for the replacement parts? We now have the answer to that question, I believe, but, because of all the negative publicity *Hubble* has gotten so far, it's awfully difficult to get it across fairly and clearly above the tumult and shouting. As one of the first actual users of the device, here are the solid facts, in my opinion. First and foremost, what we have really lost is sensitivity not resolution which is the ability to separate two or more objects from each other. This simply means that we can still resolve stars or quasars with the same accuracy as designed provided they are bright enough. Nothing has changed here and we can still do at least ten times better in this regard than we can from the ground, hands down. That's reason enough for rejoicing. We can and have started to resolve and understand the inner workings of a number of complex objects which were just a blur from the ground.

That's the good news; the bad news is that you have to pay a steep price for this gift by sacrificing sensitivity. This means that we simply cannot reach as deep or as far into the Universe as we would have desired. The loss here is real and painful and no amount of fiddling with the data or hocus-pocus with computers is going to get us out of this pickle. It's particularly serious because many of the science progams that *Hubble* was specifically designed for depended critically on its ability to see individual objects in very distant galaxies to establish the size of the Universe or witness the birth of a galaxy, for example. It also means that *Hubble* almost certainly will not be able to discover planets around other stars beyond our own solar system. Those kinds of breathtaking results are not likely to happen any time soon, unfortunately, or at least not until some of the current instruments are replaced in orbit in several years time.

. . . All this bounty produces an inevitable bitter-sweet taste to scientists like myself taking the first images of the cosmos with *Hubble*. We look at these images and marvel at the new and beautiful natural phenomena that are revealed to us for the first time on the terminal at the control center. But, at the same time, we cannot repress the urge to imagine what the image could have revealed had the mirror been what it should have been; what incredible richness of detail would have unfolded. It is rather heartbreaking, but this is the way it goes sometimes, and it is certainly not as bad as it has been depicted in the past few months. I am certain this feeling of mine will be shared by more people as the results begin to trickle down from *Hubble*. In many respects, this endeavour has been a tremendous success and we should not, by any means, lose faith or patience in the process as we struggle to come to grips with the complex reality behind the hype that we have allowed *Hubble* to become.

By contrast, the media did pick up on a sound bite more suited to their agenda when Paresce later uttered what turned out to be one of the most widely quoted statements of *Hubble*'s commissioning period: "The telescope is working good enough to show you what you're missing."

Images like this one of R Aquarii, as well as many more to be taken during the lifetime of the *Hubble* telescope, are expected to revolutionize our ideas about such stellar "volcanoes." Future observations will also shed new light on how nature redistributes the products of nuclear fusion from deep inside stars back into interstellar space. Nova events such as those described here are of more than just passing interest to astronomers and laypeople alike. For this is one known way—in addition to the truly titanic but extremely rare supernovae—to release chemical elements heavier than hydrogen and helium into the wider, colder realms of the Universe beyond the hot islands of gas we call stars. Elements like carbon, nitrogen, and oxygen are critical building blocks of planets like the Earth and of life-forms such as our own.

Following this stunning success, the law of averages took hold again. Pointing malfunctions on the part of the telescope and unfortunate planning on the part of human beings caused the next two observations to yield nothing much useful.

A highly irregular and compact galaxy, named IZwicky1 (after the first object listed in volume I of a catalog compiled decades ago by the late Swiss-American astronomer Fritz Zwicky), was imaged with the wide-field camera. This is a distant galaxy, on the order of a billion light-years away, and the intent was to resolve its peculiar shape—resembling two frosted light bulbs colliding—better than ever before. But the guide stars specified to keep *Hubble* steady during several three-minute exposures were never acquired properly, and IZwicky1's image was blurred so badly that it appeared no better than in ground-based telescopes.

We were beginning to realize that the main mirror's aberration was definitely affecting the guide stars seen by the fine-guidance sensors, an issue that project engineers had long denied. These sensors see blurrily, too—making it harder for them to sort out the field of stars, which now hardly resemble the expected pattern of pin-point-sharp navigational beacons. The result was that the FGSs could acquire fine-lock on a pair of guide stars only if they were sufficiently bright—at least the thirteenth magnitude rather than the design (Level-1) spec of 14.5 magnitudes—thus limiting, quite considerably, the availability of guide stars in many parts of the sky.

Consequently, depending on the magnitude of the guide stars so chosen, and complicated by occasional crowding of comparably bright stars within the sensors' field of view, sometimes guide stars would be acquired and sometimes not. And when they were, it was never clear whether fine-lock would hold, or merely coarse track. What's more, all of this "go, no-go" regarding guide stars was compounded by spacecraft jitter twice each orbit, which further tended to unlock those beacons needed for precise aiming.

The other unprofitable exposure that yielded little excitement was taken, ironically, toward an especially exciting region. The Crab Nebula is a remnant of a star that was observed to explode in the year A.D. 1054. Records show that Chinese, Arabs, and possibly American Indians noticed the appearance of a new light that year in the nighttime sky. This was a supernova—the near obliteration of a massive star some 6,000 light-years away. The result was a potpourri of many phenomena studied in modern astronomy: hot plasma, expanding gases, magnetic fields, a core remnant that pulses—a pulsar, or neutron star—and a region that emits radiation of all kinds, clear across the electromagnetic spectrum, from radio waves to gamma rays. Accordingly, many astronomers regard the Crab Nebula as another kind of Rosetta stone, in that it has much to tell us about many aspects of astrophysical events. The abstract of the proposal that led to this early-release observation did not disguise our enthusiasm: "Images . . . are bound to provide new insight into the physical processes occurring in the nebula in the aftermath of the explosion that gave rise to it."

Our intention with *Hubble* was to image the filamentary structure of the Crab's dying remnants. Although these stringy gaseous wisps can be seen with ground-based telescopes, the finer resolution of *Hubble* should enable us to see more clearly in and around the expelled debris. Since the apparent size of the Crab Nebula is a few arc minutes, its image would fit nicely within the viewing area of the wide-field camera. But, owing to a little astropolitics between the two camera teams—Who had observational priority for this source?—this region was in the end imaged with the narrower field of view of the faint-object camera. The result showed several of the Crab's southern filaments, but the exposure time—twenty-five minutes, and my mistake—was too short, the image too dim, and nothing terribly interesting was learned.

Despite our calling as scientists who supposedly take little stock in hunches, we pleaded with thin air that now was the time for the pendulum of fortune to swing in the other direction. And swing back it did, delivering one of *Hubble*'s first unexpected findings—nearly a genuine discovery.

We commanded the spaceborne observatory to slew again to the Large Magellanic Cloud. Now we would examine another region not far from the star-forming 30 Doradus complex—but this one displayed evidence of recent stellar death. Several years earlier, on February 23, 1987, a bright supernova had appeared in this region among a group of mostly young stars and diffuse matter. It was one of the most spectacular and unexpected events of this century—the great supernova of 1987. SN1987A, as it is called for short, is the first supernova to reach naked-eye visibility since the one studied by Johannes Kepler in A.D. 1604, a few years prior to Galileo's pioneering observations with his newly constructed telescope.

Although SN1987A is estimated to be more than a hundred times fainter than any of its predecessors in the last millennium, it has been observed in such detail and with such accuracy that we can define this event as a "first" in many respects: a burst of subatomic neutrinos was detected approximately coincident with the initial flash of light, a gamma-ray flux was observed shortly thereafter, its progenitor star was identified on photographic plates, and its recent evolutionary history is reasonably well known from the study of those archival photos. Surely, SN1987A is the best-studied supernova in all of history.

Supernovae are crucial in the cosmic scheme of things. Almost all the matter of which we are made can be traced to these cataclysmic events. All the elements except for a handful of the lightest ones, such as hydrogen and helium, have either been produced within nuclear furnaces deep inside massive stars (and subsequently released into space by supernovae) or created by nuclear reactions during the supernova itself (or in later events of radioactive decay). Because life, as well as the Earth from which life arises, is made partly of heavy elements, we know that the matter of which we are formed once resided inside stars that perished as supernovae even before our solar system materialized, some 5 billion years ago.

During the several years following the discovery of SN1987A, astronomers meticulously monitored its evolution, using both ground-based and space-based instrumentation. Over a few weeks, the supernova increased in brightness by a factor of about a hundred, and it has since dimmed until it is nearly a million times fainter than it was at peak intensity. This dimming has enabled astronomers to take a closer look at both the supernova proper and its surroundings.

From the presence of hydrogen in the ejected matter and from the conspicuous flux of neutrinos, researchers deduced that the exploded star—called Sk −69° 202—had been quite massive, perhaps containing as much as twenty times the mass of our Sun. (Astronomers classify SN1987A as a type-II supernova—the violent self-destruction of a huge star at least ten times as massive as

This view, taken through a ground-based telescope, shows the Large Magellanic Cloud with a bright supernova within it. The 30 Doradus star-forming complex is at left, the supernova of a dead star at right. The photograph was made in April 1987, more than a month after SN1987A had appeared in the sky, the result of a star having exploded there some 170,000 years ago. [AURA]

our Sun; type-I supernovae occur only in binary-star systems, their progenitors thought to be smaller, white dwarfs.) What puzzled astronomers most was the observation, made and published numerous times during the last century, that Sk −69° 202 was a blue supergiant; by contrast, red supergiants are thought to be the progenitors of supernovae in current theories of stellar evolution.

In mid-1988, observations made with a small satellite operating in geosynchronous orbit, the *International Ultraviolet Explorer* (IUE), indirectly suggested the presence of matter surrounding the supernova. Apparently, that matter had been expelled in the form of a gentle wind by the progenitor star when it *was* a red supergiant, approximately 10,000 years ago. Thus, the theories are not so wrong after all, but it was unclear how the transformation happened and in what pattern the matter might now be distributed.

Even before *Hubble* took a look, we knew from theoretical studies and observational (mostly spectroscopic) evidence that SN1987A must be expanding rapidly. The outer regions of the exploded star are racing outward at speeds as great as 65 million miles per hour (30,000 kilometers per second). Hence, in the time span of three and a half years, the supernova debris should have mushroomed to a size of 0.1–0.2 light-year in diameter. Given its 170,000 light-year distance, the angular size of SN1987A's litter as seen from Earth would be about 0.1–0.2 arc second in diameter; it could not be resolved or studied with ground-based telescopes (or from the small IUE satellite, which has no imaging capability). Thus, our objective was to observe SN1987A with *Hubble*'s faint-object camera, intending in particular to study the structure and properties of its surrounding environment. Everyone agreed that this was a perfect early target for *Hubble* to examine, and it did not disappoint.

The first observation, deep in the ultraviolet at 1,750 angstroms, was only mildly interesting. In the OSS area on the evening of August 23–24, we watched the monitors display a moderately long, sixteen-minute exposure. Two well-known stars, labeled merely "star 2" and "star 3" by astronomers who had first studied SN1987A, were clearly separated thanks to *Hubble*'s superb resolution—an amusing advance, knowing how hard two colleagues at the institute had earlier struggled for months to examine ground-based photography while trying to distinguish these stars from the brightly lit supernova. Although the two stars are unrelated to the supernova, they lie close to the line of sight and are difficult to separate on older photographic plates with 1–2 arc-second seeing. In the *Hubble* ultraviolet image, those two stars are easily distinguished from one another and in turn from a third blob between them, which is the supernova wreckage itself. (The fact that the supernova had by then dimmed about a million times from its peak intensity also helped in distinguishing it from adjacent stars.)

Through *Hubble*'s eye, the supernova remnant is resolved to nearly 0.2 arc second in diameter—just about that expected for a fading cloud of debris several years after its initial burst of light. Significantly broader and clearly extended relative to *Hubble*'s image of a normal star, the remnant's physical size is nearly 0.2 light-year across, or about a hundred times the diameter of our solar system. Although the acuity of the faint-object camera operating in

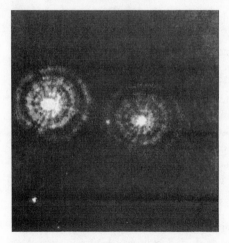

This ultraviolet image, recorded toward the supernova SN1987A, at 1,750 angstroms by *Hubble*'s faint-object camera, shows two bright (fifteenth-magnitude) blue stars astride the supernova itself near the middle of the frame. This is a raw image, not deconvolved or enhanced in any way; the circular features, or "halos," around the stars are therefore not real, but are due to spherical aberration. *(See also color insert section.)* [NASA and ESA]

the ultraviolet—*Hubble*'s sharpest viewing combination—is yet too coarse to reveal much detailed structure, this was the first time that the exploded debris of SN1987A was photographically resolved—a feat that clearly demonstrated the power of the spaceborne telescope, spherical aberration notwithstanding.

The next two observations, also in the ultraviolet but at different wavelengths, likewise showed nothing especially exciting. But throughout the night, an Italian theoretical astronomer, Nino Panagia, who had studied SN1987A a great deal, kept saying excitedly, "Just wait until it's imaged at five double-oh seven." He was referring to 5,007 angstroms, the specific wavelength at which doubly ionized oxygen radiates in the middle of the visible spectrum. Convinced that something intriguing, perhaps a shell, would be seen in the light of this specific element, Nino had calculated that the luminous flash of the supernova ought to have been just right to excite nearby oxygen atoms, after which those atoms would radiate enough for us to see some surrounding matter. For technical reasons having to do with our earlier failure to see much in the Crab Nebula supernova debris, I was skeptical of finding much in SN1987A, and we debated the issue at length that evening. But I am pleased to report that I was dead wrong!

A lengthy, twenty-eight-minute exposure was taken during a single orbit while the spacecraft cruised on coarse track, during which it experienced a terminator crossing, jittered somewhat, yet held its pointing sufficient to continue the observation. When the data were relayed to us, the displayed image took everyone except Panagia by surprise. In the light of twice-ionized oxygen, the supernova had a curious luminescent elliptical ring around it more than a light-year in extent. Even in the raw and aberrated image, the ring is very narrow and well defined, and on that memorable evening it seemed to jump out of the computer screen as brightly as the two flanking stars. But, unlike the pointlike stars' images of which *Hubble* has recorded innumerable numbers, the near-perfect ellipse—not a shell—is quite different from anything else we had ever seen in the sky.

I glanced across the control room and caught sight of a Swiss astronomer exclaiming something in French, and immediately thought of the "ring of light" that the French had proposed placing into Earth orbit a few years ago in order to celebrate the centennial of the Eiffel Tower. Fortunately, owing partly to an outcry of the world's astronomers who argued that our modern civilization had already polluted the sky with enough glowing street light, orbiting debris, and radio interference, the French backed off and the bright human-made ring was never deployed in space. Now, here was a hugely larger ring, stretching more than a light-year across—but visible with only the most advanced telescope ever built by humans.

Hints of the existence of a circumstellar nebula surrounding the site of the blast had been gleaned from previous ground-based and ultraviolet space-based observations. Only now, however, was it seen as a well-defined, thin ellipse—a genuine ring and not a brightened limb of a shell or bubble, much as for some planetary nebulae. We reason it is a genuine ring both because of its elliptical shape and because the brightness inside the ring is much lower than expected if we were looking down through the surface of a shell. Interestingly, the ring could not have been produced by the supernova itself. The expanding debris from the explosion has not had nearly enough time to fill the volume of space cicumscribed by the ring. Rather, the eerie ring must be a relic—a ghostly footprint—of the hydrogen-rich stellar envelope that was ejected, in the form of a gentle "stellar wind," by the progenitor in its red-giant stage several thousand years prior to the explosion.

For millennia the ring was diffuse gas, cold and dark. Slowly moving away from the star, it was subsequently swept and compressed into a narrow, high-density ring by a rapidly moving stellar wind as the red supergiant shed its outer envelope to become the recently observed blue supergiant. But the outlying ring was still invisible. Then, within the first year following the supernova explosion, the former star's formidable ultraviolet flash plowed into the ring, heating it sufficiently to ionize the gas and causing it to glow as a "halo" in the *Hubble* image. In fact, the brightness of the ring indicates that the gas is still at rather high temperatures—about 20,000°C—and is still glowing, several years after being irradiated by the blast of the supernova.

Why does the ring appear elliptical and not circular? And why did the mass lost from the star not pile up in the form of a three-dimensional shell? Perhaps the progenitor star had some considerable spin and thereby expelled its outer envelope from its equatorial region, in the shape of a rotating, flattened ellipsoid. Directionally oriented stellar winds could presumably also help to produce a ringlike structure rather than a shell. Additional observations with *Hubble*—especially using its spectroscopic equipment—should be able to focus onto parts of the ring and test some of these ideas directly, provided the telescope's pointing and eyesight can be properly repaired.

This new *Hubble* image provides important insights into the nature, history, and evolution of massive stars and their catastrophic deaths in supernovae. Continued *Hubble* viewing of this object will show cosmic evolution in action—something that is very difficult to witness during the lifetime of a given human being. A relatively short-lived structure, the slowly expanding ring will be overtaken by the swiftly moving ejecta from the supernova within a decade or two. The fastest, outermost parts of the supernova's debris are moving outward at nearly one-tenth the speed of light and therefore should hit the ring in about the year 2004. The resulting collision will likely heat the ring enough to cause it to glow brightly in ultraviolet and especially X-ray radiation.

Within a few more decades, the ring will disintegrate as it is engulfed by the supernova remnant, which itself will remain visible for centuries as a dimming clutter of ash in the southern sky. Said Nino Panagia in his own enthusiastic way, "In all, these are *direct* measurements of the evolutionary times for an *individual* star. Such results have never been obtained before in any other case on time scales so diverse."

About a month after astronomers had finished analyzing the *Hubble* image of SN1987A and the initial findings were rushed off for publication in the *Astrophysical Journal*, another group of researchers realized that these new data could be combined with earlier data to make a significant advance in our efforts to map better the neighboring realms of the Universe. The first step was to computer-enhance the *Hubble* image of the supernova's ring, thereby correcting it for faulty seeing. Aside from revealing that the gaseous matter is heavily clumped, the deconvolved image enables a precise measurement of the *apparent* size of the ring—1.66 arc seconds across, which is equivalent to locating and resolving a burning automobile tire at a distance of about a hundred miles. The next step required an accurate measurement of the ring's *intrinsic* size. But how could that be done?

Since *Hubble's* observation of its elliptical shape implies that the ring is inclined by about 45° to our line of sight, we might expect that light emitted from the far edge of the ring arrived at Earth nearly a year after light arrived from its forward edge. Although the IUE satellite in geosynchronous orbit cannot see the ring, reanalysis of the waxing and waning luminosity buried in the previous three years of data collected by the satellite does indeed show evidence for such a delay in light-travel times. And this in turn enables astronomers to derive an extremely good measure of the ring's intrinsic diameter—1.37 light-years.

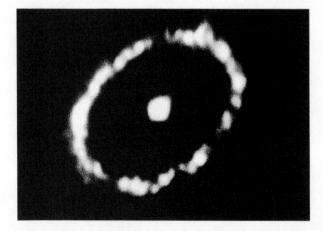

This image of SN1987A is seen in the light of doubly ionized oxygen, at 5,007 angstroms. It was obtained with *Hubble's* faint-object camera and then computer reconstructed to bring out additional detail. Resolution here is an exquisite 0.07 arc second, close to the theoretical "diffraction limit" of Space Telescope. The clumped and knotted structure in the ring is clearly evident, as are the expanding remnants of the exploded star itself that *Hubble* is continuing to track. [NASA, ESA, and N. Panagia]

By comparing the ring's apparent and intrinsic diameters, we can use simple trigonometry—in fact, the "skinny-angle approximation" will do nicely—to calculate the distance to the Large Magellanic Cloud more accurately than for any object outside our Milky Way Galaxy. Given the measured values of the two diameters, we now know that distance—169,000 light-years—to an unprecedented precision of 5 percent. By contrast, our previous estimates of the cloud's distance were hardly known to within 15 percent uncertainty, typically ranging from 140,000 to 180,000 light-years. That we now have a better-calibrated cosmic yardstick was summed up haltingly by astronomer Alan Sandage, often considered to be Edwin Hubble's successor: "This result is an important recalibration of the first step on the scale of intergalactic distances."

Unfortunately (and this was the source of Sandage's hesitancy), given *Hubble*'s much lamented aberration and thus its inability to measure well the brightness of Cepheid variable stars in far-flung galaxies hundreds of times more distant, as explained earlier, we shall not likely be able to take the next leap forward in our quest to better the cosmological distance scale of the entire Universe—at least not until *Hubble* is properly repaired on-orbit or its successor has been built and launched.

The imaging campaign, still under way, was already an unqualified success. We had struck a lode of fascinating, often unique, scientific data. No outright discoveries had yet been made, but the findings to this point exceeded what even the most optimistic among us thought possible when *Hubble*'s aberration was first uncovered. The early imaging results became so robust, in fact, that it is doubtful that *Hubble* would have achieved such stunning science results merely four months after launch even if it had worked as designed with perfect optics.

Given these telling *Hubble* accomplishments, the grave need for positive publicity, and the political reality of congressional appropriations meetings then about to commence, it was clearly time for a press release. The plan was to announce a few science results just prior to the reconvening of the Congress from its summer recess, followed by another of the institute's Science Writer Workshops to celebrate the success of the imaging campaign by releasing a barrage of images into the public domain. The workshop would not be a NASA-style press conference with a bunch of bobbing heads lined up awkwardly behind a long dais, but would be a series of lively, illustrated reports and roundtable discussions stressing pedagogical explanations and led by scientists having a direct stake in the outcome.

The fact that the Science Institute was taking the lead and making decisions regarding all aspects of the conduct of the mission must have pained NASA greatly, but the institute had the momentum now and there was no turning back—at least not until the imaging campaign had run its course. Admittedly, I felt a little out of character when besieged officials at NASA Headquarters would call to make suggestions and we would say something curtly to the effect, "No, that's not a good idea," or, "We need to change the strategy on this issue," and they would surprisingly and immediately back off.

Even so, with this project there seems to be a conservation principle that operates constantly: somebody or some group will always emerge to challenge what's being done with *Hubble,* and in this case it was a disgruntled scientist.

The neat *Hubble* observations of the ring around SN1987A and the potential black hole in NGC 7457—one result equitably chosen from each of the camera's findings to date—entered the public domain as paper press releases for the print media and as "B-roll" for their TV counterparts, topped off by a well-done press conference, featuring astronomers Nino Panagia and Tod Lauer, and televised over NASA's cable channel. This was practically the first good news to emanate from the project, and all the elements of a successful public airing had gone according to our specifically orchestrated plan. Now we would reap the rewards in the print and television media worldwide—or so we thought.

Almost coincident with the end of the press conference, we began taking calls from the Associated Press about a story that had just appeared in the morning edition of the *San Jose Mercury News,* headlined "Hubble Shots to Be Released with a 'Spin'," and claiming that the new *Hubble* results were nothing more than a public-relations stunt. It so happened that one of the members of the wide-field camera team was not happy to see the *Hubble* data released publicly—especially not wide-field camera data. These new findings clearly proved that Space Telescope could still do fine, indeed peerless, scientific research even in its disabled state, and they undermined her agenda to have much of *Hubble's* science program postponed until after the telescope might be fixed. So, the day before the press conference, this astronomer publicly ridiculed the memo I wrote triggering the imaging campaign; prematurely released (and downplayed) the new *Hubble* results twenty-four hours before the end of a previously agreed-upon news embargo; and gave an interview critical of our efforts to disseminate *Hubble's* early findings. But she didn't grant the interview to just any journalist; she critiqued our efforts before an investigative reporter who had been hostile to the Space Telescope project since well before launch—indeed, one who had on at least two occasions illegally entered the Science Institute seeking controversy, once while impersonating a technician.

The result, naturally, was a largely negative article that raced across the news wires and that badly colored many of the subsequent press accounts of the new SN1987A and NGC 7457 science results. As the journalists were emerging from one of the first upbeat press conferences since launch, they were met with accusations by a fellow science writer that they had just witnessed a PR facade blown all out of proportion. Perplexed, some journalists who otherwise would have written positively about the *Hubble* findings elected not to write anything at all, claiming they had been scooped by the California paper.

Many others who did report on the new results did so skeptically, suspecting that we had somehow "manufactured" the good news. For example, *Newsweek's* lopsided coverage of the mission caused its writers to make three mistakes in a single sentence: "The two photos, which confirm that *Hubble* sees farther than ground-based telescopes [wrong], but not with the sharpness it should have [wrong], may yet yield some worthwhile astrophysics; for now, they're colorful PR [wrong]." An Associated Press editor was especially livid on the other end of my telephone: "Why the hell should I put anything about your science on the news wires, if this two-bit paper on the West Coast is saying it's all a setup?" Thus, *Hubble's* early treasure trove did little to dispel the perception that it was broken. Most people never saw the images or heard of the telescope's new findings.

By contrast, my incoming e-mail log lit up that night, with *Nuncius* repeatedly pulsing me. Professional astronomers were unquestionably spreading word of the new results and a flood of electronic messages, little of it addressed to me but intercepted and copied my way, began entering my computer workstation from all nodes of approach around the world. Excitement among much of the scientific community, at least, was beginning to build.

Science Institute personnel did all the legwork for the late-summer press releases—coordinating with the principal astronomers, writing the prose of the releases themselves, electronically preparing the data, printing thousands of copies of the images, and building computer animations—"video bites"— to accompany each release. When the product was complete and ready for distribution, NASA contributed but one comment—a complaint that we had placed the logos of the Science Institute and of the European Space Agency aside the NASA crest. These releases, the agency said, must be solely attributed to NASA—this from the Goddard office that had earlier and repeatedly accused the institute of not being a team player. And while the agency was quick to blame others (usually contractors and usually incorrectly) for all manner of problems, when it now became apparent that a fragment of positive news was about to be announced, NASA argued that no one else should share in the credit. So much for teamwork.

Late the following evening while in the OSS area monitoring more incoming *Hubble* data, I was asked to join a small meeting under way in the director's office at the institute. There I found ESA's chief scientist, Roger Bonnet, an uncommonly polite Frenchman whom I had last seen jogging on Cocoa Beach on the eve of launch. Tired from his flight directly from Paris, he was uncharacteristically livid at NASA's total disregard for his space agency. Between fuming statements and emotional mixtures of French and English, he informed us that he intended to upstage NASA the following week by releasing exclusively from Paris several more *Hubble* science results taken with ESA's faint-object camera. Considering NASA's singular preoccupation with the need for credit, this was the programmatic equivalent of a declaration of war. Accordingly, and in the presence of the ESA man, I recommended to Giacconi that the Science Institute remain neutral, for if it looked as though we had allied with ESA in the secret release of *Hubble* results, the institute would be damned.

As things turned out, ESA completely botched their attempt to rebuff NASA. While analyzing the data and debating the words interpreting their findings for a press release, the ESA *Hubble* contingent fragmented badly, breaking up into various nationalistic factions, with Italians feuding with

British, Germans disagreeing with most, and a key Dutch scientist seemingly unable to get along with anyone. As for the critically important images chosen to accompany the press release, the Parisian-based contingent produced pictures so poorly that the science results explained in the figure captions couldn't even be seen in the images.

The faint-object camera team at the institute had indeed shipped electronically (and clandestinely) the relevant digital data to ESA Headquarters, but technicians there and elsewhere in Europe were unable to transfer cleanly the high-quality images glowing on their computer screens onto paper, film, negative, or whatever. The *Hubble* data became a prisoner in their machines, another comedy of errors—all the while they frantically raced to get the new results out the door before NASA.

Steadfastly determined to present the magnificent pictures to French President François Mitterrand lest NASA deliver copies to the White House first, someone in Paris took quick-and-dirty, hand-held snapshots of the exquisite *Hubble* images displayed on computer monitors, and quick-and-dirty they looked. Actually, it really didn't matter, for when it came to releasing the science findings into the public domain, someone else in the Paris office prematurely leaked to the Reuters News Agency copies of the poorly rendered images without any descriptive words at all. Since reporters had only fuzzy images whose contents were undecipherable, hardly any media coverage was won. ESA's attempt to circumvent NASA had summarily backfired, as I had visions of Keystone cops running around Paris like fools.

In the end, in the spirit of Thoreau's civil disobedience, we at the institute ignored NASA's directives regarding logos when announcing new science results. All press releases issued by the institute throughout the commissioning period were regularly accompanied by institute, ESA, and NASA logos equally emblazoned across the masthead. As for NASA, the agency each time took our complete press package, rekeyed the prose into their own word processors, and reprinted everything we had provided on a masthead with only the NASA logo. Astute science journalists who received in the mail both versions of the same press release knew exactly what was happening, with some even noting privately that this kind of heavy-handed search for credit was precisely that which was losing for the space agency the very credibility it so desperately sought.

Each of these incidents—an astronomer's attempt to deflate the public release of Hubble science results and revengeful feelings of some Europeans toward NASA (and conversely)—pale in comparison to the salient issue of a stultifyingly bureaucratic U.S. space agency that day in and day out mounted roadblocks along the path of an honest public discussion of *Hubble's* mission status, but they do serve to exemplify numerous deviant episodes and the many varied, esoteric factors necessarily juggled while trying to navigate the *Hubble* program through troubled times. Literally scores of incidents like these hobbled the project throughout its commissioning period because, without strong leadership, managerially or technically, the *Hubble* project was hostage to the subjective whims and hidden agendas of many individuals and groups, as well as to ill-informed perceptions.

*   *   *

As the imaging campaign ramped up again, I scrambled back into the trenches, trying to escape from those who worry daily about logo credit, petty politics, and bureaucratic control. We quickly experienced a string of four unsuccessful observations, all of which served to reinforce the notion that we were still learning psychologically how to observe with the strangely behaving *Hubble*. Two of these objects mimicked earlier failures—for the nebular region HH1, even a long, forty-minute exposure captured too few ultraviolet photons to analyze its knots and filaments credibly (much as we had found earlier for the Crab Nebula), confirming that *Hubble*'s throughput was much diminished owing to aberration; and for one of the nearest known quasars, Tonanzintla 256, the weak and fuzzy light patterns of its host galaxy matched all too well the scale of the aberration surrounding the central quasar (confirming our earlier disappointing results toward AP Librae), making it impossible to decipher the hints and clues about the quasar hidden within its surrounding environment.

The third failure occurred when trying to capture data from Eta Carinae, an extraordinarily colorful nebula and the highest-priority EROs target. As we attempted to acquire multiple exposures with the wide-field camera through red, green, and blue filters, *Hubble*'s fixed-head star trackers were puzzlingly unable to identify a few bright stars in the Carina constellation, and its onboard computer rightfully refused to allow the camera's shutter to open. A few of us immediately appealed to NASA officials to allow us to retry this observation in a week or two, but they were unaccommodating at best—even if most of us were convinced that this image would be nothing short of spectacular.

The fourth observation failed all right, but tantalized us badly. The target was a globular cluster, a region containing typically hundreds of thousands of stars tightly packed into a relatively small ball-shaped region ten-ish to a hundred light-years across. Such clusters are so densely congested with stars that no ground-based telescope has ever been able to see clearly into their cores. And it is in such cores that many theorists reason there ought to be black holes. Before launch a prime question often asked was, Will *Hubble* be able to discriminate among the central stars of a globular cluster, thus seeking a characteristic signature of a black hole hidden within their midst? Now the question was, Could *Hubble* do it given its faulty optics? This would be a severe test for Space Telescope and for the sophisticated image reconstruction algorithms then being coded—we were "pushing the envelope"—since globular clusters represent some of the most crowded fields of stars known anywhere. The target chosen was M15, the fifteenth object in the Messier catalog, a well-known globular for which we have excellent ground-based images for comparison.

In all, *Hubble* was commanded to take ten exposures (with the planetary camera for nearly maximum resolution), all but one of which aborted. Again, the culprit was the star trackers, which had clearly become the weak link in *Hubble*'s pointing-control system. The final exposure was effectively forced, without guide-star acquisition and while *Hubble* was pointing only with its gyroscopes. The resultant image would therefore show stars as linear streaks rather than round points, caused by the drifting of the telescope during a thirteen-minute exposure.

Accordingly, and discouraged by this latest string of failures, I didn't even bother to display the single image of M15 on the monitors at the Science Institute's OSS area. Later that Saturday evening—it was now August 25—the camera's principal investigator, Jim Westphal, called me, yelling into the phone that for curiosity's sake he had taken a look at that one image of M15. And what an image it was! I eventually brought it up on my *Nuncius* and it was immediately clear that the core of the cluster had been nearly penetrated, gyroscopic trailing notwithstanding. A globular cluster had been nearly resolved. One could see right into its core and again a rush of "eureka" swept past. But the lack of accurate pointing made it tricky to do any meaningful science and especially impossible to address the issue of a central black hole. One of the most vexing issues in modern astronomy would have to wait a while longer.

The next morning, Sunday, August 26, after a series of delays through the night concerning continued troublesome operation of the telescope's star trackers, the imaging campaign resumed. It would be the last and most spectacular day of *Hubble*'s inaugural science observations.

Originally, of all the planets available, we had chosen Jupiter as one of the highest-priority EROs targets. We knew that *Hubble* observations of this giant gas ball would be spectacular, showing its turbulent and colorful atmosphere with resolution seen only briefly as the *Voyager* spacecraft approached the "big banded one" in 1979. Besides, there was some symbolism here; Jupiter was one of Galileo's first telescopic sights, in fact the one that most convincingly bolstered the Copernican worldview that Earth revolves around the Sun.

But by late summer, Jupiter's position on the sky had drifted within 50° of the Sun, a region that *Hubble* normally avoids lest it be potentially harmed—much as Galileo was partly blinded when he looked through a telescope directly at the Sun early in the seventeenth century. This large "solar avoidance zone" can be manually overridden, allowing the spacecraft to "creep" ever closer toward the Sun, but we were unwilling to increase the risk to *Hubble* so early in the mission. We had enough problems on our hands as it was.

Instead, we reactivated an earlier EROs proposal to observe the planet Saturn, another favorite target of Galileo's first lens. We knew that the image would not be quite as awesome for two reasons: Saturn is farther away than Jupiter and it does not display nearly as much structure as its neighboring gas giant. Still, the public would respond to an image of Saturn, perhaps like none other.

Planets, of course, appear to move on the sky at different rates than stars. That is because in addition to Earth's spin, which causes distant stars to seemingly trek across the sky, nearby planets have their own motions in their orbits about the Sun. While *Hubble*'s novel pointing system that relies on guide stars works reasonably well for relatively fixed objects such as stars and galaxies, a more sophisticated version of the pointing system must be employed to track "moving targets"—some of which, like Mars, occasionally loop back and forth during a so-called retrograde maneuver.

Computer software needed for planetary tracking had been a topic of much concern and some effort among the programmers in the operations areas at the Science Institute during the four-year delay since the *Challenger* tragedy. But, given the press of other tasks in the multifarious arena of telescope planning and scheduling, the coding of "polynomial tracking commands" had never been deemed of great urgency. Even several months after launch, the pointing system was still not robust enough to accurately track planets, moons, comets, and other solar system objects. Consequently, images of planets could suffer from blurring caused by faulty pointing, thus robbing us of the promised advance in resolution.

At any rate, as anyone knows who has looked through a small backyard telescope once or twice, Saturn is a relatively bright object. Provided the Hubble exposure was kept to less than a second—an "aim-and-shoot" observation while under gyroscopic control—then the image should be adequate to resolve interesting features. Accordingly, *Hubble* was commanded to take three observations, each for a few tenths of a second through red, green, and blue filters, which when combined would yield a true color photograph. Although we realized that the planetary camera (with its smaller field of view) would be better able to image fine details on the planet (hence the camera's name), the pointing concerns noted above caused us to use the wide-field camera. Its larger field of view would have a better chance of capturing all of Saturn and its rings which—at 860 million miles (or 1.4 billion kilometers) from Earth—measured at the time about 40 arc seconds across.

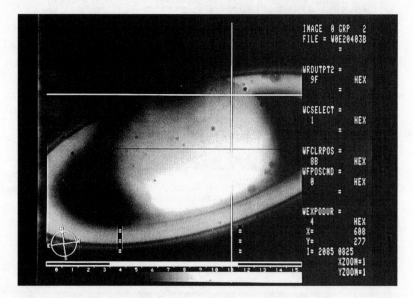

This raw and uncorrected image of Saturn, just as we saw it at the blue wavelength of 4,390 angstroms on the monitors in the Science Institute's OSS area, was captured by a single CCD chip of the wide-field camera mode. Here, the field of view has been clipped (or effectively zoomed) to about 20 arc seconds across to display fine details. *[NASA]*

As the first exposure (with the blue filter that we saw in black and white) appeared on our monitors at the Science Institute, we were startled. The image of Saturn was absolutely magnificent. After trekking through hell for the past few months, we were suddenly glimpsing heaven. Despite *Hubble*'s poor vision, this new view of the ringed planet—for many, even astronomers, the most spectacular object in the sky—seemed to engulf first the oversized computer monitors, then the whole operations room! We had sighted our first planet with *Hubble* and it was the ammunition that I needed.

Shawn Ewald, a skilled WF/PC team member in permanent residence at the institute, immediately went to work on the image in a business-like manner—"flat fielding" it to correct for unwanted instrumental effects, interpolating pixels to remove blemishes caused by cosmic rays, and deconvolving it (forty times) to minimize the corruptive stigma of aberration. The latter task was not trivial as the extended (non-point-source) Saturn consists of thousands of smeared and strongly overlapping point-spread functions, each with halos about one-fifth as wide as the disk of the planet itself. Ewald did this for each of the red, green, and blue images individually, whereupon he sent the images to my computer workstation, *Nuncius,* and then on to the high-tech workstations, *Paris* and *Cassandra,* in our visualization lab. There, we registered (that is, aligned) the images atop one another and synthesized them into a single picture—one that approximates true physiological color.

The result was breathtaking. The classic features of Saturn's vast system

This computer-enhanced *Hubble* image of Saturn shows clearly the band structure on the ball of the planet, as well as the Cassini gap halfway out in the rings. Despite *Hubble*'s aberration, the narrow Encke division in its outermost "A" ring can also be discerned in this deconvolved image. Because the trajectories of the *Voyager* space probes were programmed to sweep past Saturn's rings in 1981, the planet's north polar cap was not seen then as well as in this photo; in particular, a polar "hexagon" structure seen by *Voyager* is still there, but some of the details of the belts and zones have changed. Resolution is about 420 miles (670 kilometers) per pixel. Saturn appears here much as our naked eyes would see it if it were only twice as far away as the Moon. *(See also color insert section.)* [NASA and J. Westphal]

   (Note that these published images of Saturn, as well as of all the other cosmic objects in this book, are "twice removed"—that is, they are not seen as clearly here as on our computer monitors, for they must be first converted from the electronic medium to print medium (as slides or glossies) and then printed in this book; both processes degrade the resolution of the displayed images.)

of equatorially orbiting rings were clearly present. From the outer to the inner edge, the bright A and B rings were divided by the so-called Cassini gap, followed by the extremely faint inner C, or crepe, ring that appears in the image due to the wide-field camera's large dynamic range—both intense and dim light can be simultaneously captured with charge-coupled detectors (CCDs), unlike with photographic film. (Not solid structures, Saturn's thin rings are composed of chunks of rock and ice, ranging in size from microscopic specks to house-size boulders.) What's more, we could see greater spatial detail in the planet's ring system and cloud belts than can be achieved with any ground-based telescope. And the Encke division, a dark narrow gap near the outer edge of the outermost A ring, which to my knowledge has never before been photographed from Earth, was clearly evident. This was good stuff.

Word spread throughout the institute and across the Johns Hopkins campus, and soon we had a parade of people wishing to see the final product. Distinguished scientists, leading academics, and assorted hangers-on trotted into my office and our imaging lab, practically half out of breath and begging a peek. For not only was this a "pretty picture" but it also contained some new science in Saturn's northern polar cap region that was not well seen by the *Voyager* spacecraft while reconnoitering the planet in 1981. Most importantly, this extremely brief observation demonstrated great potential for doing state-of-the-art science in the solar system, since with *Hubble* we could achieve this kind of clarity virtually any time we wanted. Gazing across my office at *Nuncius* and seeing Saturn's image brightly aglow, it was as if the planet illuminated a neon sign proclaiming: *Hubble* is not broken! "Good stuff indeed," said Giacconi. "Keep it coming." By evening, President Bush was passing around a glossy image of Saturn at a White House cabinet meeting, proprietary issues notwithstanding.

My euphoria quickly turned to anger as I wondered aloud why the project hadn't permitted several months earlier a few quick "aim-and-shoot" observations of this kind. The total exposure toward Saturn in all three filters was less than one second. We literally could have acquired such a raw image of Saturn even before we had discovered the telescope's aberration—with our "hopelessly blind" Space Telescope. It was particularly galling that the scientist most opposed to the early release of exploratory images had, shortly before the start of the imaging campaign, testified at yet another Capitol Hill hearing by saying, "Mr. Congressman, I have been here many times in the past to promise you beautiful pictures and great discoveries, and I've come today to tell you about the heartbreak that we all have in not being able to provide those to you." This distinguished astronomer had become so obstinately opposed to sharing images with the public that he seemed unable to understand that statements like these were fueling the notion that the project should be terminated, a possibility that most definitely was being advocated in some influential congressional circles at that time.

I was further angered a day after our observation when a member of the wide-field and planetary camera team blurted out, "That Saturn image is worse than what I could see in my backyard with a 6-inch telescope. I do not want my name associated with an image that amateur astronomers could criticize." This charge was largely repeated a few days later in the presence of high officials at NASA Headquarters when I briefly displayed a slide of *Hubble*'s deconvolved Saturn image, which by then I had been carrying around in my pocket, prepared to show anyone and everyone. This time the camera's principal investigator, Jim Westphal, somewhat more impressed with the image now but still unsure on what side of the fence to sit, amended his teammate's comment yet refused to take a close look at it, saying, "I could probably do as well with a small telescope on Mauna Kea."

Clearly, while one part of the wide-field and planetary camera team was thrilled with the Saturn results, another faction—about half the wiffpickers— was still posturing politically, still refusing to sign up to the benefits of computer enhancement and still trying to mortgage the telescope until a new WF/PC might correct the telescope's faulty optics—all the while seeking to prohibit others from using the current camera to examine sources of interest to them.

Only a month later, toward the end of September, after heavy lobbying changed Westphal's mind, would this team admit that *Hubble*'s photo of Saturn was superior in quality to any ground-based telescopic view, but by then additional damage had been done. For meanwhile, we were denied public use

Later in the commissioning period, astronomers managed to secure some quick looks at a high-priority EROs target, Jupiter. Among the first views of the planet seen by *Hubble*, this computer-enhanced image was taken with the planetary camera exposed to green light for mere seconds. Amidst a wealth of detail in the clouds covering this huge gas ball, the centuries-old Great Red Spot is rotating out of the picture in the lower right (southeast), while near the equator at right the moon Europa is just about to disappear behind the limb of the planet. As with the *Hubble* images of Saturn, those of Jupiter are as sharp (approximately 0.15-arc-second resolution) as the pictures taken by the *Voyager* spacecraft about five days before its closest approach a decade earlier. Note the computer cursor mistakenly left on the image at the planet's midsection; fortunately the media overlooked it, and we saw no wild stories in the tabloids about a monolith on an alien world. *(See also color insert section.)* [NASA and J. Westphal]

of the image at a critical political juncture. By the rules of the game, the Saturn image was taken with the wide-field camera and thus its digital data were the property of that team; even EROs images were part of the camera team's "cosmic turf" for thirty days. Once again, NASA's own policies governing proprietary rights were hurting the space agency and lingering concerns about data ownership were jeopardizing the *Hubble* project.

The next observation that last Sunday in August called for *Hubble* to execute a brief series of only a few command lines on the institute's observing calendar uplinked to the spacecraft. The objective was to image the core of an AGN— astronomers' slang for an active galactic nucleus, the central region of a violent galaxy that many scientists suspect is the site of a supermassive black hole. The wide-field and planetary camera was used in an attempt to resolve knots of emission near a suspected hole, especially those knots that might be radiating strongly in the light of doubly ionized oxygen at 5,007 angstroms. This is the same wavelength at which the ring around SN1987A had been so spectacularly seen a few days earlier, and because some of the thermal properties of nebulae are thought to be found in black-hole environments, this narrow-line feature might be a good place in the spectrum to image otherwise dim regions. The result was a magnificently detailed view of highly energetic events in the core of an active galaxy.

The target galaxy chosen is called NGC 1068, a rather famous AGN among working astronomers, if only because it is the closest active galaxy of its kind. Located some 30 million light-years away, or roughly half the way to the benchmark Virgo Cluster of galaxies, NGC 1068 looks on photographic plates like a normal barred-spiral galaxy—that is, a galaxy whose spiral arms originate at the ends of a linear array of gas bisecting the galaxy's nucleus. But looks can be deceiving, even among galaxies.

Since the year 1909, the core of this object has been recognized as the source of unusual activity—activity now evidenced by the presence of extremely hot, fast-moving clouds of gas in the vicinity of the galaxy's nucleus. The clouds are ionized enough to emit bright, broad emission lines and they have (presumably rotational) speeds of up to 2 million miles per hour (1,000 kilometers/second). Similar galactic fireworks have been detected at the heart of other galaxies, and they are the telltale sign of a class of objects known as Seyfert galaxies, named after the late U.S. astronomer Carl Seyfert, who first studied these galaxies in the 1940s.

The most powerful AGNs emit as much energy as twenty Milky Way Galaxies combined—the equivalent of a trillion stars like our Sun. By contrast, NGC 1068 is an energetically weak member of the Seyfert class, with its core displaying about one-fiftieth of the total luminosity of our Milky Way Galaxy. Even so, NGC 1068's core is several hundred times more energetic than the core of the Milky Way, and shines with the equivalent energy of about a billion suns. What's more, all AGNs, including NGC 1068, vary their brightness on the scale of weeks, implying that the diameter of the region producing their enormous luminosity is no more than a few tens of light-*days* across, or at most a hundred times the size of our solar system.

This in fact is the dilemma for astronomers trying to understand all types of active galaxies in the Universe: Huge amounts of energy seem to arise from relatively small regions at the centers of active galaxies. Thermonuclear reactions that power the Sun and stars cannot generate such prodigious amounts of energy in such small spaces. The favored mechanism is a massive black hole, wherein energy is released as matter plunges toward the hole's deep "gravity well." And in the case of NGC 1068, it seems that a black hole having vast quantities of matter would indeed be required to account for its nuclear activity.

NGC 1068 was chosen for exploration because, despite its comparatively low power among AGNs, it is relatively close and therefore bright. We reasoned that images taken with *Hubble* would enable us to pinpoint the location of the suspected black hole and to determine the effects of jets, winds, and ionizing radiation from the black hole on the innermost, central few hundred light-years of the surrounding galaxy. This then was not an ordinary science assessment observation. We were going for the jugular, for a supermassive black hole in the heart of a distant galaxy.

Two planetary camera images of NGC 1068 were acquired in sequential exposures totaling eighteen minutes, thus proving again that this camera is by no means limited in its use toward planets. In addition to an observation with a narrow-band filter that allows the light of stars and of ionized oxygen gas to reach the CCDs, a second observation was made using a broad-band filter that passes light mostly from stars while excluding light from ionized oxygen. The two pictures were then subtracted to obtain an image that shows only the clouds of ionized gas near the galaxy's nucleus. Finally, the *Hubble* image was

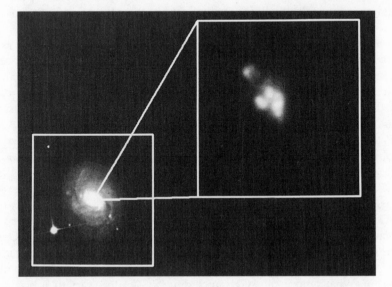

The bottom left image of NGC 1068 was taken with a ground-based telescope at the Kitt Peak National Observatory outside Tucson, and clearly displays the galaxy's barred spiral-arm features. The upper right image shows the unenhanced observation of its core made with the Hubble Space Telescope. The ground view measures about 250 arc seconds on a side, or 40,000 light-years across; *Hubble*'s view is about a hundred times magnified, measuring a few arc seconds, or 500 light-years across. [AURA and NASA]

In this *Hubble* view of the core of NGC 1068 (which measures only a few arc seconds or a few hundred light-years across), more details are evident than have ever been seen from the ground. This image was made by subtracting a broad-band (continuum) observation from one taken through a narrow-band filter (centered at 5,007 angstroms) with the the planetary camera. The resulting image was then computer enhanced to reveal structure on the order of 0.1 to 0.2 arc second. Clouds of ionized oxygen as small as 10 light-years across are clearly resolved in the galaxy's central 150-light-year radius. The cone was artistically superposed on the image to suggest how the clouds are energized by radiation beamed from the hidden nucleus. *[NASA and H. Ford]*

enhanced, using standard deconvolution techniques to correct for the telescope's aberration.

The resulting *Hubble* image of NGC 1068 resolves the inner sanctum of the galaxy and reveals far more detail than is seen in any pictures taken from the ground. Although its precise core is not directly visible—not surprising if it really is a black hole (which would also be obscured by a dusty accretion disk)—its effect is clearly evident on numerous gas clouds in the vicinity of the galaxy's nucleus. Several clouds of gas, glowing because of radiation beamed like a searchlight from the hidden "central engine," are clearly resolved in the galaxy's innermost few hundred light-years; some of the clouds are as "small" as 10 light-years across, which is testimony to *Hubble's* exquisite resolution, given the considerable distance to this galaxy. Three of the clouds correspond closely to the brightest radio sources previously observed near the core. At least one of those clouds shows signs of being compressed by a jet of plasma ejected from the nucleus. Joked an institute astronomer examining the raw image of NGC 1068: "It looks like a clutch of frog eggs."

Briefly, the favored interpretation of these new findings is that the spatial arrangement of the clouds, even on the smallest observed scale, appears to trace a cone of ionizing radiation beamed from the invisible central engine. The *Hubble* image confirms an earlier theory that the suspected black hole in NGC 1068 is shrouded in a dense ring of dust—or perhaps more like a torus resembling a fat doughnut. Energy released from the nucleus escapes through the ring's narrow hole to produce a brilliant "searchlight" of radiation, in which the clouds are illuminated like moths in a flashlight beam. Reminiscent of the much more collimated extragalactic jets flowing from even more powerful galaxies in deeper space (such as PKS 0521-36, imaged earlier), this jet is a little different in that it lies in the plane of NGC 1068, and therefore plows through the interstellar medium to carve out a cavity of ionized gas, more so in this case like a blow torch through butter. All things considered, a black hole containing about a hundred million times the mass of our Sun would be sufficient to account for the nuclear activity in NGC 1068.

In follow-up observations, astronomers intend someday to use this remarkable new view of NGC 1068 much like a road map in order to aim *Hubble's* spectrographs at specific clouds that might be reflecting light from the vicinity of the suspected black hole. Of perhaps greatest interest, the apex of an imaginary cone superposed on the *Hubble* image evidently points the way toward the site of the suspected black hole, whose accretion disk might be examined in subsequent *Hubble* observations. Ongoing analysis of the light from deep within this galaxy's nucleus should help us better understand the dynamics of the central engines that power this unworldly class of violent galaxies.

Partly in an effort to balance the early imaging observations equally between the wide-field and planetary camera and the faint-object camera, we next trained the latter on a well-known globular cluster. Much like the dense star cluster M15, whose observation had tantalized us a few days earlier, the globular cluster M14 (also known as NGC 6402) is a tight-knit swarm containing hundreds of thousands of (mostly) old red stars gravitationally bound together within a ball-shaped region a few light-years across. Some 70,000 light-years distant and in the constellation Ophiuchus, M14 resides well up out of the plane of our Milky Way Galaxy.

Our principal objective was to test to what extent *Hubble*'s cameras could resolve faint stars in the extremely crowded fields typical of globular clusters, a task that no ground-based telescope has ever managed very well. A secondary objective—which quickly became a fascinating detective story—was to try to identify a historical nova, last seen decades ago about halfway between the edge and the center of the cluster. The hidden agenda, then, was a "needle in a haystack" search for a single star, now lost amid more than 100,000 other stars bunched together almost clear across the Galaxy.

Most novae are spotted as a sudden eruption in the brightness of a star where, in just a few days, the star flares thousands of times brighter than normal. (The brief flare-up does not cause the star to destroy itself, as in a much more violent yet rarer supernova explosion.) Over the next few months, the outburst fades and the star returns to its normal light level. A few novae undergo repeated outbursts, typically separated by decades. The most widely accepted theory today is that novae occur when two stars are locked in a close orbit by their mutual gravitational attraction. One member of the pair is a normal star similar to our own Sun. Its companion is a white dwarf, a dense, nearly dead-end star that has collapsed to roughly the size of Earth.

Although R Aquarii, imaged by *Hubble* a few days earlier, is by no means a classical nova, the salient features of the M14 nova are similar: Hydrogen gas stripped off the outer layers of the normal star by the strong gravitational pull of the white dwarf gradually accumulates and is squeezed into an ever denser shell on the white dwarf's surface. Eventually, the temperature and density of the hydrogen in the shell rise to those values needed to ignite the gas spontaneously in a thermonuclear explosion, much like a hydrogen bomb. The immense energy released explains both the sudden brightening and the rapidly ejected shell that reaches speeds as high as thousands of miles per second.

Throughout the past two decades, astronomers have searched for a nova that drew attention to itself a half-century ago in M14—a phenomenon so rare for a globular cluster that only one other is known to have displayed it in

Ground-based observation of M14, a tenth (integrated)-magnitude globular cluster in the halo of our Milky Way Galaxy. Although by no means the best pre-*Hubble* image of M14, this picture typifies the problem confronting ground-based telescopes while trying to separate (resolve) any of a globular cluster's densely packed stars from one another. The region imaged with *Hubble* is the northeast (upper left) quadrant. [AURA]

the history of astronomy. However, the nova's stellar neighborhood, even in the suburbs of M14, is heavily blurred and confused when viewed from the ground, due to the high crowding of stars in the cluster. This, then, was a perfect region in which to test *Hubble's* precision imaging, to see if the telescope could uncover the vanished nova amid the cluster's myriad other stars.

The nova that became the target of the *Hubble* observation—officially called Nova Ophiuchi 1938—was accidentally photographed that year with a moderate-size (72-inch) telescope at the Dunlap Observatory, near Toronto. However, it wasn't until twenty-five years later that the nova was actually recognized, when a new generation of researchers began systematically studying numerous photographic plates of M14. The series of plates for 1938 revealed the telltale presence of a bright star where none was previously, nor subsequently, visible. Apparently, the nova had gone unnoticed in 1938, but it probably erupted sometime in May of that year, shortly before the serendipitous photograph. Other plates show that by 1939 the nova had disappeared into the background of thousands of neighboring, brighter stars in the cluster. Today, the nova is presumably in its quiescent dim state as an ordinary binary-star system.

Before the age of *Hubble,* only a few such binaries had been sighted in globular clusters, mainly because they are hard to detect amid the clutter of so many other stars—but more on this later. Such binaries are thought to be less numerous in globulars than they are in the more open star clusters in our galactic neighborhood. However, if even a small number of binaries exist in globular clusters, they can, at least theoretically, drastically influence both the motions of other stars and the behavior of the whole cluster. Thus, the issue of binary stars inside globular clusters is an important one if we are to understand the dynamical evolution of these intriguing celestial objects.

The *Hubble* image acquired toward M14 is not centered on the core of the cluster but rather in its northeast quadrant. Even so, the image reveals literally hundreds of individual stars in a tiny region of the cluster where only dozens are distinguishable on ground-based images. Numerous faint puffs of light, appearing as single objects in high-quality ground-based photos, are clearly resolved into multiple stars in the *Hubble* image. These space-based

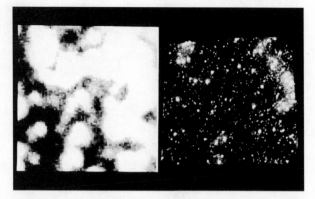

These two images, both having a field of view of approximately 22 x 22 arc seconds (approximately 7 light-years square) centered some 30 arc seconds from the core of M14, vividly demonstrate the exquisite resolving power of the Hubble Space Telescope. The image on the left, taken with the 4-meter telescope at the Cerro Tololo Inter-American Observatory atop the Chilean Andes, has a resolution of about 1 arc second. Its graininess results from the charged coupled device (CCD) used to detect the light. [AURA]

The image on the right, taken with Hubble's faint-object camera, yields stellar (core) diameters of only 0.08 arc second; their median brightness is about the twenty-second magnitude. This is a forty-two-minute exposure taken at a (blue) wavelength of about 4,300 angstroms in an attempt to spot in a field of ultraviolet-weak stars an ultraviolet-bright object, namely a nova. As this image is raw—that is, not deconvolved or computer enhanced in any way—the small concentric halos around each star clearly introduce a kind of aberrant "fog" that deprives us of a crystal-clear view of the cluster. Even so, the Hubble image on the right is demonstrably sharper than the ground-based photo on the left. (The dark diagonal strips at the upper left and lower right corners of the Hubble image result from occulting "fingers" designed to block light from bright sources.) [NASA and ESA]

images were taken in the blue and ultraviolet parts of the spectrum and are reasonably long (forty-minute) exposures. Short-wavelength light was chosen in an attempt to develop high color contrast by suppressing the reddishness of the giant stars that dominate globular clusters while enhancing the emission from blue/ultraviolet objects such as bright, hot accretion disks often associated with even dormant novae.

Globular-cluster expert Bruce Margon, of the University of Washington, to whom the Hubble image of M14 was shipped electronically for analysis, sent me several increasingly enthusiastic electronic messages about the M14 data, and he finally talked to me the old-fasioned way—on the telephone, exclaiming, "These Hubble data provide humanity's best view ever of a globular cluster, clearly a project that would be hopeless from the ground." To say that he was delirious with excitement is an understatement—which was a breath of fresh air, given all the doom and gloom then effusing from the U.S. astronomical community.

Nature continues to outwit us, however, as our detective story is not yet finished. The faint-object camera found five and perhaps six separate stars at the candidate nova's suspected position, where previously only one cottony blob of light was seen in ground-based pictures. Presumably only one of these half-dozen stars is the old nova, although it is also possible that the culprit is too faint to be seen even in the *Hubble* image. In any event, we had quickly learned something new from these most recent science findings—namely, that the brightness of the nova remnant today must be much less than suggested by the ground-based data, which we now know to have summed at least five separate stars. "Nature has decided to be particularly devious in this case," said Margon gleefully.

The cosmic hunt for Nova Ophiuchi 1938 will continue, although it would appear that this version of the haystack's needle can be addressed only with *Hubble*'s superb resolution. The next step in pinning down the nova's location in M14 requires further analysis of these and additional images taken with either of *Hubble*'s cameras. Because images have already been obtained in two different colors (at 3,420 ultraviolet angstroms and 4,300 blue angstroms), astronomers were able to search the dense star field for an object with unusual colors in comparison with its hundreds of neighbors. Strong ultraviolet radiation is one such signature of many old novae. And although two of the six stellar candidates do seem to have slightly peculiar colors—an "ultraviolet excess"—the observed oddity is neither extreme nor perhaps even significant. We still do not know whether this fact alone represents a bonus find—for example, that old novae in globular clusters differ from the kind outside such clusters.

We should obtain yet another clue to the puzzle once spectra of the candidate stars are in hand. Both spectrographs on board Space Telescope are capable of seeking the spectral signatures characteristic of novae, a task that is all but impossible from the ground given the crowding of nearby, unrelated stars in M14. Provided *Hubble* is repaired, such spectroscopic observations will go forward during the next few years as part of the guaranteed telescope time granted to the faint-object spectrograph team.

The final observation of the imaging campaign led to a dramatic finding—nothing less than a confirmation of Einstein's General Theory of Relativity.

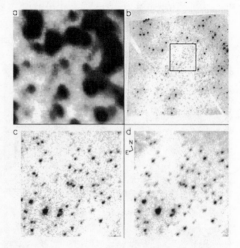

As an example of how astronomers prefer to work with negative images—the contrast seems better—this series of four photos illustrates the hunt for the elusive nova in the suburbs of M14. Frames (a) and (b) are merely the same as the previous figure, though shown here as black stars on a white background—(a) is a good ground-based view of a part of M14 and (b) is the *Hubble* target-acquisition image taken with the faint-object camera, both of them 22 x 22 arc seconds across. Frame (b) also has superposed on it a 5-arc-second square area inside of which resides the nova in question. Frame (c) shows an ultraviolet (or "U" filter) image of that small 5-arc-second square area, and frame (d) shows the same piece of celestial real estate as observed through *Hubble*'s blue or "B" filter. The nova is suspected to be at the center of frames (c) and (d), but there at least five stars seem present, whereas from the ground only one such star was spotted. *[AURA, NASA, ESA, and M. Shara]*

Buried in the unassuming proposal title "Verification of Point Sources in Backgrounds" was a measurement of a "gravitational lens" in space.

Gravitational lensing occurs when light from a distant source is bent while passing virtually through or close to a massive foreground object. Depending on the detailed alignment of the foreground and background objects along the line of sight from Earth, multiple images of the background source can be seen. No magic is involved here, nor is this merely an illusion, implying a trick of some sort. Rather, the multiple images comprise a mirage, much like the pool of water one sees over the horizon in a desert. In a cosmic setting, as Einstein taught us, the structure of space and time is warped by matter to produce the effect of gravity. The presence of a large mass can cause space to be curved in much the same way that a heavy rock lying on a thin rubber membrane such as a trampoline causes the fabric to bend. The difference is that the rubber sheet is two-dimensional, whereas space is three-dimensional; in fact, the nature or "fabric" of the Universe—spacetime—is four-dimensional.

Just as a curved piece of glass can bend light, curved space can also make light appear to bend. That's why, in an analogy with lenses made of glass, the bending of light by matter is called "gravitational lensing." Hold a plain, inexpensive wine glass up to a single light bulb and you can often see multiple representations of the bulb; this is a mirage. The exact number and arrangement of the multiple images of an object seen in a gravitational lens depend on the particular geometry and alignment of the foreground and background objects; sometimes several small points or blobs of light are seen, in other instances arcs of glowing light. (The possibility that one is not merely seeing several actual sources can be disproved only by taking spectra of each light source, thus proving, if they really are part of a mirage, that all the sources have identical spectra.)

That gravitational lenses are important to astronomy is evident in a pair of simple words: *dark matter*. Astronomers estimate that at least 90 percent of the Universe consists of matter that does not emit any radiation detectable by current instruments. Although dark matter—also known as "hidden matter" or "missing matter"—cannot be seen directly, its existence has been inferred from its gravitational influence on the motions of stars in galaxies and the motions of galaxies in galaxy clusters. The phenomenon of gravitational lens-

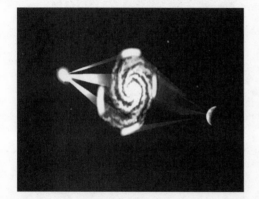

This piece of art illustrates the concept of a gravitational lens. Light from a distant source (left) is bent while passing near or through an intervening, massive galaxy or cluster of galaxies. Depending on the geometry and positioning of each of the objects, we at Earth (right) can see one or more images of the single distant source arrayed as blobs or arcs along a ring around the intervening galaxy. In a special case not yet observed in the cosmos, the entire ring could conceivably be seen completely surrounding the intermediate lensing source. In the case illustrated here, light from a distant quasar is imaged or "miraged" four times. [D. Berry]

ing, which partly utilizes unseen objects, then provides a powerful probe in the ongoing search for dark matter.

For this observation, the faint-object camera was directed toward an object having the peculiar designation G2237+0305, a name that derives from its galactic coordinates. This object is known from ground-based studies to be a gravitational-lensing source in which light from a quasar 8 billion light-years away is seen miraged several times by an intervening (fourteenth-magnitude) galaxy estimated to be about twenty times closer, or 400 million light-years away.

The objective of this science-assessment observation was to check to what extent astronomers could do imagery (map a source) and photometry (measure its brightness accurately) toward faint objects (the quasar) adjacent to a relatively bright background (the intervening galaxy). But the hidden agenda on the part of Duccio Macchetto, who was leading both this camera team and this observation in particular, was to make a discovery, an attitude that once again severely torqued the hysterical opponent of the imaging campaign, fearing that someone else was going to demystify the quasar quandary before he could get a crack at it.

The initial attempt to bring home the image suffered from underexposure and poor telescope pointing, but a second exposure was made on coarse track, using a yellow-green filter, for twenty-five minutes. The result, once again including the 5,007-angstrom spectral emission of doubly ionized oxygen that had led to earlier successes, was better than anyone expected. Owing to an extremely fortuitous alignment, the light of the distant quasar is well bent in its path by the gravitational field of the intervening galaxy, resulting in four distinctly separate images of the same quasar arranged as points of a rectangle, but with the intervening galaxy (and possibly a fifth mirage) in the middle, looking more like a symmetrical cross. The angular separation between any two adjacent mirages of the distant quasar is only about an arc second, clearly testifying to *Hubble's* superb resolution.

Gravitational lenses, such as G2237+0305, are useful probes of many types of phenomena occurring in the cosmos. In addition to implying the presence of dark matter, gravitational lenses make possible the "weighing" of

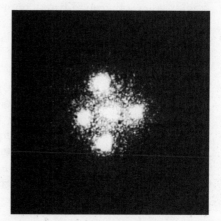

This *Hubble* image, which is not computer enhanced, shows the most detailed observation ever made of the gravitational lens system, G2237+0305—sometimes referred to as the "Einstein Cross." The four external patches of light comprise mirages of a single, very distant quasar that has been multiply imaged by a relatively nearby galaxy. The quasar is about 8 billion light-years distant, whereas the galaxy is "only" 400 million light-years away. The fifth, diffuse splotch of light at the center is probably the core of the intervening lensing galaxy. The angular separation between the core and any one of the outlying blobs is less than 1 arc second; the separation between opposite blobs is 1.6 arc seconds. [Courtesy NASA, ESA, and D. Macchetto]

the foreground galaxy by measuring the relative positions and the brightnesses of the mirages of the quasar. This should be possible to do more accurately given the exquisite view of the images obtained with the faint-object camera. Analyses of this and similar images scheduled to be acquired throughout Space Telescope's mission are expected to provide a wealth of information on the often misconstrued details of lensing galaxies, as well as on the eerie process of gravitational lensing itself. Further probes of the early Universe, of the elusive dark matter, and of Einstein's grandest accomplishment would all seem possible by means of gravitational-lensing observations. Whether *Hubble* is up to the task in providing new and significant insight regarding these fascinating issues remains to be seen.

Two additional observations were made in September to supplement *Hubble's* inaugural imaging campaign. Both were solar system objects, and one of them demonstrated *Hubble's* ability to make moderate exposures of moving targets. Of course, all celestial objects travel through space, but planets, moons, and their neighboring comets, asteroids, and meteoroids all appear to move at faster rates in the sky than do stars—hence the curious term "moving targets."

One such target was our solar system's most distant and enigmatic object—the planet Pluto. The ninth and last real planet known, and the only planet not yet visited by a robotic space probe, the elusive Pluto was discovered in 1930 by the American astronomer Clyde Tombaugh, who was searching for the source of slightly irregular motions of Uranus in its orbit around the Sun. In fact, we now know that his discovery was serendipitous, since Pluto is far too small to affect Uranus's orbit (the source of irregularity still being unknown). Pluto, with a diameter of only 1,400 miles (or 2,300 kilometers), is the tiniest planet in our solar system, smaller even than Earth's Moon.

That Pluto is a peculiar object is an understatement: Its orbit is more inclined and more elliptical than those of any of the other known planets. It rotates "upside-down" with its North Pole below the plane of the solar system (the ecliptic). Unlike the gassy Jovian worlds in the outer solar system, tiny Pluto is made entirely of solid rock and ice. Perhaps its most fascinating property was uncovered in 1978, when a huge companion satellite was detected from studies of ground-based photographs. This moon, which is named Charon, revolves around Pluto in a mere six days, compared to the month-long period of the Earth-Moon system. Subsequent investigations have shown that Charon is about half the size of Pluto, making it the largest known satellite relative to its parent planet. For this reason, these two worlds are often collectively referred to as the "double planet."

Actually, *Hubble's* observation of Pluto was ordered by "higher authorities," although with some prodding from the Science Institute. In late July, the number-two man at NASA Headquarters—J. R. Thompson, a rough-and-tumble Texan who speaks his mind squarely—had visited the Science Institute. During the course of a tour and briefing we gave him on the status of the project to date, I expressed my concern that we keenly needed some science images from *Hubble* if we were to reverse the widespread impression that the spacecraft was a dud. "My friend, we need more than pictures—we need a dis-covery," he drawled while looking me straight in the eyes. And then he put it to me gruffly, further substantiating official NASA's lack of appreciation for the scientific endeavor: "Can you order me up a discovery? And make it one that the public can understand."

I told him just as bluntly that no one could "order" a discovery, but that our on-again, off-again list of early-release observations had one target that might come close. *Hubble* should be able to resolve Pluto and its moon Charon, and to separate the two of them clearly—something that had·not been done from the ground, and of added benefit something that might well appeal to the public. Moreover, a possible sighting for the first time of surface features on Pluto would attract much attention worldwide. (Had *Hubble* worked properly, it could have probably seen Pluto just about as well as our naked eyes see the Moon from Earth.) Ignoring the chain of command, I repeated my argument with a handwritten note that I sent Thompson the next day. A few days thereafter, we received an electronic message from God-dard that orders had come from Washington to have *Hubble* image Pluto pronto: "NASA HQ wishes to do this in a way that is understandable to the 'man in the street.' . . . This activity has the personal interest of NASA's deputy administrator . . . [but] we want as close to zero advance publicity of this attempt as possible."

An observation of Pluto could not be achieved during the ten-day marathon of early science observations in late August. Pluto is a great deal fainter than Saturn and thus a longer exposure was needed to collect enough light from it. Since *Hubble* was not at the time capable of tracking planets automatically, the observation had to be scheduled manually to execute a methodically planned series of operations known colloquially as "parallax, step, aim, and shoot"—a little like ambushing the planet by timing the shutter opening to coincide with the planet's entry into the telescope's narrow eye-sight. To avoid smearing the images, ground controllers at the institute had to preprogram *Hubble* to track Pluto extremely accurately and to compensate exactly (by periodically "stepping" and minutely adjusting the pointing) for the parallax introduced by the combined motions of Pluto, Earth, and *Hubble* in their respective orbits. (Parallax is the apparent motion of an object caused by the motion of an observer—which in this case was the *Hubble* spacecraft rapidly moving around Earth.)

The first time we attempted to image Pluto, the observation mostly failed, largely due to pointing errors. The faint-object camera was used and, as noted before, its narrow field of view is not forgiving. If you miss the target, then you miss the target. A small amount of data was accumulated, but a combination of *Hubble* drifting on gyros and problems with the makeshift parallax-plus-step technique prevented a clear registration of several short exposures. The resulting image was heavily blurred and was no better than that achievable from the ground—which is not very good at all.

Previously, when an early-release observation failed for some reason and we petitioned NASA to let us try it again, we would be met, often in writing, with the flat-footed response from Goddard, "No EROs repeats." But since

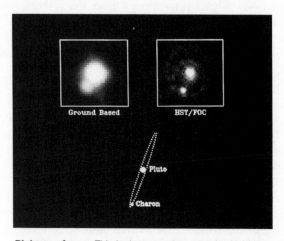

Right top frame: This is the second attempt by *Hubble* to image Pluto and its moon, using the faint-object camera in the f/288 mode at 4,300 angstroms. This image has not been computer enhanced to minimize aberration; hence, the unreal rings around the planet. Pluto is at the center of the frame while Charon is the fainter object at the lower left. The registration of numerous short exposures stacked together is good and the resolution measures approximately 0.1 arc second, fully ten times that typically achievable at the best observing sites on the ground. For the first time, these two icy worlds are clearly distinguished. At the time of observation, Charon was near its maximum angular separation of about 0.9 arc second—equivalent to the angular separation between a pair of car headlights about 300 miles away. *[NASA, ESA, and R. Albrecht]*

Left top frame: For comparison, this is a one-minute exposure of the Pluto-Charon system, taken with the Canada-France-Hawaii 3.6-meter telescope atop Mauna Kea, Hawaii. Seeing at the time was a superb 0.5 arc second, virtually the best that any (civilian) ground-based telescope can manage. The graininess of the image is caused by the rather small number of pixels in the CCD array attached to that telescope. In all such ground-based photos, the pair appears blurred into a nearly single egg-shaped blob of light. *[Canada-France-Hawaii Telescope Corporation]*

Bottom frame: Charon's orbit around Pluto is a circle seen nearly edge-on from Earth, with a radius of nearly 12,000 miles (20,000 kilometers)—a span of about 1.5 times the diameter of Earth.

the Pluto observation had attracted higher-level involvement, Goddard, looking out for its own preservation, this time aimed to please. A second attempt was made, and the success was immediately apparent. *Hubble* had indeed managed to take a long-duration (ten-minute) photograph of a moving target. Pluto and its moon were resolved for the first time. The superior resolving power of the faint-object camera was clearly evident, *Hubble*'s optical flaw notwithstanding. And although the new image of the double planet was not really a discovery, it was an important milestone—for planetary scientists and for the *Hubble* project.

Currently, at a distance of approximately 3 billion miles (or 4.5 billion kilometers), Pluto is near its closest approach to Earth in its 248-year journey around the Sun. In fact, during the interval 1979 to 1999, Pluto is actually inside the orbit of Neptune, making Neptune temporarily the most distant planet, a point of heavenly trivia nearly guaranteed to win bets at cocktail parties. Even so, Pluto averages forty times farther from the Sun than Earth and therefore we should not be surprised that Charon is so faint. Charon is smaller than Pluto and its surface is probably covered with water ice, whereas Pluto is covered mainly with methane frost or snow, which is more reflective.

Further *Hubble* observations of Pluto and Charon will be important in elucidating the nature and origin of these fascinating and frigid worlds, where the average ambient temperature approaches –215°C—only 58°C above absolute zero. Although the "fog" surrounding the image, caused by the aberration of *Hubble's* primary mirror, severely reduces the chances of spotting surface features on Pluto, several other critical pieces of information can be extracted from continuing observations. For example, detailed analysis of brightness (albedo) variations of both Pluto and Charon should provide a wealth of knowledge about their surfaces and atmospheres impossible to obtain from the ground.

Precise measurements of the orbital parameters of the Pluto-Charon system are also now possible and this will enable astronomers to derive accurately the individual masses and densities of the two objects, thereby providing vital clues to their origins. One idea is that myriad objects—planetary embryos—similar to Pluto and Charon accreted from loose gas and dust in the outer fringes of the primordial solar nebula some 4.5 billion years ago, but were either expelled from the system or gobbled up by the giant gas planets, Jupiter, Saturn, Uranus, and Neptune, with Pluto and Charon the sole survivors. Lots of answers are needed here among theories that are far out in front of the data. Continued monitoring of these two fascinating objects at the outer edge of our solar system throughout the lifetime of the Hubble Space Telescope will contribute to astronomers' quest to understand the nature of the nebula from which all of us originated.

We ended the summer's imaging campaign by commanding *Hubble* to take a quick look at the brightest comet in the sky at the time. Repeatedly, since well before launch, a young planetary scientist at the institute, Hal Weaver, had urged me to include a comet in our early observations. To his credit, he kept track of comets that came and went during the several months since launch, and regularly updated a proposal for ready execution in the event an opportunity arose. And arise it did.

By late summer, it had become clear that Comet Levy, discovered earlier in the spring by an amateur astronomer, might become the brightest comet since Comet Halley's apparition during 1985–1986. Technically termed comet 1990c, Comet Levy is a small ball of dirty ice that was then picking up speed while wending its way toward the Sun. And because ground-based telescopes were detecting variability in its light output, suggestive of jets expelling gas and dust from volatile pockets on the comet's nucleus, *Hubble's* observations had the potential to provide important insights into the nature of cometary morphology.

However, most NASA officials, despite the success with Pluto-Charon, were reluctant to risk imaging a "speeding target"; comets in the inner solar system move across the sky even faster than planets, and such an observation would be widening *Hubble's* envelope still further—something the scientists very much wanted to do to assess the observatory's true capabilities, yet the conservative project engineers were loathe to try, lest something else break.

By this time, the Science Institute had had a lot of practice working the NASA bureaucracy. We arranged for Weaver and his team of cometary observers to petition that the comet be declared a "target of opportunity," for which director's discretionary time could normally be used—although we were bluffing since the director of the institute did not officially own any such discretionary time during the orbital and science verification phases of the project. When there was no response from NASA, we simply inserted a brief observation onto *Hubble's* calendar and waited for flak from the space agency. It never came, probably because project officials were unaware of what was really happening in the *Hubble* trenches on a daily basis.

Because of the lingering immaturity of *Hubble* to point accurately and because the comet was traveling so fast, the telescope's widest field of view was used for this observation. We literally caught Comet Levy on the fly. A series of short infrared exposures with the wide-field camera successfully imaged the comet's coma, a region of loose gas and dust surrounding its icy nucleus. As with the Pluto-Charon image, these were "aim-and-shoot" observations, involving no tracking. Exposure times were kept to less than a few seconds each, in order to minimize the trail of the comet through the camera's field of view. In the first *Hubble* image, the comet landed within about 3 arc seconds of its expected position, but this pointing discrepancy degraded significantly in subsequent exposures. Even so, the resulting images showed considerably more detail than could any ground-based telescope.

The comet's image surprised us at first. It seemed to have a short tail pointing *toward* the Sun. Comet tails are caused by the "solar wind" of radiation and matter escaping from the Sun; this wind blows volatile gas and dust from the comet, thus, their tails always point away from the Sun. After a few moments of puzzlement and a reminder that this image had captured infrared radiation, someone suggested that most of what we were seeing must be sunlight reflected by grains of cometary dust, carried outward by gas boiled off the leading edge of the icy nucleus as it was warmed by the Sun. The resulting coma is brightest (approximately fifth magnitude) in a fan-shaped region on the sunlit side of the nucleus, but the computer-enhanced images show no clear evidence for a jet. The nucleus itself was unresolved; at a distance of about 100 million miles from Earth, the resolution is about 50 miles per pixel, which is some ten times greater than the estimated size of the nucleus itself.

Comet Levy is interesting because its orbital track suggests that it comes from the outermost reaches of the solar system, nearly as far away as the hypothesized but never confirmed Oort Cloud of millions of comets thought to girdle our planetary system well beyond Pluto. If this is so, its nucleus is

7

More Early Science Results

On the 7th day of January in the present year, 1610, in the first hour of the following night, when I was viewing the constellation of the heavens through a telescope, the planet Jupiter presented itself to my view, and as I had prepared for myself a very excellent instrument, I noticed a circumstance which I had never been able to notice before, owing to want of power in my other telescope, namely, that three little stars, small but very bright, were near the planet; and although I believed them to belong to the number of fixed stars, yet they made me somewhat wonder, because they seemed to be arranged exactly in a straight line parallel to the ecliptic, and to be brighter than the rest of the stars, equal to them in magnitude. The position of them with reference to one another and to Jupiter was as follows.

On the east side there were two stars, and a single one towards the west. The star which was furthest towards the east, and the western star, appeared rather larger than the third.

I scarcely troubled at all about the distance between them and Jupiter, for, as I have already said, at first I believed them to be fixed stars; but when on January 8th, led by some fatality, I turned again to look at the same part of the heavens, I found a very different state of things, for there were three little stars all west of Jupiter, and nearer together than on the previous night, and they were separated from one another by equal intervals. . . .

At this point, although I had not turned my thoughts at all upon the approximation of the stars to one another, yet by surprise began to be excited, how Jupiter could one day be found to the east of all the aforesaid fixed stars when the day before it had been west of two of them; and forthwith I became afraid lest the planet might have moved differently from the calculation of astronomers, and so had passed those stars by its own proper motion. I therefore waited for the next night with the most intense longing, but I was disappointed of my hope, for the sky was covered with clouds in every direction. . . .

Accordingly, on January 11th I saw an arrangement of the following kind, namely, only two stars to the east of Jupiter, the nearer of which was distant from Jupiter three times as far as from the star further to the east; and the star furthest to the east was nearly twice as large as the other one; whereas on the previous night they had appeared nearly of equal magnitude. I therefore concluded, and decided unhesitatingly, that there are three stars in the heavens moving about Jupiter, as Venus and Mercury round the Sun; which at length was established as clear as daylight by numerous other subsequent observations.

Sidereus Nuncius
GALILEO GALILEI, Venice, 1610

As the leaves began to fall, we retreated—literally. The Science Institute's directorate and division heads (mostly all astronomers) crossed the Chesapeake Bay and held a reflective postmortem for a few days in what seemed like a winterized safe house on Maryland's eastern shore. With the engineers driving *Hubble* once more, we sought to escape the chaos and to understand better where we stood in the light of all that had whirled around us since spring. We had known that the *Hubble* project would likely be a focus of public attention during the months following launch, and so it had—but, again, for the wrong reasons. We clearly needed to get out of the petri dish for a while—to recover our equilibrium, to assess how each of our divisions had weathered the postlaunch crises, and to examine where we might be individually and institutionally headed with this project.

Although almost everyone dressed more informally than I had ever seen among Institute personnel—mostly jeans, old sweaters, and sneakers—we addressed many concerns with an intensity that surprised me. We first looked back, replaying in our collective minds a nearly worst-case scenario that had transpired before us, yet I personally worried about the extent to which we were self-congratulatory. It was true that each of the institute's divisions had led the way in virtually every postlaunch category, while our NASA counterparts seemed paralyzed: Our scientists had discovered the telescope's aberration, had exploited computer-enhancement techniques to recover the intended resolution, had lobbied to trigger the imaging campaign, and had pushed hard for a more robust repair of *Hubble* on orbit. But were we missing the forest for the trees, as had so many others on this project who failed to examine the bigger picture? Despite these many accomplishments, we were still nursing along an injured bird and it was not at all clear what might be the long-term prognosis of this celebrated space mission.

Especially worrisome to us was the attitude of the scientific community regarding the conduct of the mission that they were watching so closely. Astronomers all over the world were irate at NASA (and I suppose ESA, too), and we were concerned that they, whom we at the institute were supposed to represent, would soon take aim at us. On the other hand, we had seen the initial anger of our own institute staff give way to guarded euphoria as the imaging campaign took hold. We were confident that in time, with a steady stream of new science results from *Hubble,* the international community of astronomers would realize that the Space Telescope, however impaired, was able to do certain kinds of science that simply could not be done with ground-based instruments.

As the astronomers closest to the mission, we debated whether we should lambaste NASA and its industrial contractors for having built a $2-billion machine riddled with bullet holes and accumulating band-aids quicker than we could have imagined. Some argued that the astronomical community was now waiting for leadership from the dozen of us at this very retreat. Perhaps a statement of no confidence in the space agency was called for. Or some sort of break in diplomatic relations. But how can you break relations with your sole source of funding? Some of us felt like hostages; others' deep-seated frustrations emerged as in a group therapy session; all of us were confused about what posture to assume toward NASA.

One participant asked, "Are we wasting our lives continuing to hitch our wagon to NASA?" "Look," said another, "there has been no evil intent or fraud, just human stupidity and frailty. A lot of stupidity and frailty, but no skulduggery." In the end, we reached a weak consensus that the degree of programmatic inertia at the agency was so overwhelming that nothing we could publicly say or do would likely make a difference, and we decided not to speak out as a group. In hindsight, our reluctance to adopt a strong public stand might have been a mistake.

We also looked critically at ourselves. Whether in small, formal working groups that weightily debated the issues, or among a few of us during equally serious self-scrutiny while informally hiking around an abandoned, wind-whipped golf course, we wondered about what our Science Institute had become. We reexamined the 1976 National Academy report that had recommended the institute's creation, and we found amusing how naive that report had been regarding our topsy-turvy relationship with NASA: our creators had not recognized the extent of bureaucratic arteriosclerosis at NASA and had underestimated the amount of daily trench work needed to operate such a complicated robot. While the academy had envisioned the institute as a small, sleek, efficient group of several dozen (mostly astronomer) professionals, we had become an order of magnitude larger, including well the majority of the staff involved in telescope operations, not pure science.

The institute's prelaunch growth owed largely to our unexpected (but necessary) involvement in the ground system needed to command and control *Hubble,* and especially in the calibration of the telescope and the analysis of its downlinked science data. Yet, to our horror, one of the astronomers present blurted out, "Are we becoming another Goddard?" At which point, Giacconi, clearly becoming uncomfortable with all this retrospective flagellation, suggested impatiently that we treat the original charter of the institute as history: "The STScI is built. It's different from what we thought, period. Let's move on!"

We resolved, among other things, to redouble our efforts to ensure *Hubble's* success in the short term and its potential recovery in the long term. We also decided to formally assume a more active role regarding scientific oversight of both hardware and software components for future optical-ultraviolet space missions. "We can't wait for Daddy to build our instruments and then just expect to use them," Giacconi told us, becoming ever more upset the more he spoke. "Instead of worrying whether our lives are wasted, we scientists need to take greater personal responsibility, and even put our careers on the line, to be sure things turn out better next time."

For a variety of reasons—including both scientific curiosity and the desire to maintain a kind of drumbeat of public releases—several of us very much wanted to redo a number of observations that had failed during the imaging campaign. There was the globular cluster M15, which had tantalized us with a quick peek into its heart, a type of astronomical source never before clearly penetrated. And there was Eta Carinae, our highest-priority EROs target and perhaps the most fascinating object in the southern sky. A few more good images, we reasoned, as well as some early spectroscopic observations, would go a long way toward achieving our dual objectives—to prove to the world that *Hubble* could do significant science on a regular basis and to assess which kinds of science programs were best rendered in light of the telescope's problems.

However, the engineers had now resumed their many varied tests, and a request for even a few more orbits of science time might be regarded as the equivalent of a palace revolution. Despite the considerable success of the imaging campaign, the *Hubble* project office, which is populated almost exclusively by engineers and administrators, had seemingly lost track of the reason Space Telescope was built in the first place. The issue of the importance of conducting science was, ironically, settled with the help of Mikhail Gorbachev's former science advisor.

Visiting the Science Institute one crisp, fall afternoon, Roald Sagdeev, the head of the USSR's Institute for Space Research, sat with astronomers Giacconi, Norman, and a few others, all of us staring at a large wall of my office where I had mounted an evolving display of images surrounding a huge map of the Universe. Throughout commissioning, I had been charting our progress on the science front on a wall-size status board—a colleague called it a "battle map"—complete with color-coded push-pins signaling those objects observed and those yet to be targeted. After a thorough discussion of the success and significance of the imaging campaign, and a few jokes comparing the

A wall of the author's office at the Science Institute became an informal status board for the *Hubble* imaging campaign in the late summer and fall of 1990. Here, around the perimeter of a large, oval shaped map of the nighttime sky were displayed finding charts and guide-star plots in the vicinity of each of the early *Hubble* targets; as soon as the newly acquired *Hubble* images were in hand, they too were posted. This was major-league exploration, and great fun too—the way science ought to be—for a variety of cosmic phenomena were being examined much more clearly than ever before.

Soviet agricultural system to the American space shuttle program, the Russian declared in a clear and serious voice, "Imaging campaign is most important thing you can do right now. It should continue until fire is out and citizens feel good about Space Telescope mission." Quite independent of the concept of *glasnost* then sweeping his homeland, he was genuinely explicit on the need for clear and open dissemination—both good and bad—of the *Hubble* project's status. Otherwise, he chuckled, *perestroika*—"restructuring"—would apply to the U.S. space program as much as to the Soviet economic system. He promised to communicate his sentiments to the administrator of NASA with whom he was having dinner a few days later.

By means not too unlike those that sparked the imaging campaign, we eventually did manage to have several more EROs and SAT targets examined that fall. Among the most spectacular was M15: the hunt for an immense black hole was on again.

Residing some 42,000 light-years away in the constellation Pegasus, M15 (also known as NGC 7078) is a rich globular cluster high above the galactic plane, in the halo of our Milky Way Galaxy. Globulars such as M15 differ from open (or galactic) clusters in that they are old (approximately 12 billion years), tight-knit spherical aggregates of uncounted hundreds of thousands, and sometimes millions, of stars bound by their mutual gravity within a relatively small region of, typically, a few hundred cubic light-years. The density of stars within such clusters is among the highest known in the Universe— roughly a million times that in our part of the Galaxy.

M15 is so compact, in fact, that many scientists suspected before *Hubble* took a look that its core must have gravitationally collapsed to form a huge black hole containing roughly a thousand solar massses. Others disagreed, suggesting that the motions of individual stars could help to buoy the core, effectively prohibiting its catastrophic collapse in much the same way that the motions of individual gas molecules keep Earth's atmosphere from falling to the ground. Binary stars, in particular, some of which orbit each other at a million miles per hour, can act like a storage battery of kinetic energy; loose stars moving past such a binary system can rob it of some of its energy and get a boost in speed, like a pinball bouncing off a bumper, thereby inhibiting the cluster's stars from coalescing at the core.

Those holding this view do admit that something peculiar must reside within the core of globular clusters like M15, including some unspecified dark matter, but they are unwilling to concede that it need be anything more exotic than an exceptionally dense cluster of ordinary stars swarming, like bees around a hive, within a region of loose gas and dust at the center of the cluster. At any rate, although M15 is easily visible in a small backyard telescope, not even the most powerful scopes on Earth have been able to penetrate the bright ball of light to reveal the central engine at the core of the cluster.

Throughout the years leading up to launch, M15 was considered a high-priority target for the Hubble Space Telescope. It was, in fact, the celestial source that had most often triggered the volcanic reactions of the now resigned seventh-team member, and I therefore steered the EROs program clear of it. The Science Assessment Team, however, had no such reservations, and charged on, knowing that even the most powerful ground-based telescopes cannot resolve stars well enough to visually probe any globular's mysterious core; here was a case where Space Telescope observations might force a rewriting of the astronomy textbooks.

Despite *Hubble's* aberration, it could still resolve features to some 0.1 arc second, which, at the distance of M15, equals about 0.02 light-year, or about twenty times the size of our solar system. This is roughly the scale suggested by the best theoretical models for the angular size of M15's inner sanctum. If the core really has collapsed into a black hole, it would be a massive one, and vast quantities of light would be emitted just outside its "event horizon" (or maximum radius) as matter, under phenomenal stress and tidal distortion, fell into it. Such a characteristic black-hole signature should, paradoxically, form a bright "cusp" of light rising to an unresolved point, like the one that *Hubble* found earlier for the galaxy NGC 7457, which indeed harbors a good candidate for a supermassive black hole. (All black holes contain much mass, but unwritten convention among astronomers has it that "ordinary" black holes contain tens of solar masses, "massive" black holes harbor of the order of a thousand solar masses, and "supermassive" black holes have upward of a million or more solar masses.)

Recall that *Hubble's* initial attempt to observe M15 had excited us with a view that seemed to reach well into the cluster's core. That earlier attempt was judged a failure, because of lack of guide-star acquisition, but that brief glimpse made us want to return to explore this object as soon as possible. As with the previous effort, Westphal's postdoctoral assistant, Tod Lauer, again used the planetary camera, this time acquiring four 13-minute exposures in violet light. This highest-resolution mode of the wide-field and planetary camera was deemed necessary if we were to have a chance of separating the innumerable stars crowded together in the central region of the cluster.

Imaging in violet color would also be helpful in suppressing the glow of the red-giant stars abundant in the suburbs of M15. Previous observations of globular clusters, made with ground-based telescopes, had all displayed vast fuzzy balls of pinkish light, much of it dominated by red giants, which greatly outshine everything else. In violet light, those red giants would be less prevalent, allowing us a much clearer view of the cluster's enigmatic core. Thus, every reasonable advantage was utilized to penetrate the core of the cluster. And this time, thank goodness, guide stars were acquired and held the spacecraft in fine-lock for all the observations.

The results showed us that the *Hubble* telescope can measure, and not just detect, celestial objects as faint as the nineteenth magnitude in a single orbit's exposure. Furthermore, observations taken over a span of several orbits can be accumulated (provided the spacecraft's roll angle has not appreciably changed) to analyze sources in crowded fields as faint as the twenty-third magnitude. The original hope and intent, however, had been that *Hubble* would reach at least the twenty-sixth-magnitude objects in the highly crowded star fields of globular clusters. Sadly, the M15 observations confirmed that although *Hubble's* designed resolution was mostly recoverable, its sensitivity was irretrievably reduced (short of a fix on orbit).

Even so, thanks to the unmatched clarity of the *Hubble* image and computer massaging of the data, the cardinal mystery of at least this globular cluster has apparently been solved. By resolving red giants in M15's outskirts for the first time, Lauer and colleagues were able to subtract them from the image, effectively peeling away M15's outer envelope. The remaining group of thousands of faint stars, many still unresolved because of tight crowding near the cluster's center, form a diffuse background with a surprisingly large radius, about 0.4 light-year—much larger (by more than an order of magnitude) than that predicted for the event horizon of a classical black hole. The absence of any sharp spike, or cusp, of light on sub-arc-second scales bolsters the idea that M15 probably does not have a black hole in its midst. In fact, the precise nucleus of the cluster does not coincide with any of the observed bright points of light in the *Hubble* image. Just as we had expected before launch, *Hubble* was indeed beginning to cause the textbooks to be rewritten.

For me, this rather mundane development was welcome news. While teaching college astronomy courses over the years, I had resisted the temptation to endow the heart of virtually every poorly understood object in the Universe with a black hole. The "bandwagon" appeal among astronomers who would have black holes lurking in darkened nooks and crannies practically everywhere—in the centers of galaxies, star clusters, exploding stars, even at the core of our Sun—was unconvincing to me, especially since there is no unambiguous evidence that even one such black hole actually exists anywhere. They probably do, but they could just as well be figments of our imagination.

In the case of M15, the evidence—genuine data acquired with humankind's most powerful telescope—reveals a core region that is doubtless unearthly but not one that requires us to conjure up a pinhead-size source of

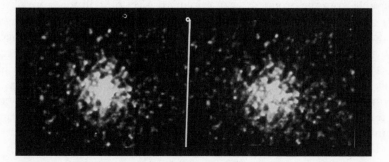

The image at left shows a nearly one-hour exposure of the core of M15 made with *Hubble*'s planetary camera in deep violet light (or near-ultraviolet, at 3,360 angstroms). The area displayed is 17 arc seconds (approximately 3 light-years) on a side and represents only the central portion of the star cluster. The image has been computer massaged twice—once to reduce the glare by subtracting the bright red giant stars, and again to clean the image of the degrading halos surrounding each star. The light that remains, now resolved to about 0.1 arc second, arises from the combined emission of thousands of faint (dimmer than nineteenth-magnitude) stars swarming in and around the core. By contrast, the panel on the right shows a computer simulation of what the image would look like if a massive black hole were present at the cluster's center. In this case, the faint stars would be much more tightly concentrated into a bright nucleus, which would have been easily visible with the *Hubble* telescope. The true image on the left clearly lacks such a strong central concentration of light, or "luminosity cusp," implying that M15 does not likely have a massive black hole in its midst. *[NASA and T. Lauer]*

gravity capable of swallowing entire stars. Though the media took little note of it, the true power of Space Telescope was now emerging—especially its ability to test conclusions drawn from a limited, ground-based perspective. And not disappointingly, it was compelling astronomers to return to a more sensible appraisal of nature. Emerson would have been amused.

In late September, a remarkable apparition took shape in the sky. A huge bright spot was sighted just north of Saturn's equator, evidently a major atmospheric storm brewing on the planet. As is often the case, this celestial discovery was made by an amateur astronomer, this one in New Mexico. (The great supernova of 1987 was also first spotted by an amateur astronomer in Chile, and Comet Levy by an amateur stargazer in Arizona.) Unlike Jupiter's famed Red Spot, the Saturnian spot was white. And it was immense, at its maximum some 35,000 miles (50,000 kilometers) across, or several times the size of Earth!

In the days following the discovery, many professional astronomers at ground-based telescopes around the world departed from their prepared observing sessions in order to track the storm, which was seen to be spreading along the planet's equatorial region. The effects of shadowing seen in some of the pictures suggested that it was climbing higher in the cloud deck. But none of the ground-based telescopes could see the spot much more clearly than its discoverer's homemade 6-inch telescope. (Again, large ground-based telescopes collect more light and are therefore more sensitive to faint sources, but being inside Earth's atmosphere they do not resolve phenomena in the Universe much better than does a backyard telescope.)

The wide-field and planetary camera team, by now having come around to the notion that the earlier *Hubble* image of Saturn was superior to anything achievable from the ground, began cajoling project officials for more observations of the ringed planet. We at the Science Institute joined in, arguing that the development of "The Great White Spot" was an event of rare scientific importance and that we should depart from the scheduled set of engineering tests then under way. Lots of discussion ensued, back and forth, again like a tennis match, but, once more, none of the NASA managers wanted to take responsibility for ordering the observation; the media might find out about it, and if it failed the helmsmen would be plastered against the wall by screaming administrators farther up the NASA chain.

The issue was further complicated, for time was of the essence: Within a couple of months, Saturn would enter the area around the Sun that *Hubble* must normally avoid lest its optics be harmed. When the planet emerged, a half-year later, the spot might be gone. Finally, after weeks of haggling with the space agency, Giacconi let loose with another of his operatic roars—"In the interest of science, *just do it!*"—thereby declaring another "target of opportunity," for which he would grant some of his (still nonexistent) director's discretionary time.

Starting on November 6 and extending over half a week, Bill Baum of the Lowell Observatory joined Jim Westphal and others in using the planetary camera to capture forty-eight exposures in blue (4,390 angstroms) and near-infrared (8,890 angstroms) light—colors chosen to aid monitoring of the billowing White Spot and its associated clouds in Saturn's upper atmosphere. The hope was that we might even have a chance to track the vertical growth of the vast cloud, since the upper decks of the cloud might be seen best in short-wavelength blue light, whereas the lower decks should be penetrable by the longer-wavelength infrared radiation.

The resulting *Hubble* image clearly depicts a very turbulent atmosphere, similar to the cloud system that trails the Red Spot on Jupiter. And it was clear that Coriolis forces and wind shear had smeared the spot into a broad white equatorial belt. Much as NASA launches and deploys canisters of glowing chemicals to study the dynamics of Earth's jet streams, the spreading Saturnian White Spot resembled a milky dye in a stream of virtual fluid, allowing astronomers a rare opportunity to trace the cloudlike disturbances in Saturn's otherwise bland atmosphere.

Similar large-scale outbursts had occurred in 1876 and 1933, but no other Saturnian event (or "burp," as it is called) since then had erupted on such a wide scale. Now, a half-century later, we could study the fine structure and complex turbulence of a Saturnian storm, witness the sudden outbursts of minor spots swirling around the primary one, and test the proposition that atmospheric bands detected on the Jovian planets may have had their origin in the dispersal of such spots. If so, why has Jupiter's Red Spot remained more or less intact for at least three centuries, while Saturn's White Spot was already dispersing into an equatorial band within a few weeks? "With observations of this superb quality," declared an excited planetary expert, "we are into the realm of meteorology." At least as regards planetary observations, *Hubble* was living up to its prelaunch billing, enabling us to trace synoptically such evanescent phenomena as cloud detail and variable wind speed on other worlds.

As this first set of observations caused considerable excitement among both planetary astronomers and atmospheric scientists, official NASA warmed to the task, and we took advantage of the thaw to schedule no less than 160 images (in red, green, and blue filters) to be executed later in the month, over the course of more than fifty nearly consecutive orbits. By the time Saturn entered the solar-avoidance zone, we had acquired a vast volume of telemetered bits—in fact, enough data to make *Saturn: The Movie* in color.

Snapping pictures of Saturn almost continuously around the clock for more than three full days truly was like drinking from a firehose. Given the fact that Saturn rotates once on its axis about every ten hours, and that *Hubble* revolves around Earth about once every hour and a half, we had sufficient data to follow several rotations of the planet. With the finished movie in hand—initially assembled by Ed Groth at Princeton and then refined by Dana Berry in the astronomy visualization lab at the institute—we were able to watch the spot spread almost completely around the planet, in what was probably a once-in-a-lifetime event. Major-league exploration, to be sure!

We now know that the White Spot had been raging on the ringed planet

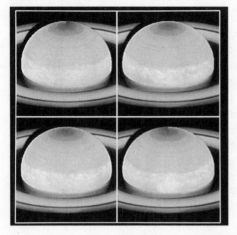

These four computer-enhanced images of Saturn were frame-grabbed from a time-lapse sequence of numerous images acquired with *Hubble*'s planetary camera throughout the course of a complete planetary rotation. These and nearly a hundred other frames allowed astronomers to follow the scalloped remnants of the White Spot across the face of the planet, watching them evolve while interacting with other gases in Saturn's upper atmosphere. The level of detail is unprecedented for this giant planet, for even the *Voyager* fly-bys in 1981 did not record this kind of tumultuous activity. *(See also color insert section.)* [*NASA, J. Westphal, and E. Groth*]

when the first *Hubble* image of Saturn was acquired months earlier. The storm was at the time confined to a small location on the opposite side of the planet. As Saturn's rotation period is less than half an Earth day, *Hubble* had missed making its first genuine discovery by only a few hours. Not even Lady Luck was with us.

Early that fall, the two teams for the spectrographs on board *Hubble* had been making a serious effort to address their science agendas. As the imaging cameras became increasingly productive, the principal investigators for the high-resolution spectrograph and faint-object spectrograph naturally sought opportunities to use their instruments, which had been partially calibrated during late summer. One thing was certain: the spectrograph's more esoteric data—bright and dark lines or bands in an object's spectrum—would be trickier to understand than the picturesque images.

Spectroscopy is an observational technique enabling scientists to study the nature of matter by the way its constituent atoms (and sometimes molecules) emit and absorb radiation. This technique has become an indispensable tool of modern astronomy, enabling scientists to identify and analyze the physical and chemical properties of objects much too far away to study directly. Such experiments are done with the aid of an apparatus known variously as a spectroscope, spectrometer, or spectrograph. This equipment can be as simple as a prism, which disperses light into its component colors, or it can be a more complex device (a radio receiver, or an X-ray diffraction grating) capable of dissecting invisible radiation. Basically, the modern research spectroscope comprises a telescope to capture spectral features, a dispersing device to display them, and a detector to record them.

Hubble's high-resolution spectrograph concentrates entirely on ultraviolet spectroscopy: its two 512-channel Digicon electronic detectors are deliberately blind to visible light, enabling astronomers to study subtle ultraviolet features from bright stars. One tube is sensitive to wavelengths ranging from 1,100 to 1,700 angstroms, and the other to wavelengths from 1,150 to 3,200 angstroms. In addition, this spectrograph can detect light pulses as rapid as 100 milliseconds (a tenth of a second) apart.

The high-resolution spectrograph has three resolution modes—although here we are speaking of the ability to discern the fine details of spectral lines, not of any imaging capability. This spectrograph's poorest resolution is better than the best resolution attainable by its companion, the faint-object spectrograph; the HRS's low-resolution mode can measure spectral features having widths of about an angstrom. For example, at 1,200 angstroms, a spectral feature deep in the ultraviolet from a nineteenth-magnitude object is designed to be resolved to about 0.6 angstrom. Medium resolution achieves about 0.1 angstrom; thus, the same spectral feature can be examined to a precision of 0.06 angstrom, but in this case the object must be brighter than the sixteenth magnitude. The high-resolution mode can map spectral lines with superb detail, enabling that same feature from a fourteenth-magnitude object or brighter to be studied down to 0.01 angstrom. This latter resolution is unprecedented in space astronomy missions.

The instrument's various modes are actuated by prismlike gratings mounted on a carousel and controlled by *Hubble*'s NSSC-I computer; the HRS is the only science instrument on board that does not use its own microprocessor. The carousel resembles that on a home slide projector, though an expensive one, for it divides its circular drum into 65,536 slots for accurate optical alignment—further testimony to the complexity of *Hubble*'s science-instrument payload.

The numbers just discussed are design goals specified by the high-resolution spectrograph's builder, Ball Aerospace. But some of them necessarily changed, owing to the aberrated performance of the telescope. In brief, *Hubble*'s inability to focus most of its captured light into a small, 0.1-arc-second

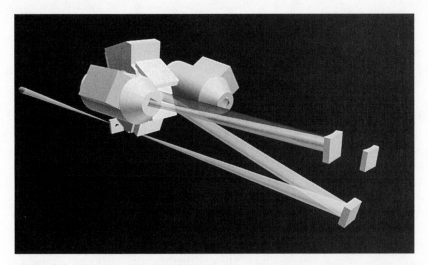

A sketch of the optical innards of the high-resolution spectrograph aboard the Hubble Space Telescope. Light enters from left, hits one or more mirrors at right, is reflected back to a diffraction grating (which like a prism splits light into its component colors), and is eventually captured by one of two canister-like detectors. Actual device records radiation only in the ultraviolet part of the spectrum. *[D. Berry]*

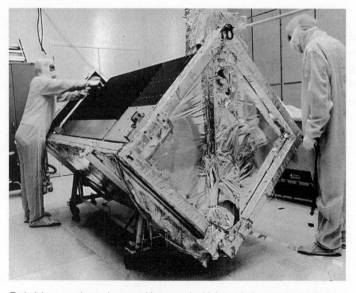

Technicians are shown here working over the high-resolution spectrograph prior to installation aboard Space Telescope. The telephone-booth-size instrument measures 3 x 3 x 7.2 feet and weighs on Earth nearly 700 pounds. *[NASA]*

domain meant that the HRS was less sensitive than expected. As with the cameras, the principal loss was to sensitivity, but for technical reasons aberration also degraded the effective resolving power of the spectrograph, some of which could be recovered by means of spectral-deconvolution computer algorithms. This "hit to sensitivity" was especially severe regarding the narrower of this spectrograph's apertures (0.25 arc second); since much of a star's light was blurred over several arc seconds, a good deal of the light was scattered outside the aperture and was therefore not detected.

In addition, *Hubble's* pointing problems made it hard to place the spectrograph's small entrance slit precisely on the specific desired target. After all, this tiny entrance aperture is only about one-seventy-fifth of an inch across, 0.25 arc second being the apparent size of a dime viewed from a distance of 8 miles.

Hubble's first spectroscopic observation of a science target was made with the high-resolution spectrograph early in the morning of October 4. This was the night before we hosted our first postlaunch Science Writer Workshop at the Science Institute. Although all the presentations that day highlighted early results obtained with *Hubble's* cameras, I thought it might be nice to let the assembled journalists know that the spaceborne observatory was beginning to do spectroscopy too—and to show them the historic first spectrum.

Again, my hidden agenda was to create an atmosphere conducive to the reporting of warm, human-interest stories, something that the journalists might write about anecdotally to enliven their factual accounts. I pulsed the NASA public-affairs system for permission to allow the reporters to visit the Science Institute's OSS area for a quick look at the first spectrum, which had

been displayed hours before on our monitors. "No, no, the press has no business there. Keep them out. Your OSS area remains off-limits to the media,"
was typical of the response from Goddard management.

But, by then, most of us at the Science Institute were feeling more confident, given that we had led the imaging campaign to an unqualified success.
We were especially flush that day, when a barrage of early science results were
being presented to all the world. I conferred with Giacconi and our operations chiefs (all of whom welcomed my media initiative), and then approached
Goddard scientist Dave Leckrone, a key member of the HRS team, and a rare
man who can communicate well with both technical and nontechnical people.
I explained to him my motives for allowing the press to see the first spectrum
and he immediately agreed, saying that he would be happy to explain the significance of the observation to the journalists. I dared not tell this NASA
employee that his own space center had ruled against the visit, as I had
resolved to treat NASA's admonition, once more, with benign neglect.

When I announced during the morning workshop session that there
would be an opportunity to tour the institute's operations areas and to see
Hubble's historic first spectrum, several senior NASA officials present were not
amused. Still, the workshop was generating a momentum of its own, and they
could hardly object without looking openly bureaucratic and perhaps a little
foolish.

For some, the tour of the OSS area was the highlight of the workshop. A
few dozen science writers were finally being allowed to see the innards of some
of *Hubble*'s control rooms and to talk with the people who work the telescope's central nervous system. The journalists' wide-eyed grins were an interesting contrast to the glum expressions on the faces of the NASA officials who
accompanied us on the tour. These officials stuck to me like glue, listening
attentively to every word I uttered, and I figured that I would soon be taking
some heat for my enterprise. But in the end, nothing came of it. By this time,
NASA had apparently become accustomed to the unexpected behavior of the
mavericks at the Science Institute. Or maybe the agency, too, was beginning to
realize the value of a more open and honest relationship with the press.

Frankly, what we saw on the monitors that day was not terribly exciting
scientifically—as was true also of the cameras' first observations of innocuous
star clusters, back in May and June. But it was first light for *Hubble*'s spectrographs, and for that reason the sight of a rich set of spectral features spread
across the computer consoles was gratifying. The target was a lukewarm star
named Chi Lupi, a calibration source known from previous observations with
the *International Ultraviolet Explorer* satellite to display a complex and puzzling absorption spectrum abundant in the element mercury. (By "lukewarm"
is meant a surface temperature of about 10,000°C, which is still hotter than
our Sun's surface of nearly 6000°C.)

A single exposure was taken for 5.5 minutes, centered at the ultraviolet
wavelength of approximately 2,000 angstroms. The spectrograph's wider
aperture, of 2 arc seconds, was used, in the hope that Chi Lupi would appear
somewhere within its field of view. (A subsequent attempt to observe the

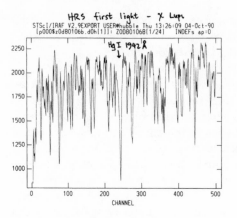

HRS first light - χ Lupi

STScI/IRAF V2.9EXPORT USER●hubble Thu 13:26:09 04-Oct-90
[p000$z0d801066.d0h[1]]: ZODB0106B[1/24]   INDEFs ap:0

Hg I 1942 Å

CHANNEL

This is a tracing of the first spectrum acquired by *Hubble*—a milestone "first light" for spectroscopy—just as it was telemetered on October 4, 1990, to the Science Institute's OSS area. The spectrum was obtained through the high-resolution spectrograph toward the star Chi Lupi, centered near an ultraviolet wavelength of 1,942 angstroms. As with all spectra in this book, intensity is plotted vertically, wavelength horizontally. Hg is the chemical symbol for the element mercury. *[NASA and D. Leckrone]*

spectrum of Chi Lupi through the 0.25-arc-second aperture failed because of erratic pointing.) The resulting observation provided the highest-quality spectrum of any star (save the Sun) ever obtained in the ultraviolet part of the spectrum, displaying clear evidence for such rare elements in the star's atmosphere as platinum, arsenic, and ruthenium. Although it had been a long time coming, the tracing of jagged curves and bumpy lines was a notable beginning for the spectroscopists who were used to seeing their data take a back seat to the more glamorous results of the imaging scientists.

As Leckrone briefly interpreted for the journalists the spectrum on the OSS monitors, it was clear that this small goodwill gesture had gone a long way toward sharing with the media a peek at the process and methodology of big science. Although I can't recall any journalists emphasizing the spectral results in the stories they filed, the many upbeat articles recounting the telescope's improved performance and new science images were clearly colored by the hospitable reception of their authors in the institute's operations areas. The project's siege mentality was beginning to lift, if forceably, and the positive achievements of *Hubble*'s science results would hereafter be balanced against the negative aspects of its continuing engineering ills.

A few weeks later, the high-resolution spectrograph acquired its first exciting science result. The target was a star named Melnick 42, an alluring celestial object in the Large Magellanic Cloud less than an arc degree north of 30 Doradus and SN1987A imaged earlier. This rare star is thought to be young and massive—perhaps only 2 million years old, compared to our Sun's estimated age of 5 billion years, and housing perhaps as much as a hundred times the mass of our Sun. It is also known to have a surface temperature of about 48,000°C (nearly 90,000°F)—some eight times hotter than our Sun—and is one of a class of stars known as "torrid blue supergiants." Melnick 42 shines about a million times brighter than the Sun, and in fact is one of the brightest stars known anywhere, making it an excellent candidate to explode as a supernova within the next few million years. Given the 170,000 light-years' distance, of course, this star's pyrotechnics might already be racing across the skies.

One scientific objective of the observation was to study how the chemical makeup of hot stars influences the way in which they evolve on their inex-orable road to catastrophic explosion. Knowing that many stars in the Large Magellanic Cloud, in contrast with those in the Milky Way, seem to have low abundances of elements heavier than helium, we thought prior to this *Hubble* observation that such stars must have weak stellar winds. But these new data seem to contradict this expectation.

A second, general objective of the high-resolution spectrograph team is to understand a star's atmosphere in intimate detail—a daunting task, for which astronomers have not had much success to date. Computer modeling of stellar atmospheres is plagued by uncertainties regarding the amounts of heavy metals, which play crucial roles in the thermodynamic and radiative properties of atmospheric gases. Melnick 42 is surrounded by other stars in the cloud having only about a tenth of the heavy-element abundance of our Sun, and is regarded as a prime touchstone for testing theoretical models of stellar atmospheres.

Before obtaining a spectrum of Melnick 42, Jack Brandt and Sally Heap, the co-principal investigators for the HRS, ordered up a target-acquisition image of the star with the wide-field and planetary camera. This image con-firmed that the star was indeed a single star, not a member of a multiple-star cluster—a possibility that lingered since its mass was so large. The image also specified the position of the star with greater accuracy than then known. By repositioning the spacecraft so that the image of Melnick 42 was at the very center of the HRS's field of view, and further checking that maneuver with the instrument's 2-arc-second aperture, the device was then stopped down to its 0.25-arc-second aperture, thereby focusing the bright core of the star's fuzzy image into the tiny entrance hole of the spectrograph. This minute field of view excluded light from stars near Melnick 42, which have always ham-pered ground-based observations and which confuse the spectral analysis.

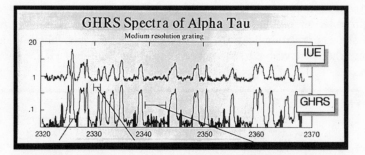

A twenty-minute exposure with *Hubble's* high-resolution spectrograph, shown at the bottom centered at the ultraviolet wavelength of 2,345 angstroms, is seen to be far superior compared with a nearly ten-*hour* expo-sure (at top) taken with the *International Ultraviolet Explorer* satellite now operating in geosynchronous orbit. Both were obtained toward the relatively cool, red-giant (4,000°C-surface-temperature) star Alpha Tau, popularly known as Aldebaran, during early checkout phases of *Hubble's* spectroscopy experi-ments. Despite the telescope's faulty optics, the *Hubble* spectrum is clearly more sensitive, showing many new and fully articulated spectral features, among them many emissions of the element iron from the star's chromo-sphere, or inner atmosphere. Although only an early test, this science assess-ment observation of Alpha Tau produced the highest-quality ultraviolet spec-trum ever obtained of a cool star. *[NASA and K. Carpenter]*

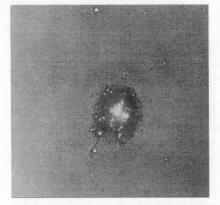

This is a raw, target-acquisition image of the area surrounding the thirteenth-magnitude star Melnick 42 inside the Magellanic Cloud, taken in the ultraviolet with *Hubble*'s planetary camera. The larger cluster of stars toward the bottom of the frame is the 30 Doradus region discussed earlier. The smaller, isolated point of light, about 8 arc seconds to the north (top), is Melnick 42, and it was toward this star that an ultraviolet spectrum was obtained during one of *Hubble*'s first spectroscopic studies. [NASA and J. Brandt]

The resulting spectrum, exposed for about twenty-nine minutes, was wonderfully rich in the kind of detail never before seen by astronomers. The most obvious spectral features observed (in absorption, since some of the light from Melnick 42 is blocked by gases between it and Earth) are those of triply ionized carbon (i.e., three of its six electrons have been stripped away). The shape of the absorbing features suggests a surprising finding: Melnick 42 is apparently shedding its hot gases at a furious rate. This "stellar wind," derived directly from the spectral observations, is estimated to reach speeds of several million miles per hour. Knowing both the rate of mass lost and the amount of starlight absorbed by the gas in the stellar wind, we find another surprising result: Melnick 42 is radiating away an amount of gas equal in mass to our Sun every hundred thousand years. Hence the reason why this supergiant star is not likely to last too much longer, at least in the cosmic scheme of things.

The most spectacular of the early-release observations with the high-resolution spectrograph enabled astronomers to probe indirectly a nearby star system thought to be surrounded by a protoplanetary disk—a thin plane of loose matter possibly coagulating to form planets. Beta Pictoris is a fourth-magnitude star (and therefore visible to the naked eye) in the Southern Hemisphere, esti-

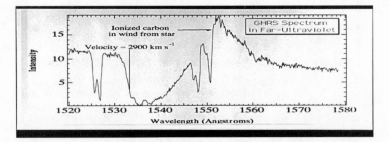

This graph shows a portion of the ultraviolet spectrum of the star Melnick 42, taken with the high-resolution spectrograph on board *Hubble*. The deepest and widest absorption feature (at about 1,535 angstroms) is caused by triply ionized carbon gas, which is flowing from Melnick 42 at a speed of more than 6 million miles per hour (or nearly 3,000 kilometers per second). It is seen in absorption because the streaming carbon blocks some of the light emitted by Melnick 42 from reaching the Earth. [NASA and S. Heap]

mated to be a billion years old, or about one-fifth the age of our Sun. Approximately 53 light-years away, it is an ordinary dwarf star, much like the Sun.

In the early 1980s, an unusual excess of infrared radiation, suggestive of circumstellar material, was detected by the *Infrared Astronomy Satellite* toward Beta Pictoris, called "Beta Pic" for short. Subsequent ground-based observations revealed direct evidence for a highly flattened and symmetrical disk surrounding the star and reflecting the light from it. The disk is at least 40 billion miles (65 billion kilometers) across, or some 300 times the distance between the Earth and the Sun. By using special occulting devices that diminish much of the star's glare, astronomers have in recent years been able to explore the inner parts of the disk, as close as a few arc seconds to the star itself; the structure closer to Beta Pictoris is blocked by the occulting device. The disk dramatically increases in brightness toward its center and then seems to drop off—a gap in the disk where, just perhaps, planets might have swept up the loose matter.

We had anticipated the day when *Hubble* would probe regions of the disk that were much closer to the Beta Pictoris central star, and also detect fine structure in the disk itself, possibly even the faint glow of protoplanets. Alas, given *Hubble*'s problems with its pointing system and especially with its optical system, any imaging of Beta Pictoris would be worthless; the scattered light from the telescope's spherical aberration would strongly interfere with the reflected light of the disk, making any interpretation highly suspect. The

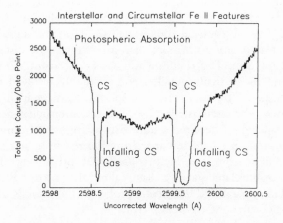

This is an expanded plot of a small part of the ultraviolet spectrum sampled by *Hubble*'s high-resolution spectrograph toward Beta Pictoris. The broad absorption feature stretching from one end of the plot to the other is caused by trace amounts of hot ionized iron gas in the star's photosphere, just inside its lowest atmosphere. Superposed on this wide feature are much narrower lines of cooler ionized iron gas. Some of this cooler gas is in the circumstellar disk (labeled CS) surrounding Beta Pictoris, and the rest resides in interstellar (IS) clouds somewhere along our line of sight. It is the circumstellar gas that is of greatest interest; this spectrum contains evidence that some of this gas is falling toward the star. Note that the entire spectrum spans only 2.5 angstroms, which is the typical width of many stellar spectral lines; the resolution and richness of detail in this spectrum are unprecedented for space astronomy. *[NASA and A. Boggess]*

search for planets around other stars simply could not be done with an unhealthy *Hubble*. Even so, the orbiting observatory was able to contribute to our knowledge of this fascinating object in the nearby sky.

Conjecture has it that the disk around Beta Pictoris is made of coagulated particles at least a micron in size, which is a little bigger than a millionth of an inch. If the disk's matter were extremely fine dust, like that commonly found in interstellar space, it would scatter light differently, making it bluer, like our daytime sky. By contrast, the agglomeration process is apparently in an advanced state. As the clumped particles plow their way through the disk around Beta Pictoris, they accrete increasing amounts of matter, much like a snowball thrown through a blizzard; the fine interstellar dust sticks gradually to become sand, which in turn becomes pebbles, and eventually planetary-size bodies.

In this way our own solar system almost certainly accreted from the thick dust grains that formed a circumstellar nebula accompanying the birth of our Sun, some 5 billion years ago. Ground-based studies were unable to determine if protoplanets had already formed in the disk around Beta Pictoris. Nor had they been able to specify the composition of the particles, though they likely contain silicates, carbonaceous materials, water ice, and some trace amounts of metals. And this was where Space Telescope could make its mark.

In order to take the cleanest possible spectrum of the Beta Pictoris system, astronomer Al Boggess and colleagues used the high-resolution spectrograph in its highest resolution mode; to further minimize confusion in this complex celestial environment, they also used the smaller aperture (0.25 arc second across). The observation thus pertains only to the inner part of Beta Pic's disk, an area spanning no more than a few times the distance between the Earth and the Sun. The resulting ultraviolet spectrum, centered around 2,600 angstroms, is so richly detailed and so highly resolved that one can clearly distinguish, for the first time, the spectral features attributable to the star's photosphere, the circumstellar disk, and the interstellar medium. "It's a spectroscopist's delight," declared a university astronomer from the Midwest. Most of the spectral lines detected arise from ionized iron, and all are seen in absorption; the light from Beta Pictoris itself is absorbed both by its circumstellar disk and by unrelated galactic clouds along the line of sight. Subsequent analysis provided unparalleled diagnostic probes of the velocity and density of the gas in the disk surrounding Beta Pictoris.

Among the most interesting findings is evidence that some of the circumstellar gas is falling in toward the star, whose formative stage is apparently still under way. Most interesting of all, the spectrum seems to change with time. This is especially true of the shape and intensity of the absorption lines arising in the circumstellar disk. As the disk revolves about Beta Pic, the absorbing matter between us and the star naturally changes. If the spectrum displayed little change with time, then we would conclude that the absorbing matter must be smoothly spread throughout the disk. But the fact that it changes in less than a month, with some absorption features appearing and disappearing altogether, implies that the disk is made of clumped matter. Drawled Boggess as excitedly as he could, "This is the closest thing we've seen anywhere in the

This is an artist's conception, not a Space Telescope image, meant to interpret *Hubble*'s spectroscopic observations of Beta Pictoris. This peculiar star was long believed to have a disk surrounding it with new planets in the midst of forming, but the new *Hubble* data suggest that this illustration is more the case. Here, we see a stable disk of gas and dust with huge, comet-like objects falling in toward the central star. *[D. Berry]*

Universe to what we think our own solar system looked like billions of years ago."

Might there already be kilometer-size protoplanetary chunks, or maybe even fully formed planets, orbiting Beta Pictoris? Now that *Hubble*'s high-resolution spectrograph can provide us with exquisitely detailed spectral features, astronomers will undoubtedly continue to monitor the ultraviolet spectrum of this intriguing star system. But the acknowledged discovery of planets beyond our own solar system will probably have to await better imaging capability than what *Hubble* could provide, for many people—including astronomers—will not likely be convinced that other planets have been found until a camera actually takes a picture of one.

Hubble's faint-object spectrograph, as its name implies, is designed to examine fainter cosmic objects than the high-resolution spectrograph, and it can study them across a much wider portion of the electromagnetic spectrum—from deep in the ultraviolet (around 1,150 angstroms) through the visible spectrum and a little beyond (to about 8,000 angstroms). The trade-off, again as implied by their respective names, is that the FOS cannot resolve spectral features as clearly as can its counterpart instrument, the HRS.

The faint-object spectrograph uses two Digicon (silicon diode) sensors to make simultaneous measurements at 512 closely spaced wavelengths. The "blue" tube is sensitive from 1,150 to 5,500 angstroms, while the "red" tube covers 1,800 to 8,000 angstroms; tube selection is made by telescope pointing. On-orbit tests show that oxide contaminants had built up within the instrument, apparently during the half-dozen years when it was hibernating inside Space Telescope awaiting launch. Accordingly, ultraviolet sensitivity shortward of about 2,200 angstroms was greatly degraded—"the price paid for delay," according to Rich Harms, the instrument's principal investigator.

Light can enter the faint-object spectrograph through any of eleven different apertures, from 0.1 to about 4 arc seconds in diameter. There are also two occulting devices that can block light from the center of a source, while allowing its peripheral light to pass through. These occulting devices permit, for example, the analysis of the shells of gas around giant stars or of the faint galaxies suspected to host quasars. This spectrograph can also detect in its targets light pulses as rapid as 40 milliseconds (or one-twenty-fifth of a second).

Built by the Martin-Marietta Corporation, the faint-object spectrograph has two modes of operation. At low-spectral resolution, it is designed to reach twenty-sixth magnitude objects in a one-hour exposure with a resolving power that can trace out spectral features no narrower than about 10 to 20 angstroms. In such a mode, it would miss, for example, much of the detail just discussed for the Beta Pictoris system. In its high-resolution mode, it is supposed to reach only about twenty-second magnitude in an hour exposure, but here it has a resolving power capable of tracing spectral features as narrow as 2 to 3 angstroms. These numbers aside, the FOS's great advantage is that it can acquire spectra toward celestial objects much fainter than can the high-resolution spectrograph. Thus, like *Hubble*'s two cameras that have complementary capabilities—one to image great detail and the other to take in the "bigger picture"— the two spectrographs on board *Hubble* also complement each other nicely.

However, owing to *Hubble*'s faulty vision, there was a loss in sensitivity,

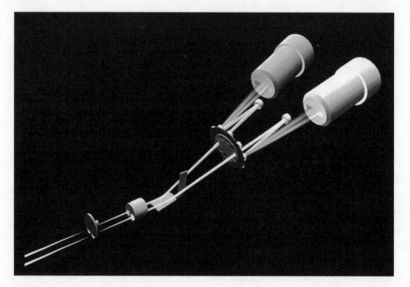

Schematic diagram of the salient features of the faint-object spectrograph aboard Space Telescope. Light enters at bottom left, passes through collimating devices, reflects off mirrors, through a filter wheel, off more mirrors, and is dispersed into its component colors before finally entering twin red and blue detectors. The instrument measures 3 x 3 x 7.2 feet and weighs on Earth 680 pounds. *[D. Berry]*

or "throughput," amounting to two or three magnitudes, and the spectrograph's design specifications must be adjusted accordingly. In addition, some of the earliest tests showed that its sensors, especially the red detector, had not been adequately shielded by the manufacturer. Consequently, the Earth's magnetic field interferes with the spectrograph's magnetic focusing device (much like that on an ordinary television), causing the observed spectrum to exhibit a cyclical drift back and forth in wavelength during each orbit of *Hubble* around the Earth. This in turn degraded resolution and complicated calibration, making some planned high-resolution observations difficult at best.

The target object for the FOS's inaugural science spectrum was a faint (seventeenth-magnitude) quasar, catalogued dandily UM675, billions of light-years distant from our Milky Way and receding from us at more than 80 percent of the speed of light. Thus, the quasar launched its radiation when the Universe was about a fifth of its present age; alternatively stated, it must be about 12 billion light-years away, assuming a universal age of 15 billion years. UM675 can be seen in large ground-based telescopes only as a dim starlike object, some 6 million times fainter than the bright star Vega. This truly distant celestial source was chosen for two reasons: First, it would provide a realistic science assessment of the telescope's ability to acquire spectra of celestial objects near the limits of the observable Universe. Second, if successful, the effort would provide insight about the chemical composition of the truly remote Universe. And given that "looking out is looking back," we would be using *Hubble* as a time machine more powerfully than at any moment in the six months since launch.

Specifically, by measuring the strength of a particular spectral signature arising from helium, the second most abundant element in the Universe, seen in absorption as UM675's light passed through clouds associated with the quasar as well as through unrelated intergalactic clouds along the line of sight, we should be able to tell if most of the helium we see today was made in the immediate aftermath of the Big Bang or more gradually, in fusion events at the cores of stars ever since that creation event. This experiment could not have been performed before the *Hubble* mission, because the spectral feature sought normally occurs in the largely unobservable deep ultraviolet, at nearly 600 angstroms, where very few scientific instruments have successfully operated.

Although wavelengths shorter than about 1,000 angstroms are also beyond *Hubble's* range of detectability, the great speed at which UM675 is moving away from Earth would cause the helium line in question to be Doppler shifted substantially, enough to put it into the viewing range of *Hubble* at around 2,000 angstroms. (This redshift—a z of 2.1 in astronomers' slang—is commonly attributed to the expansion of the Universe, and thus the distant quasar should not be visualized as traveling through space as much as moving along with the expanding "spacetime fabric" of the Universe.)

None of this kind of work, incidentally, can be accomplished from the ground, regardless of how large the telescope. Earth's atmospheric ozone absorbs all the ultraviolet radiation reaching the outskirts of our planet, making it flat-out impossible to detect any such highly energetic photons on the surface of Earth.

Because UM675 is so faint, a long exposure was required in order to capture enough photons of light to see the expected pattern. In fact, this would be the longest observation to date with *Hubble,* and would be assembled from three 33-minute exposures. The red detector on the faint-object spectrograph was chosen for the observation, and was used in the low-resolution mode to guarantee enough signal to warrant analysis. Fortunately, *Hubble* was first commanded to display the results of a short (approximately two-minute) target-acquisition observation made using a large 4-arc-second aperture; this observation indicated that UM675 was about 1 arc second off-center; thus, *Hubble* was ordered, via real-time link to one of the TDRSS relay satellites high above, to adjust its sights by the required amount to center the object. The FOS team then began blindly taking data through a smaller 1-arc-second aperture for the main observation. Several orbits were dedicated to this task, and when the data were integrated into a single set of observations, the total exposure time exceeded one and a half hours.

To everyone's delight, the results seem cosmologically significant. A helium feature was possibly detected just where predicted, amid numerous spectral features caused by the absorption of UM675's light by various elements in many clouds along the line of sight to the quasar. The observation bodes well for one of the key projects of the *Hubble* mission: the "Lyman-alpha forest" program,

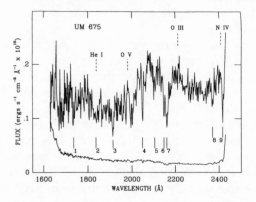

Ultraviolet radiation from quasar UM675 is here dispersed into its component wavelengths by *Hubble's* faint-object spectrograph. Although the spectrum is noisy—"groddy" in astronomer's lingo—several elemental signatures are identifiable, including possibly that of helium at 1,836 angstroms. This absorption feature normally occurs in the unreachable deep ultraviolet (584 angstroms), but because the quasar and its associated helium clouds are receding so fast, some lines in the spectrum are red-shifted (Doppler) enough for the light from the background quasar, robbed from our line of sight while passing through the intergalactic helium clouds, to appear within *Hubble's* observing range. Possible absorption lines are numbered below the spectrum, while predicted positions of emission lines (none clearly detected here) are shown by vertical dashes above the spectrum. *[NASA and M. Burbidge]*

whereby studies are planned of the rich absorption spectra often observed toward galaxies that light up the far away and the long ago.

The pursuit of this project is crucial, if we are to crack one of astronomy's foremost problems: the birth of galaxies. These grandest of nature's structures are scattered everywhere throughout the cosmos, by some estimates upward of 100 billion of them in the visible Universe alone. Yet, embarrassingly, we don't know where they came from. By using the remote quasars as probes of ancient epochs we may be able to decipher the evolutionary steps that led to galaxies. On its way to us, the light from a quasar will inevitably pass through youthful clouds laden with hydrogen, whose principal spectral line (at 1,216 angstroms) is called Lyman-alpha. And since the multitude of such primordial clouds along any quasar's line of sight gives rise to a wealth of absorbing features spread out along a spectrum—to some resembling a "forest" of lines—the so-called Lyman-alpha forest is thought to contain clues regarding dark, or missing, matter in the Universe.

This initial science test of the faint-object spectrograph was not meant to address directly the Lyman-alpha forest issue. In fact, by searching for a helium feature originating in the early Universe, it was intended to be even more challenging. If the helium-line identification is confirmed by more stringent observations (which will require longer exposures), then we can confidently conclude that the cosmic helium abundance early in the Universe was nearly the same as now. Thus, an innocuous set of wiggly lines would seem to be important for cosmology—the study of the origin, structure, and fate of the Universe: Most of cosmic helium was produced practically all at once shortly after the Big Bang. Apparently only token amounts have been forged in the hearts of stars since the Big Bang. This is a result that most astronomers anticipated, but it's nice to see measurements validating our long-held ideas.

"We've waited a long time, but I'm just ecstatic," declared Margaret Burbidge, one of the world's most eminent astronomers. Seeing this elderly woman and leading quasar expert so eager to share her team's results with everyone—she gave me a spontaneous and wonderful twenty-minute pep talk about her passion for this inaugural observation, displaying the infectious enthusiasm of a child—was a breath of fresh air in a project still exuding gloom and doom. Had NASA any savvy, the space agency would have utilized her to highlight the exciting UM675 results, for Burbidge has an innate sense of good cheer as well as a brilliant mind. She would have been an outstanding ambassador for the *Hubble* project.

Instead, the bastion of male conservatism that is NASA sent out a botched press release—half of the UM675 report stapled mistakenly to half of another, completely unrelated release about NASA's ill-fated aerospace plane. The result was much confusion among journalists who received the announcement, another comedy of errors on the public-relations front. Since Goddard had also neglected to inform the Science Institute of its announcement—so much, again, for teamwork—we were unsure what to tell the dozens of reporters who called the institute seeking clarification, lest we be accused of divulging proprietary data. A few days later, new press releases were hurriedly issued to amend for the Goddard slip-up, but by then these science findings were no longer fresh news, and the reporters were again snickering at the lack of professionalism. To my knowledge, this important science result received no media coverage.

In addition to its two cameras and two spectrographs, *Hubble* has on board an extremely sensitive photometer, which detects and counts photons of light one at a time. Somewhat like the light meter of any household camera, this instrument was added to the design of the orbiting observatory mostly as an afterthought, largely because virtually every telescope ever made on the ground has had a device dedicated to the accurate measurement of light intensity. The main advantage of *Hubble*'s photometer is its rapid-response circuitry, which allows measurement of light pulses occurring as frequently as every 10 microseconds. Since the turbulence in Earth's atmosphere masks the rapid light pulsations of certain types of celestial objects, it is usually not possible to detect from the ground brightness fluctuations with periods shorter than one second. The high-speed photometer should be able to examine objects as faint as the twenty-fourth magnitude, during a typical thirty-minute, half-orbit observation. Wavelength sensitivity is broad—from 1,200 angstroms in the ultraviolet, to 8,000 angstroms in the red end of the visible spectrum.

The high-speed photometer is the simplest instrument on board *Hubble;* it has no moving parts. To take advantage of its many filters and tiny entrance apertures (some as small as 0.4 arc second), the telescope itself must be aimed precisely, so that light can pass through the desired aperture-and-filter combination. More than 150 such combinations are possible, each selected solely by telescope pointing. Needless to say, given *Hubble*'s considerable shortcomings in this respect, use of the HSP became something of a challenge. During periods of jitter, it clearly could not be used effectively, since spacecraft fluctuations caused variations in a star's light streaming through a small instrument aperture, hopelessly confusing it with any intrinsic variation in the star's output; even during the quietest periods of orbital night, when the spacecraft stability met or exceeded design specifications, the photometer was usually affected adversely by the aberrated light reaching its detectors.

Partly because of these problems, and partly because Bob Bless, the photometer's principal investigator, rightfully became disgusted with the confused state of *Hubble*'s early life (which he had generally foreseen well before launch), no significant science was achieved with this instrument during the commissioning period. Bless, to my mind the elder statesman of the *Hubble* project, put it this way at the end of commissioning while addressing the Space Telescope Institute Council, one of our premier advisory bodies: "All GTOs carry around so much baggage—we are all tired of it. I want to put it all behind me. Most distressingly, there are too many lawyers and litigiousness lurking underneath astronomers' clothes."

Throughout the fall and into the winter, as *Hubble*'s space trials played out, several other fascinating science images were acquired under the aegis of the early-release observations program. Some of these efforts resulted in mundane findings, more useful for calibrating various filters, gratings, and observing modes than for the advancement of science. We also began to interweave some of the first guaranteed-time observations reserved for the principal investigators, who had led teams building *Hubble's* science instruments. These data for the most part would be governed by proprietary policies and thus would be held closely for at least a year. However, a few EROs images percolated through the data-reduction pipeline at the Science Institute, and we kept the pressure on to have them released as quickly as possible. I remained tenaciously convinced that the only way to recover *Hubble's* good name was to maintain that steady drumbeat of science results.

One of the earliest targets of the late-summer imaging campaign had been the active galaxy, PKS 0521-36, whose spectacular jet was discussed in the previous chapter. However, given this object's considerable distance of some billion light-years, even *Hubble* can see it only so well. Since many active galaxies are known from radio-astronomy studies to have similar jets protruding from their cores, we were eager to train *Hubble* on a much closer object of this type, so that we might better discern any intricate structure in the jet. As noted, such jets are taken by some astronomers to be signatures of, indeed to be pointing directly at, supermassive black holes, and these are among the cosmic denizens that *Hubble* was expected to probe with great fanfare. An alternative view is that black holes are not needed to power such jets; bursts of star formation could conceivably be their trigger. Our hope was that *Hubble* might be able to distinguish between these two very different origins for the active-galaxy jets.

The closest powerful extragalactic jet to us is perhaps the most famous of them all—the archetypical M87 active galaxy, located at the center of the Virgo cluster of galaxies some 50 million light-years away. Beginning well more than a year before launch, some of us had discussed imaging this most interesting "neighbor" as early as possible after launch. M87 became one of a handful of prime EROs targets—those sure both to fire the imagination of the public and to yield results that might significantly advance our knowledge of the Universe. But since three of *Hubble's* key players each wanted first dibs on this object, we were confronted with multiple conflicts of interest.

The principal investigator for the wide-field camera, Jim Westphal, wanted the initial peek, knowing that his camera's wide field of view and high resolution would likely show structure well beyond that visible with any telescope on the ground. The P.I. for the faint-object camera, Duccio Macchetto, likewise argued that he deserved first look, in that he is one of the world's leading experts regarding jets in active galaxies and much of his planned science program with *Hubble* concerned such objects. And the P.I. for the faint-object spectrograph, Rich Harms, maintained that black-hole physics would not likely be definitively addressed in M87 until a spectrum was obtained of the bright, inner parts of the jet, before which he needed an image for target acquisition purposes. Each of these distinguished scientists had a meritorious argument, yet each suspected that the very first image of M87 would be so spectacular that none wanted to give way to his colleagues.

The frontiers of science are awash in fears (but few examples) among scientists that they will be scooped by their colleagues, and such fears were certainly operating in this case. To settle the issue, for *Hubble* would have to explore M87 eventually, the three P.I.s did something highly unusual for scientists: they signed a gentleman's agreement (no, not in blood, as was rumored) that no team would publish an initial paper on M87 without consulting the other team's members, and preferably they would release their results simultaneously.

At a hastily called gathering a few months before launch in the lobby of a hotel on Cocoa Beach, outside the Kennedy Space Center, they informed me that they would support my contention that M87 should be a high-priority EROs target whose initial image should quickly be shared with the public. As things turned out, all the astropoliticking about a minute speck of light in the heavens was moot, since by the onset of the imaging campaign M87 had entered the zone surrounding the Sun that *Hubble* must normally avoid. No one, so early in the mission—not even I as "principal instigator" for the EROs program—wanted to manually override *Hubble*'s onboard safety algorithms to explore an M87 that was drifting on the sky ever closer to the Sun.

Since M87 would remain within the zone of solar avoidance until the spring of 1991, we opted to return to 3C66B, the active galaxy that we had tried to image three days into the imaging campaign only to be thwarted by *Hubble*'s entry into safemode for reasons that we do not understand to this day. At a distance of about 270 million light-years, 3C66B is farther away than M87 but closer than PKS 0521-36. Accordingly, it should display some noticeable structure within its jet, at least enough to whet our appetites for the day, months thereafter, when *Hubble* would be able to put M87 in its sights. The new data obtained toward 3C66B were not disappointing, as they provided intriguing new detail to aid our understanding of the central engines within such powerhouse galaxies.

The faint-object camera was brought to bear on the issue since this camera yields the highest angular resolution possible with *Hubble,* especially when used at the shortest wavelengths of its sensitivity spectrum. 3C66B was imaged in ultraviolet radiation, and the resulting picture clearly reveals intricate patterns within one of nature's most awesome phenomena. The narrow beam of glowing plasma ejected from the galaxy's heart displays structure heretofore unseen in any such jet. In general, these observations resemble PKS 0521-36, but there are significant differences in the details, suggesting that a different mechanism is at work.

Two kinds of computer algorithms were used by Macchetto and his group to enhance the view of the jet in 3C66B. First, the core of the host galaxy was electronically subtracted, to allow the jet itself to be seen more clearly; this can be accomplished with a high degree of confidence because such jets are known to be relatively bright in the ultraviolet compared to their host galaxies. Second, the image of the jet was subjected to reconstruction

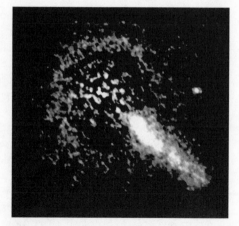

Hubble's faint-object camera was used to observe the 10,000-light-year-long jet of plasma emanating from the active galaxy 3C66B. In this computer-enhanced ultraviolet image, the host galaxy has been (mostly) sub-tracted to allow details in the fingerlike jet to be traced nearly all the way into the galaxy's nucleus. Resolution here is about 0.1 arc second, which is more than ten times better than previous ground-based optical images and three times better than any radio maps made of the jet. *[ESA, NASA, and D. Macchetto]*

techniques to minimize the effects of spherical aberration; in so doing, we were (and we remain) reasonably confident that the process of deconvolution did not unacceptably amplify noise peaks, which might be taken for real structures. The resulting image shows a 10,000-light-year-long jet containing gaseous filaments, bright knots, and a pair of unusual kinks. Many of these features had already been suggested by ground-based radio observations at much longer wavelengths. However, the *Hubble* image reveals a peculiar braided, almost double-helical structure, much like a twisted pair of wires. The helical strands of plasma are separated by about 500 light-years, and exemplify the kind of detail that simply cannot be discerned with even the most powerful optical telescopes on the ground.

The radiation in such cosmic jets is produced by electrons spiraling through magnetic fields at velocities close to the speed of light. As these electrons spiral outward along magnetic-field lines, they lose energy in proportion to their frequency and to the strength of the magnetic field. In a time span of only a few hundred years, the electrons responsible for optical and ultraviolet emission dissipate much of their energy. By contrast, the electrons that produce radio emission lose energy less rapidly and thus can survive in the same magnetic field for tens of thousands of years. Accordingly, most galactic jets that extend for thousands of light-years are detected at radio wavelengths. In only a few cases to date have optical counterparts been observed for such lengthy jets. What's more, since the optically visible parts of the jets are younger, we should be able to trace them back closer to the central engines that give rise to such cosmic "roman candles."

Among many puzzling aspects of the 3C66B observation, perhaps the most troubling is the following: How do the electrons remain energetic enough to radiate ultraviolet signals throughout their 10,000-year journey along the jet? One possibility is that the electrons experience repeated boosts in energy while propagating outward, perhaps by means of instabilities at the edge of the plasma flow, which might produce shock waves. Another possibility, which the jet's braided appearance suggests, is that the electrons spiral through a channel or along the edge of a nearly hollow "tube" having a much lower magnetic-field strength, and hence experience lower energy loss. Two sharp bends, or kinks, in the strand (at 3,000 and 8,000 light-years out from the nucleus) are especially hard to explain. At the least, they suggest that the galaxy's central engine does not release energy at a steady rate but rather "hiccoughs," or fluctuates in output. The kinks might also be produced by a complex magnetic field along the jet, or by collisions with dense regions in the intergalactic gas through which the jet is plowing.

Further Space Telescope observations, especially of the nearby M87 object, are virtually guaranteed to provide fundamentally new information about the nature of galaxy jets—in particular, how energy is transported from the central engine along such jets, and what role magnetic fields play in channeling matter from the core of a galaxy into intergalactic space. If we are on the wrong intellectual track in attributing the powerhouse cores of galaxies to supermassive black holes that act not just as sinks but also as sources of unworldly power, then we shall likely need to invent mechanisms so exotic as to require wholly new physics. As we bear down on one of nature's most challenging puzzles, the Hubble Space Telescope might well be the instrument that will lead us to that new physics.

Much closer to home, during early autumn we were able to take a quick look at part of the Orion Nebula—also known to working astronomers as M42 or NGC 1976. One of the most beautiful gaseous nebulae in the sky, this colorful object was originally among the high-priority prelaunch EROs targets, if only because we knew that almost every citizen has seen or at least heard of the constellation Orion and thus would be able to relate to a *Hubble* image of it more readily than to many of the other, more distant galaxies and star clusters.

However, early plans to observe Orion had been axed by the Science Working Group well before launch, for fear that even a single image might give away some of the nebula's secrets prematurely. Now, we were returning to it, partly to help further calibrate ("flat field") *Hubble*'s wide-field camera and partly to help us gauge *Hubble*'s science capabilities in a region where we thought we knew more or less what to expect. At a distance of 1,500 light-years, Orion is the nearest bright nebula, and this relative proximity enables astronomers to see its gaseous structure in "up-close" detail. We felt we understood this region pretty well—at least until *Hubble* took a look.

Even after all we had been through during postlaunch, one astronomer who had earlier complained that I had "gored his ox" by proposing Orion as an EROs target was now worried that the wide-field camera team was "poaching on his territory." Claiming proprietary rights to the Orion Nebula, he wanted to defer his planned and extensive observations until after *Hubble* was repaired several years hence, while prohibiting others from taking a quick look at a cosmic region that cries out to be examined with Space Telescope. In the end, the objector was ignored during the observation, yet was invited to join in the subsequent data analysis and interpretation.

The Orion constellation is one of the most recognizable star fields in the Northern Hemisphere's winter sky. Even with the naked eye, the three bright stars in the Hunter's belt can be readily discerned among other easily recognized stars, such as the old red giant Betelgeuse, to the north, and the young blue star Rigel, to the south. Below the belt lies the Hunter's sword, at the tip of which is a fuzzy patch of light that appears, even through a small telescope, a little like a glowing wad of pinkish cotton candy. This is "The Great Nebula in Orion," a genuine stellar nursery.

Largely on the basis of radio and infrared studies, astronomers reason that the Orion Nebula originated roughly a million years ago when a cluster of stars—notably four in the nebula's core, known collectively as the Trapezium—were born at the edge of a much larger, cool, dark interstellar cloud. These four stars are a good deal hotter than the Sun and therefore emit vast quantities of highly energetic ultraviolet radiation, which ionizes atoms of interstellar gas throughout several tens of cubic light-years. The resulting plasma glows continuously as its constituent electrons and protons recombine to form atoms (after which they ionize again, recombine, and so on).

Currently, the Trapezium stars are located in an open cavity on the front face of a giant cloud rich in molecules (of mostly hydrogen yet including some organics) into which an ionization front is propagating. The Orion region is also abundant in other young stars not quite as energetic as those in the Trapezium, and the nebula itself is regarded as a hotbed of activity, as stars form like condensing raindrops in the surrounding galactic ecology.

The wide-field camera was brought into play for these observations, and the resulting image revealed structures the likes of which we had never before seen in gaseous nebulae. Three exposures of ten minutes each were made, enabling a group led by Jeff Hester of Caltech to construct a color image much as the human eye would see—if we had *Hubble*'s vision, which, even with its myopia, is still several hundred times sharper than our own. The

At the left, the constellation Orion, "The Hunter," is notable for the three bright stars in the Hunter's belt, and for the Orion Nebula, which is at the end of the sword hanging from the belt. Shown here at right as an inset, the Orion Nebula is a rich region of star formation—young stars and protostars partly embedded within a dense and dark galactic cloud. This huge interstellar cloud of atoms and molecules, which is invisible and mostly behind the nebula, is the parent region that spawned the Orion Nebula about a million years ago. Both of these photos were taken through telescopes on the ground. [AURA]

observations were made in the visible part of the spectrum, using narrow-band filters that captured blue light from ionized oxygen (3,727 angstroms), red light from ionized sulfur (6,720 angstroms), and green light from neutral hydrogen ("H beta" at 4,861 angstroms).

The resulting color composite shows a thin sheet of glowing gas in the northwest quadrant of Orion, well beyond the core of the central exciting stars. Given the thermodynamic conditions (approximately 5,000°C) in this part of the nebula—intermediate between those in the interior nebula and those in the surrounding dark cloud—the light emission from the ionized sulfur is especially striking. Intermingled throughout a diaphanous morphology of thin translucent gases, discrete nebular features can be noted down to the limit of *Hubble*'s resolution—0.1 arc second, which at Orion's distance is about equal to the radius of our solar system.

Several colleagues immediately compared the plasma's intricate structure

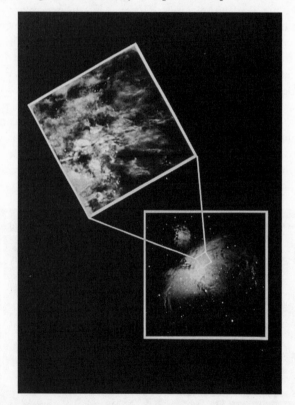

At bottom right, the image of the Orion Nebula is shown majestically as seen through a large ground-based telescope. The full expanse of the nebula spans some 10 arc minutes, which at a distance of 1,500 light-years is about 4 light-years in size. Only a small part of the parent galactic cloud from which the nebula originated is evident, as the tongue of a dark, dusty region wraps around the front of the glowing plasma to the east (which is conventionally to the left in astronomical images of nonplanets). Just to the west (right) of the dark tongue, where the nebula is brightest, lies the Trapezium—a cluster of four young hot stars that energize most of the nebula as their ultraviolet radiation streams forth to ionize the nearby gases. The superimposed square at upper left is the *Hubble* image also displayed in another figure in the color insert. *[AURA and NASA]*

to a rippled window curtain, as the most evident features seem to resemble long wispy sheets. In all, the new level of clarity reveals at least thirty knots of gas, which fluoresce under the flood of ultraviolet radiation emitted mostly by the nebula's Trapezium stars. Apparently, the light captured by *Hubble* traces the boundary between the hot, diffuse interior of the nebula and the adjacent dense, cool cloud. What we see in unprecedented—and unexpected—detail is the effect of energetic starlight and hot stellar winds causing cool, dense matter to "boil off" the face of the background dark cloud from which the stars originally formed, about a million years ago.

Of particular interest is a newly discovered jetlike feature best outlined by the emission from excited sulfur atoms. While similar jets have been detected with ground-based telescopes, the new observations reveal the structure of this jet with extraordinary clarity. Its length is about 15 arc seconds, or 0.1 light-year, and it appears to be streaming away from a young star in the late stages of stellar birth. (Such stellar jets have sizes much, much smaller than the jets observed to emanate from the cores of active galaxies.) Despite its gaseous makeup, the jet has rather sharp edges, possibly due to shock fronts at the boundary between the jet and the surrounding nebular gas. Such outflowing jets are of interest for they provide clues about how stars form and how young stars affect their environment.

Most astronomers regard such jets as a normal part of the stellar birthing process, a kind of initial howl as a newborn star initiates thermonuclear fusion and its light cuts through its placental envelope of dust and debris. Within weeks of having acquired the *Hubble* data, however, Bob O'Dell challenged this interpretation, arguing that the jet might not have any relation to the newborn star. Though the jet and the star appear aligned, there are no known velocity indicators in the area supportive of gas squirting from an infant star. The starry jet aside, one cannot deny the presence of long filamentary structures.

O'Dell compared *Hubble*'s penetration of Orion's insides to being in a cave, surrounded by vast stalactites and stalagmites of plasma. Other nebular experts urged caution, noting that some of the surprising structure might be artifacts introduced by the computer algorithm used to partly clean the image of its blurred vision. Clearly, more work is needed before we truly understand what nature has in store for us at this whole new scale of cosmic observation. But that's what science is all about and why it's so much fun. These are the kinds of unanticipated findings with which we suspected Space Telescope would make its mark. To quote again what Hubble, the man, once said while inaugurating another major telescope, on Palomar Mountain, a half-century ago: Astronomers generally expect "to find something we hadn't expected."

With this initial *Hubble* view, we have now sampled the insides of the majestic Orion Nebula at least ten times more clearly than ever before. This preliminary reconnoitering of a cosmic birthing place was a dramatic demonstration that the orbiting observatory, despite its debilitations, could still be used to obtain scientifically meaningful information with clarity far exceeding that normally possible from the ground. To be sure, it is in just this sort of morphological study of structure and geometry that *Hubble* is most valuable as a scientific tool. As someone who has observed the Orion Nebula extensively over the years, I can't wait to examine other parts of it—and especially to see the fully constructed montage of many *Hubble* images, which will give us a stupendous yet highly complex view of The Great Nebula.

Another gaseous nebula seen through *Hubble's* eyes displayed even queerer morphology. This was Eta Carinae, the number-one priority EROs target, whose planning we had labored over well before launch—a truly enigmatic patch of glowing gas about 9,000 light-years away in the Southern Hemisphere. The nebula's central star is known to be among the most massive and luminous in the Milky Way—about a hundred times as massive as our Sun and some 4 million times as luminous. Midway through the previous summer's imaging campaign, we had tried to secure some wide-field camera data of this region, but our attempts had succumbed to a string of pointing errors. Although the Goddard project office refused to authorize a revisit to EROs targets that had failed their initial observation, we managed to finagle a series of additional EROs objects onto *Hubble's* observing calendar in the telescope's "dead time"—periods when *Hubble* was otherwise idle and in between the ever-present engineering tests.

The *Hubble* image of Eta Carinae is so puzzling—so utterly unlike anything seen in the sky before—that it was nearly six months before astronomers had arrived at a consensus credible enough to release a picture and a description of Eta Carinae into the public domain. In this case, no hoarding of data or paranoid concerns held up the dissemination. Rather, the mysteries associated with this most peculiar region were deepened upon careful analysis of the *Hubble* image, as no one quite knew what to make of it.

Astronomers have long known that Eta Carinae is an extremely variable unstable star, or starlike object. Historical records show that during an outburst in 1843 it reached a magnitude of −1, flaring to become one of the brightest stars in the nighttime sky, second only to Sirius. It has since faded to the sixth magnitude (the limit of naked-eye visibility), while a cocoon of dust and gas from the 1843 outburst has expanded and obscured its emission of visible light. During the outburst, Eta Carinae also expelled copious amounts

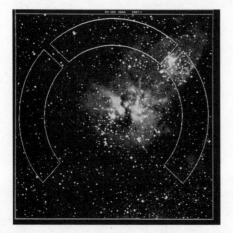

Ground-based navigational map for Space Telescope's guide stars at the perimeter of the Eta Carinae nebula. This finding chart is centered on the massive star thought to excite much of the nebula. *[Space Telescope Science Institute]*

of gaseous matter, visible from the ground today as a small, bright, oblong nebula expanding away from the star. Dubbed "the homunculus," because it has appendages vaguely resembling a head and feet, it measures about 0.7 light-year across its long axis.

Until somewhat recently, astronomers thought that the star itself might have blown to smithereens in 1843. However, infrared images taken in 1969 show that Eta Carinae is one of the brightest infrared objects in the sky, outdone only by the Sun and the Moon. This substantial glow of heat implies that the star is still intact and hidden inside a dusty cloak of its own making. (Much as red or orange light can be seen through terrestrial fog better than blue light can, infrared radiation can penetrate clouds of interstellar dust more effectively than visible light.) Apparently the central star merely "burped" some 150 years ago, the result being the odd-shaped homunculus now seen primarily as the dust-reflected light from the star.

Seven exposures were obtained with *Hubble's* wide field camera, using a filter that isolates light emitted by ionized nitrogen. The exposures were then combined and subjected to 180 passes through a deconvolution algorithm, in an attempt to make the image progressively cleaner for detailed analysis of its peculiar structure. By contrast, no other *Hubble* image in this book has been deconvolved more than fifty times. Such heavy-handed massaging of the data led some astronomers to question the extent to which some of the subtle features in the new Eta Carinae images are real or artificial. At a spring 1991 symposium at the Science Institute, the leading theorist Colin Norman challenged astronomer Jeff Hester, who had overseen this EROs experiment, and demanded to know, with some heat, why he "should spend months of my life trying to figure out the odd structures seen in this image if you cannot guarantee they're real." The controversy regarding the proper use of computer enhancement deepened when the paper announcing these results was rejected by the trade journal, *Science*, before being published in the *Astronomical Journal*.

The *Hubble* image of Eta Carinae shows that the homunculus—now looking more like a peanut shell than a man—is irregularly structured down to the limit of *Hubble's* resolution, here about 0.2 arc second. The smallest individual clumps of gas are about ten times the size of our solar system. The fact that the homunculus displays material condensations on such small scales as well as maintaining a well-defined edge suggests that it is a very thin shell,

This is among the best ground-based photographs of Eta Carinae, exposed in 1975 for one minute to the red light of the nebula and captured with the 160-inch (4-meter) telescope at the Cerro Tololo Inter-American Observatory high atop the Chilean Andes. North is at the top, east to the left, and the seeing conditions are about 1 arc second. The photograph displays evidence for material ejected from a massive, unstable star near the center; the double knot to the north, which is apparently connected to the main body of the object by a curved filament, is moving outward at 0.1 arc second per year or about 800 miles per second (approximately 3 million miles per hour). The "homunculus" is the bright inner nebula, about 0.7 light-year across its long dimension, which to some observers resembles a "little man." [AURA and N. Walborn]

rather than a filled volume, of matter. Such a dusty shell either could have been ejected in a single burst from the star or might have been swept up from matter surrounding the star. In either case, the *Hubble* image suggests that the shell fragmented—technically termed a Rayleigh-Taylor instability—as a result of the combination of radiation pressure and stellar wind.

Given the date of the explosion, and the distance from the central star to the ridge of glowing matter to the southwest (lower right), we can calculate that that part of the homunculus must be moving at a velocity of at least 3 million miles per hour (or about 1,300 kilometers per second, which equals 0.1 arc second per year). As this ejecta slams into slower-moving gas, perhaps expelled by Eta Carinae at some earlier time, shock waves heat the gas and cause it to glow. Previous studies of the spectrum of the light emitted from the southwest ridge show that this gas is rich in elements such as nitrogen and silicon known to be formed only inside massive stars. Although the ridge can be seen fuzzily in the ground-based view, the *Hubble* image clearly highlights the small knots and filaments tracing the locations of the shock fronts. The new image also suggests that this ridge forms part of a "cap" of matter, located to the southwest and behind the star.

The most striking finding that arises from these new data is the existence of a well-collimated jet of matter flowing from Eta Carinae. This jet can be

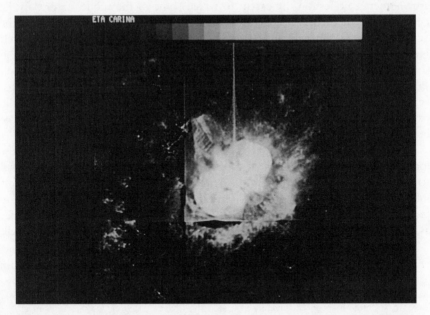

This *Hubble* image of Eta Carinae's peanut-shaped inner nebula, heavily computer enhanced, reveals surprising new features in this dusty region. These include a jet of matter flowing away from the star and a regularly spaced set of wavelike features to the northeast (upper left) as well as a complex ridge of matter to the southwest (bottom right). The image was made in the light of ionized nitrogen and displays resolution of about 0.2 arc second throughout the wide-field camera's (clipped) field of view of some 50 arc seconds—or about 2.5 light-years across. The large vertical spike (due to severe "bleeding" or overexposing of the central star), the black "peppery" spots in the nebula's midsection (due to saturation of individual pixels), and the imperfectly spliced observations made with the four CCD chips (to bottom and left of the nebula) are all instrumental artifacts and are not real structures in space. Overall, this *Hubble* image of Eta Carinae looks dirty largely because the region *is* dirty—that is, full of irregularly scattered gas and dust. *[NASA and J. Hester]*

seen as two faint parallel lines pointing to the northeast (upper left). These two lines mark the edges of the jet, a narrow tubelike structure. The jet terminates in an inverted U-shaped feature, which is probably a bow shock (much like the wake from the bow of a ship) driven by the jet into the gas surrounding the star. Such stellar jets and their associated bow shocks are often observed near young stars; they appear to be an integral part of the process of star formation. However, the existence of a jet emanating from Eta Carinae came as quite a surprise. In other stars, jets are thought to occur because the disks of matter from which stars form confine the flow of gas and dust away from the star into two thin beams. Such a disk might have accounted for the elongation of the homunculus, but it would have been perpendicular to the disk now required to explain the newly discovered jet.

Even more remarkable, the *Hubble* image also reveals what is variously described as a "ladderlike" or "riblike" protrusion associated with the jet. While this fascinating yet faint feature is by no means well understood—for nothing like this has ever been seen before in the cosmos—it seems clear that the rungs of the ladder represent some kind of wave phenomenon in the flow of matter away from the star. One possibility is that these are some sort of standing waves nearly a light-year long, much like sound waves inside an organ pipe. Another possibility is that they are ripples in the flow of matter along the jet's bow shock, much like the ripples seen in turbulent water emerging from a faucet. Exclaimed Hester, "I'm unaware of anything in astrophysics that looks like that." And to his defense, the bizarre rungs can be faintly seen in the raw *Hubble* image, so there is no denying their reality.

Prior to *Hubble*'s observation, we regarded the homunculus as a "bipolar outflow" of matter, moving to the northwest and southeast (upper right and lower left), away from Eta Carinae. But this interpretation is now questionable in the light of *Hubble*'s findings. The jet to the northeast of the star, the cap of material to the southwest, and the thinness and clumpiness of the shell suggest that the axis of the Eta Carinae system is northeast to southwest. In this new model, the bulbous homunculus is an outflowing equatorial shell, while the jet and cap comprise matter blown out of the poles of the system, perpendicular to a disk still buried in a cocoon of gas and dust.

Our newly detailed view of Eta Carinae should provide important clues about the entire class of high-mass stars called Wolf-Rayet stars. At least one other such star exhibits a similar (though much larger) oblate shell along with uneven polar blowouts. This object, called NGC 6888, might display a more evolved version of a shell much like that surrounding Eta Carinae. Both objects are regarded as strong supernova candidates; the opportunity to study Eta Carinae at a time when it is highly unstable should help us to understand the physics and evolution of the massive stars in our Milky Way Galaxy.

As we pushed deeper into the fall of 1990, an important milestone occurred some eight months after launch. This was the time frame—had the mission gone well early on—when *Hubble*'s commissioning was scheduled to end and the general scientific lifetime of the telescope begin. Long before launch, we at the Science Institute had prepared for a large and regular influx of astronomers from all over the world: these were the GOs—the general, or guest, observers, who would comprise the majority of users throughout Space Telescope's expected fifteen-year mission. But owing to the steady stream of spacecraft ills, the commissioning period's orbital verification phase, expected to last but two months after the spring launch, was still plodding along as winter approached, and the notion of "general observing" seemed to be receding faster than the galaxies.

Yet to come was the science verification part of *Hubble's* commissioning, an expected half-year process of engineering checkout of the telescope's scientific instruments and operating modes. So, with much prodding from the Science Institute, the *Hubble* project office at Goddard finally agreed that we had all had enough of the engineering tests. On November 12, 1990, NASA decided that orbital verification was over—the programmatic equivalent, as Giacconi so aptly put it, of "calling in the helicopters to the Embassy roof, declaring victory, and leaving."

We welcomed this turn of events, but our euphoria did not last long. We soon heard that NASA intended to extend the science verification period to at least a year—a decision that provoked outrage among some GOs and near mutiny at the institute. Despite the rousing scientific success of our imaging campaign, engineering managers would remain solidly in charge—"Give engineers the opportunity and they will flirt with any gadget forever," said an irked science colleague—and there seemed little prospect that they would hand over the vehicle for astronomical observations.

We knew that we had to do something to dislodge the engineers again, even if it were only symbolic. We also wanted to make an astropolitical statement: Space Telescope had been built *to do science,* and largely under the auspices of the worldwide community of astronomers—the GOs. This we felt strongly not only because we were impatient to get on with the science program ourselves but also because the Science Institute had been charged by the astronomical community to look out for its interests. After much discussion as to how the GOs might become involved, even in a token way, we concluded that the only answer was to declare, once again, director's discretionary time. Not that our director was yet entitled to any such unscheduled time during the commissioning period, but we had commandeered such time earlier. And since it had worked then, why not try again?

A group of a half-dozen, mostly Midwestern, astronomers, led by Phil James of the University of Toledo, had been preapproved for *Hubble* observing time in order to monitor seasonal changes occurring on the surface of Mars and in the planet's atmosphere. Their program was expected to extend over several years, thereby allowing better understanding of the Martian climate and the events that trigger surface changes, perhaps eventually enabling scientists to characterize global weather patterns and meteorological trends on Mars. Such studies, especially knowing how to predict dust storms, are important prerequisites for a manned expedition to the red planet.

Since Mars in December was near its closest approach to Earth (about 53 million miles, or 85 million kilometers away), the red planet would be seen more clearly than at any time in the next two years. Furthermore, within several months it would drift too close to the Sun, and it would then remain in *Hubble's* avoidance zone for more than a year. Now was the time to start the Mars monitoring program, and simultaneously obtain some science data that the public could appreciate and that might have some practical application. Once director's discretion was declared, NASA seemed in a quandary and thus stepped aside. Planning went forward to schedule several orbits of observations.

Having stood by earlier in the mission while NASA blew one opportunity after another to have the media witness the ups and downs of real science, I resolved that this historic first would be recorded—clandestinely if necessary. The institute's directorate and operations divisions telegraphed by body language alone that another expression of civil disobedience would be welcome, and Phil James, a most accommodating fellow, was quite willing to go along with whatever might help to improve *Hubble's* image publicly. I apprised him and his team of NASA's regulations that the media were *verboten* in the operations areas at the Science Institute, and we all agreed to do what we thought best, without informing NASA. After all, this was a guest astronomer—someone not previously associated with the Space Telescope project—and it was his prerogative to conduct the observational session as he saw fit. Besides, I had learned by then that whenever I asked NASA managers for anything, the answer was invariably *no;* it was best not to ask—to do what we thought appropriate, and then to file any consequent reprimand.

By this time, we were well into the filming of our weekly *Starfinder* instructional television program that was airing, mostly during the morning, across much of the nation on PBS. I invited Maryland Public Television to invade our control rooms at the Science Institute in full force, and they readily accepted, since this was an opportunity to capture on videotape a sense of what it was like to be a scientist—indeed, to be at the nerve center of the world's most sophisticated scientific apparatus. So in the dark of night, at 4 A.M. on December 13, the Mars team welcomed into the institute's inner sanctum a small army of producers, directors, videographers, camera crews, and associated PBS onlookers. The result was many hours of raw footage of the conduct of real science—and especially of the process and methodology of modern science—the real scientific method, with all its mistakes and triumphs.

Mars had been previously studied in detail by several U.S. space missions, including the now dormant *Viking* Landers, which made observations from the Martian surface for several years after touching down in 1976. Mars is a dynamic world, however; it has an active atmosphere, which changes the appearance of the planet's surface over both short and long time scales, and it also experiences seasonal change, because, like Earth, its spin axis is inclined relative to its orbit about the Sun (23.5° for Earth, 25.2° for Mars). Therefore, to understand Mars's complex meteorology and climate, the planet must be continually monitored over many of its annual cycles, much as the scanning of Earth by terrestrial weather satellites has improved our ability to understand and forecast weather on our planet.

The beauty of planetary observations made by *Hubble* is that the surface and atmosphere of a planet can be monitored on a regular basis, so that the global distribution of clouds, say (both water and dust clouds), can be studied as a function of latitude, longitude, time of day, and season. This is an impossible job for ground-based telescopes, because Earth's atmospheric turbulence blurs the telescopic image of a planet. Even under optimum observing conditions, the smallest details identifiable on Mars by ground-based telescopes are about 100 miles (160 kilometers) across. What's more, this resolution is only possible during the short period of time approximately every 780 days when Mars is closest to Earth. Most of the time, ground-based telescopes are unable to resolve features smaller than 400 miles (650 kilometers), making even major events occurring on the planet unrecognizable. (For scale, the Great Wall of China extends for about 1,500 miles.)

The planned observations with *Hubble's* planetary camera called for the use of several filters—including those at ultraviolet, blue, green, and red wavelengths—that would allow study of both atmospheric and surface features on Mars. The latter three filters would also enable us to make a color-composite image close to what we would see if we could look at Mars through *Hubble*. Observations began precisely as scheduled, at 5:34 A.M., as the spacecraft exited the South Atlantic Anomaly toward the end of orbital night during revolution #3466. A brief series of eight images were acquired, most of them exposed for a mere fraction of a second—although one of the ultraviolet sequences kept the camera shutter open for two minutes. (Mars appears about twenty-five times as bright in red light as it does in reflected ultraviolet radiation, not simply because the Sun's reflected light is brighter in the visible part of the spectrum but also because the camera's UV response is less sensitive, owing to molecular contamination inside the camera.)

In the midst of the observations, the spacecraft entered orbital day, and we watched the gyroscope telemetry indicate that the bird was experiencing its by now familiar oscillation, but since the planning sessions had specified the coarse-track guiding mode, data aquisition was not interrupted. There being no real-time contact with the TDRSS network at the time, these initial results had to be stored on one of the telescope's tape recorders. All appeared to go well, and the television crews were gathering lots of footage of scientists and operational personnel during a typical graveyard shift at the Science Institute.

While awaiting the "TDRSS dump," at which time the Mars data would be telemetered to us, we milled around the OSS area in a kind of subdued love-in, talking about the significance of these science observations and about what the investigators hoped to accomplish. It was the middle of the night, and many of us were a tad blurry eyed, but we were pleased that the *Hubble* mission had reached the point of serving these initial representatives of the international astronomical community. All the while, the crews from Maryland Public Television went about their business—remaining as they had promised, "like flies on the wall," yet picking up all that was to be seen and heard. They were determined not to be actors in the play but rather only observers, for they wanted to capture, unscripted, the true spirit of what it was like to use humankind's best telescope.

Nearly an hour later, we saw our first image of Mars on the high-resolution monitors. And we were caught with our pants down, so to speak. Only half the planet had been imaged, with the other half beyond the edge of the planetary camera's field of view. It was as though only half of your grandpa had shown up in a poorly aimed family snapshot. Testy comments were heard amidst the confusion and embarrassment. Phil James summed up his feelings smartly when he did an about-face and blurted out dejectedly and directly into a television camera, "Oh, man! Is there some coffee someplace?"

Clearly, there had been a pointing error, and most people immediately blamed it on *Hubble*—if only because the spacecraft had lost one of its gyroscopes during the previous week, owing to an electrical malfunction. But, as was so often the case during the commissioning period, the problem was not with *Hubble* as much as with a human. This time, a scheduler at the institute had made a small correction to account for Earth's precession, not remembering that the computer system in the Planning and Scheduling Branch does this automatically. Thus, the correction was executed twice, and the result was a planet half-imaged.

Meanwhile, the PBS crews continued taping exactly what NASA did not want recorded. An error had been made—an extremely obvious one—and it was now on videotape, for everyone to see. Despite all that we had been through in the *Hubble* project, NASA was still slow to admit that human beings sometimes make mistakes. In fact, errors are made all the time in science. The very process is often rife with miscalculations and false starts, many of them because of mental fatigue, or the uncertainty of working at the frontiers of science. One then strives to understand the mistake, to recover from it, to do the experiment again, and to get on with the scientific analysis. And that is exactly what we did that morning. To my delight, the TV crews continued to record everything and were thus able to document how the scientific method truly works, warts and all.

Even before we had discovered the source of the pointing error, we knew that we had to correct the orientation of the telescope. The error was minuscule—about 10 arc seconds, or well more than a hundred times smaller than the width of the full Moon—but it had to be eliminated, lest another series of Mars observations scheduled later that morning also show only half the planet. The fact that we had no real-time contact with *Hubble,* and therefore no means to uplink the correction to the spacecraft, exacerbated the problem. The confrontation in the Persian Gulf was heating up, and TDRS-East was busily occupied with priority traffic from other spacecraft then spying on the Kuwaiti theater of military operations, but our "mission-schedule events log" showed that TDRS-West might be available for unanticipated access—a "forward-link window," in operations lingo.

I informed the guest P.I. of his "rights" to make a real-time decision— namely, to arrange for an emergency uplink to *Hubble*—all of which was rather bewildering to him and his visiting team. Frankly, they seemed to have little alternative but to go along with the recommendations of the institute staff. Later, the team members confessed that they had thought we did this sort of thing all the time, whereas in reality I felt as though we were flying *Hubble* with a joystick.

By now the OSS area was full to overflowing since the daytime-shift operators were beginning to arrive and the nighttime personnel were lingering pending the outcome, along with several planetary and commanding experts who had been called in to address our dilemma. Moreover, word had spread through the institute staff that "an event" was under way in the OSS area, and technical personnel from throughout the institute came and went, though most were content to peer through the glass walls separating the OSS from the rest of the operations area. All the while, PBS continued to videotape the peculiar mix of excitement and confusion.

At 7:30, the institute's science-mission manager relayed to Goddard (with whom we had open phone lines at all times) an official "opr," or operations request, "to SIP *Hubble* through TDRS-West." SIP is another of those double-barreled acronyms standing for SOGS Invokable Procedure (where SOGS in turn stands for Science Operations Ground System), which is an infrequently used real-time contact to order *Hubble* to depart from its preprogrammed instructions and to revise the robot's task. Impressively, in a few minutes the emergency uplink was granted by the G-men controlling TDRSS, and a command was sent to *Hubble* to slew itself some 10 arc seconds east.

About noon, near the end of orbit #3471 and again while passing into orbital night, the spacecraft acquired a new set of multiwavelength observations, and when *Hubble* entered orbital day and was able to reacquire TDRS-West, the data were telemetered to the institute. The beaming faces and hearty handshakes within the OSS area as well as the applauding crowd behind the glass walls were clear testimony to our success. The full image of Mars had snapped onto our monitors and was centered precisely in the middle

A breakthrough in the public domain: *Life* magazine for May 1991. *Hubble*'s positive science findings were beginning to make their way into the everyday consciousness of American society. This image is a composite of individual images obtained through red, green, and blue filters using *Hubble*'s planetary camera. The planet's north polar cap is at upper left, where it was Martian winter at the time of observation. The spatial resolution of 0.2 arc second yields Martian surface details as small as 30 miles (or 50 kilometers) across. *(See also color insert section.)* [NASA and P. James]

of the planetary camera's field of view. And that view was wonderful, said University of Colorado's Steve Lee of the Mars team, "It's like Christmas Day—just keeps getting better here." We had made a mistake, had adjusted accordingly, and had recovered correctly. This indeed was real science, just the way it's done in the trenches most days by scientists around the world.

The resulting images of Mars show a planet at midsummer in its Southern Hemisphere. Its thin atmosphere seems relatively clear of dust; however, a thick canopy of bluish clouds obscures the icy north polar regions, as is normal during fall and winter in that hemisphere. As on Earth, the periphery of this northern polar region is the locale of intense storms, which migrate through the northern midlatitudes. Details of the cloud structures in such fronts can be seen in the *Hubble* images. Extensive thin clouds in the Southern Hemisphere are perhaps harbingers of the south polar hood, which forms as the season changes to fall in that hemisphere. The ultraviolet images show that ozone is more heavily concentrated at the planet's poles—an expected result since the water vapor that normally destroys the ozone is frozen at the poles.

The large, dark area dominating this face of Mars—officially called Syrtis Major Planitia, although to some of us more resembling a shark's fin or even an image of Goofy—was the first feature identified on the surface of the planet by terrestrial observers hundreds of years ago. It was used by Christian Huygens, the seventeenth-century Dutch master optician and telescope maker, to measure the rotation rate of Mars, namely a Martian day of twenty-four hours and thirty-seven minutes. This sloping surface region is rich in deposits of coarse dark sand, which gives it its dark hue. Relatively strong winds apparently blow upslope (from east to west), scouring the fine dust from the surface. The *Viking* images have shown the crest of Syrtis Major to be the location of one or more volcanic vents. Volcanic activity in this area of Mars probably ceased about 2.5 billion years ago, and weathering of the basaltic rock is the source of the sand.

Arabia Planitia, the lighter region to the west of Syrtis Major, is thought to be bright because its surface is covered by a reflective layer of dust perhaps as much as 10 feet deep and blown from Syrtis by the east wind. This region is heavily cratered, and several of the craters are discernible in the *Hubble* images. The equally bright region to the east is Isidis Planitia, a 600-mile-wide (1,000-kilometer) impact basin, apparently formed as a result of a large asteroid collision more than 2 billion years ago; this basin, too, is thought to be bright because it is covered with fine dust.

To the south of Syrtis Major is the bright Hellas Planitia, an even larger basin than Isidis. Perhaps formed about 3.5 billion years ago by the impact of a large asteroid, Hellas extends about 1,100 miles (or 1,800 kilometers) across and is nearly 5 miles (or 8 kilometers) deep. Normally the basin is either filled with fog and clouds or is the site of dust storms; so *Hubble* has captured an image of a rare season on Mars when the weather in Hellas is calm and its surface visible.

Our guest astronomers, who at the institute soon became known as "the Martians," continued to monitor Mars at roughly three-month intervals throughout the first half of 1991; after this, its angular separation from the

Sun was under *Hubble*'s safety limit. The visiting team returned the following year when the planet emerged from the solar-avoidance zone, and resumed their monitoring program during the height of Mars's dusty season. Among their principal findings was that the atmosphere on Mars is even colder and drier than when Space Telescope first surveyed the planet. Comparisons of such images obtained at different times are crucial if we are to be able to predict weather and climatic changes, and especially the fierce dust storms on the red planet. This will, in turn, hasten the day when we feel that we understand Mars sufficiently well to mount an expedition there—a goal considered by many to be the next great leap forward in space for our species.

We had made our point to NASA and to the world. The guest-observer era with the Hubble Space Telescope had begun.

In the spring of 1991, close to the first anniversary of Space Telescope's deployment, we at the Science Institute took stock as to where the project stood. Our Planning and Scheduling Branch had built some ninety-two weekly software calendars (SMSs) for *Hubble*—well more than the expected fifty-two because of the inability of the Marshall/Hughes/Lockheed engineering teams to cope with the early failures of the telescope optics and pointing systems. "The institute's schedulers have been very much in the hero mode," reported Giacconi to a group of visiting astronomers, "and some of our people have burned out." Once the Goddard team had taken charge, at least some science had been accomplished, although technical troubleshooting and calibration were still uppermost in the minds of those controlling the spacecraft.

Since launch, about 12,700 observations had been made, the vast majority short exposures for purposes of vehicle testing and instrument calibration. Yet we did need to archive more than a thousand magnetic tapes in our growing data bank and that represented some progress. On the downside, some of those science data were beginning to resemble the Dead Sea Scrolls, since a small fraction of them had been acquired by GOs who were then forbidden access to them by GTOs citing proprietary rights. All in all, though, when science observations were allowed to proceed amongst the engineering tests, they often provided, albeit slowly, as described in part in these past two chapters, world-class scientific results unobtainable with any other telescope.

The "science efficiency" of the spacecraft on its first anniversary was a mere 9 percent, meaning that an average of only about eight minutes of each ninety-six-minute orbit was devoted to recording meaningful data from a cosmic target of scientific interest. This was not the bird we had expected *Hubble* to be, nor was its prognosis for improvement good. If we included considerable overhead, such as the amount of time required to control the science instruments (changing filters, stopping and starting exposures, data transmissions, etc.) and to acquire guide stars prior to most science observations (up to twenty minutes to achieve fine-lock during each orbit, about twelve minutes for coarse track), then *Hubble*'s "spacecraft efficiency" rose to some 30 percent.

Telescope slews, passage through the South Atlantic Anomaly, and other periods of "idle activity" (an oxymoron for dead time) when *Hubble* had no specific observation ready to execute amounted to an additional 20 percent of its total orbital time. (Science efficiency is defined as that proportion of a full orbit that *Hubble* spends exposing one of its five main science instruments to a celestial target, whereas spacecraft efficiency equals the fraction of total time the telescope is aimed and preparing to observe, as well as actually observing, a cosmic source. All these efficiency numbers are measured assuming a half-orbit equals 50 percent, or slightly less than the maximum efficiency possible for a spacecraft in low-Earth orbit.)

By contrast, an earlier, celebrated space mission launched into low orbit in 1979—the now defunct High-Energy Astronomy Observatory, known popularly as *Einstein*—had also started out slowly, yet had managed to increase its spacecraft efficiency to some 45 percent after its first year of operations. And the small *International Ultraviolet Explorer* satellite, still operating in the more desirable geosynchronous orbit some 22,400 miles up, has achieved an impressive 60-percent science efficiency and better than 90-percent spacecraft efficiency over the past decade. These differences in efficiency are not small for major space missions. To realize *Hubble*'s trumpeted prelaunch expectations of between 20- and 30-percent science efficiency, we would effectively need two or three Space Telescopes in orbit working at the realistic 9-percent efficiency rate.

Despite *Hubble*'s inefficiency, we consoled ourselves by noting that the vehicle's complexity—as the flagship of the Great Observatories—put it into an entirely different class of spacecraft. Considering the unexpected optical aberration and pointing jitter, among a host of other technical glitches, it was not surprising that the telescope's operating efficiency had been relatively low thus far. What worried us, despite the increasing number of science observations undertaken, were indications that the spacecraft's systems failures were mounting, and that it would only be a matter of time before *Hubble*'s very survival would be in jeopardy.

Over its fifteen-year expected lifetime, the basic research and development value of the spaceborne observatory depreciates by several million dollars each week, so like an automobile, for instance, we might reasonably expect some of its components to break down. This is what the orbital servicing contingencies were supposed to be all about; the space shuttle should enable us to return to the spacecraft periodically and to fix and upgrade any of eighty replaceable parts. Yet, it was the *rate* of breakdown—and the fact that the breakdowns were systemic—that troubled us. And of course we had doubts about the ability of NASA's unreliable shuttle fleet to rendezvous with *Hubble* frequently enough.

Admitted John Campbell, one of the most capable of the Goddard engineering managers, for which he was roundly berated by his NASA superiors (as quoted in the trade journal, *Nature*), "You have to assume that it was badly built. It's like a car; if you have a lemon, it's difficult to know what to do." Although that quote clearly captured the sentiment among scientists with whom I worked, it represented a milestone of sorts because it was spoken by a leading representative of the aerospace community that had built the confounded craft.

As was the case earlier in the mission, the increasingly guarded, somewhat somber yet (I believe) realistic view widely held at the institute contrasted greatly with the "all is well" attitude among most of NASA project management. Goddard's "weekly" flight status report for April 4, 1991 (the first such report in five and a half months), noted: "At the current time (approximately three weeks short of one year in orbit) HST is performing extremely well." And in a summary briefing of Space Telescope's anniversary, on April 16 at the institute, a high-level NASA official stated flatly that, "Daily HST operations have become routine and robust." Among other pronouncements, the briefer claimed that revised software—the solar-array gain augmentation, or SAGA for short—uplinked to Hubble's flight computer had all but eliminated the largest of Hubble's instabilities (the 0.1-hertz oscillation), thus greatly reducing spacecraft jitter.

By contrast, our logs at the Science Institute had shown no clear improvement in Hubble's tracking performance; rather, we had noted continued jitter-induced blur in the science images and loss of guide-star lock on the majority of observations under way while crossing the day-night terminator. If anything, the solar-array oscillation seemed to be worsening. Scientists among other institute personnel at the anniversary meeting looked at each other with puzzlement when the NASA man projected a vugraph with the legend, "Loss of Fine-Lock Nearly Eliminated," but no one elected to say anything—partly because most of us hadn't bothered to analyze the intricacies of the software "patch" meant to fix the spacecraft vibrations, and partly because we wanted to believe that the engineers really had solved the jitter problem.

The NASA briefer also maintained that the guide-star acquisition success rate had risen from (an unheard of) 97.5 percent early in the mission to greater than 99.9 percent, all of which we found incredible, given the many science observations that had been judged worthless because of ongoing problems with guide stars. "These graphs must have been prepared for upper management—they have no relation to reality," whispered one of the principal investigators to me. When the briefer went on to claim that the nearly two hundred postlaunch engineering adjustments on the position and orientation of Hubble's secondary mirror had optimized the telescope's focus, the catcalling began. For it was our contention, although it was difficult to prove in the light of the aberration, that every science image acquired since launch was in one way or another, and aberration aside, not optimally focused.

"Continuously fooling around with the center and tilt of the secondary mirror seems like a random process," said Jim Westphal, adding pointedly, "Jerking the secondary and its actuators endlessly is very scary. Something is going to break." And Duccio Macchetto, the usually diplomatic leader of the faint-object camera team, snapped with uncommon bluntness and volume, "The engineers are just hunting, without mathematical or technical analysis. Just when are we going to have a collimated telescope?"

The cause of much of this renewed animosity between scientists and engineers was an unforeseen conflict between camera focus—a science concern—and navigational guiding—largely an engineering issue. Owing to the spherical aberration, the fine-guidance sensors work best with a certain setting of the secondary mirror, which unfortunately is not optimal for camera imaging. Acknowledged one of NASA's ubiquitous Vu-Graphs, "Best position for FGSs is unacceptable for cameras." Apparently, the peak focus settings of the two main optical systems aboard Space Telescope are mutually exclusive! And it seemed to us that NASA was attempting to maximize the performance of the FGSs rather than the clarity of the camera images—again stressing an engineering issue at the expense of science results.

NASA's Pollyanna-like attitude evaporated soon thereafter. Only days after *Hubble*'s first anniversary, the vehicle experienced what one of the institute's operations chiefs described as "the worst week since the one immediately following launch." A number of problems arose in quick succession in as many days. The carousel containing the various prismlike gratings for the high-resolution spectrograph began malfunctioning, repeatedly sending that particular instrument into safemode; a software workaround was quickly devised on paper, but nine man-months of effort on the ground would be needed to put it into practice. The high-speed photometer followed into safemode, the culprit being a hung-up systems controller (or miniature computer) inside the instrument; one of its means of routing data to the ground had effectively broken. The faint-object spectrograph missed numerous science targets for reasons unknown, although some of the failed observations might have been caused by faulty target positions uplinked to *Hubble;* this spectrograph's sensitivity also took a dive in the indispensable far-ultraviolet part of the spectrum—probably as a result of radiation damage to its innards. The wide-field and planetary camera's contamination increased (its outgassed pollution had become trapped), making ultraviolet observations impractical and exacerbating its case of the "measles," wherein splotches appeared on images taken in the visible spectrum. Spacecraft jitter and the associated loss of fine-lock on guide stars seemed to worsen noticeably, especially each time *Hubble* entered orbital day, with some of the downlinked telemetry (and direct, rare, human communication with Lockheed) suggesting that occasionally the extent of bending at the tip of at least one solar array had increased from 11 inches to several feet!

Then, in midafternoon on May 1, to a chorus of officials' beepers, *Hubble* experienced a close brush with brain death. And what a mayday call it was! A memory board on DF-224, the telescope's main operating computer, failed, causing the computer to crash and the spacecraft to enter its deepest level of (hardware) safemode. Ground controllers were uncertain for some agonizing hours what had happened to the spacecraft, and in the OSS area engineers anxiously examined the telemetry dump of *Hubble*'s memory just before the bird had become distressed. "We have entered the twilight zone," deadpanned one jaded engineer.

Insertion into deep safemode was accompanied by closure of the telescope's aperture door—only the second time that had occurred since deployment, and precisely one year since the telescope had previously so panicked. *Hubble* was "battening down the hatches," perhaps not knowing what had happened to itself. And as if these setbacks weren't enough, we also experienced the feared LOS condition—loss of signal or loss of spacecraft? Fortunately, Shuttle *Discovery* happened to be in orbit at the time, on a strategic-defense mission (to systems-test some new sensors), and ground controllers managed to induce some vital communications with *Hubble* using *Discovery* as a data link.

The brief and sluggish messages dispatched to the ground by *Hubble's* low-gain antenna indicated that while the telescope had suffered a drubbing, at least it was safe. Part of the bird's brain (a puny, last-line-of-defense safemode chip of less than 1 kilobyte memory) had correctly thrust the observatory into a comatose state, just as designed. Over the next few days, engineers managed to activate one of two spare memory boards inside the DF-224 computer and to awaken the orbiting platform gradually from its stupor.

Flatly stated, if *Hubble* has no memory, it has no brains and thus no being. What had caused the board's electronics to fail? No one knows for certain. The software amendments uplinked to dampen the telescope's jitter might have been the culprit, since various engineering tests then addressing the pointing-control problems had been overloading the small shipboard computer. In the year since launch, flight software revisions had been many; some five dozen "patches" had been sent up to the spacecraft, repeatedly altering its mental state and possibly taxing its adaptability. Or, perhaps the computer's memory had been hit by a fast-moving subatomic particle; such radiation damage had been blamed for the earlier loss of a gyroscope and some glitches in the electronic circuitry of the fine-guidance sensors. Or, the DF-224's board perhaps just succumbed to quarter-century-old technology; the failed memory board was a spare left over from the *Apollo* program, made literally to fly in a computer to the Moon.

To complicate matters further, while *Hubble* was in its deep safemode its science instruments had warmed up considerably (for the coolers were now unpowered), and the fear arose that contaminating gases inside the wide-field camera might permanently condense as a dirty, icy substance on its detectors once the camera was recooled for normal operation. These are valid and continuing examples of how, in a complex interrelated system, one problem can quickly lead to another, anticipated or not. "If the bird continues to accumulate systems failures at this rate, we are likely to reach a state of constantly having to mount shuttle repair missions just to keep it running, let alone improve it," said the institute's operations chief, Jim Crocker, at the early May director's management meeting.

Clearly, the spacecraft was limping along, the number of ailments growing, all tending to sap the spacecraft of its unparalleled versatility and power as a functioning telescope. The Associated Press newswire put it this way, on *Hubble's* anniversary date: "The Hubble Space Telescope is a 1-year-old suffering with problems of old age; it's nearsighted, it trembles in the morning, and it has trouble keeping its balance." The project had become locked in a high-stakes race against time: Could the spacecraft be fixed before it became not worth fixing?

Suddenly, in late spring, while *Hubble* was exiting the South Atlantic Anomaly, another of its gyroscopes began to falter. (The first one had shut down several months earlier, also when in or around the SAA.) Showing erratic behavior at first, the gyro finally quit after a week of intermittent operation—another victim, it would seem, of the fast-moving protons caught up in the Van Allen belts. (Several of the gyro's early electrical glitches occurred while *Hubble* was transiting the radiation-filled SAA, thereby implying a correlation, since the probability of multiple random failures while in the SAA is less than one percent.)

Intense solar-flare activity had in fact been observed a few days earlier, and had probably magnified the region's damaging effects. "Oh well," remarked a jaded colleague at the Science Institute, "another compartment flooded." Melodramatic perhaps, but it did seem ironic that the innards of the largest astronomical telescope ever lofted into space were being tripped up by submicroscopic particles, causing at an accelerated rate the same kind of "mutations" that have driven evolution on Earth for billions of years. Also ironic, all the while *Hubble* was accumulating such minute mutations, we humans—the new agents of change on Earth—were learning how to build, by trial and error, a better Space Telescope. Not unlike nature itself, for we and our man-made *Hubble* are indeed part of nature.

We were beginning to realize that the South Atlantic Anomaly was a nastier place than previously recognized. Full of high-energy elementary particles escaping the Sun, this abnormal part of the Van Allen belts out over eastern Brazil and the South Atlantic Ocean was seemingly eroding some of *Hubble's* electronics. Much of the spacecraft's vital circuitry could have been hardened—that is, shielded from the severest proton collisions—but as with so many other aspects of *Hubble's* construction there had apparently been a "design oversight," hotly denied of course by *Hubble's* makers. The problem was probably worsened as *Hubble* is normally driven in reverse, with its aft leading. This keeps debris from being scooped up by its open bow and chipping away at its mirrors, but it also means that the back-end components, such as the thinly protected gyroscopes and star trackers, are more susceptible to damage from belt radiation or orbital debris.

It will be recalled that *Hubble* went into orbit with a full complement of six primary gyroscopes: three needed for normal science operations and a fourth kept "on-line," ready to become the essential third gyro should any one of those then stabilizing the spacecraft fail. (The remaining two were initially idle and unpowered backups.) With two such gyros having quit in a single year, only four were left on its anniversary date, and—given what seemed a reasonably good correlation between gyro failure and SAA passage—it was likely that more failures were in the offing. Should *Hubble* have less than three good gyros at any time, the spacecraft would autonomously activate a special, deepest safemode state called the Retrieval-Mode Gyro Package, in which a group of three secondary (and coarser) gyroscopes would keep the spacecraft at a relatively steady attitude, in effect mothballing it until shuttle astronauts could replace the damaged gyroscope assemblies. If any one of these three coarse, emergency gyros were to fail, the observatory might well lose control, start to tumble, and thus be unapproachable by any shuttle servicing mision. *Hubble* would likely be lost altogether.

NASA, initially acknowledging in a July 5 press release that "the failed gyros' electronic circuitry appears to have been damaged ... by increased radiation within the South Atlantic Anomaly," shortly thereafter changed its tune and denied it in a July 10 press release. The agency and its industrial contractors also then refused to admit either that they had failed to properly harden *Hubble*'s gyros against radiation damage, or that the military routinely shields its orbited circuitry, or even that the military had built any such *Hubble*-class telescopes.

Instead, the project maintained that *Hubble*'s gyros had simply died of old age—"lead breaks from long-life corrosion" was how the senior manager at Lockheed put it to me in a fax sent directly to my office—an argument that surprised many of us, given NASA's earlier apoplexy whenever anyone even whispered that Space Telescope's design harks back to 1970s-vintage technology. Now project management was admitting that the multi-billion-dollar telescope embodied fifteen-year-old gyroscopes meant as spares for the much cheaper *International Ultraviolet Explorer* (IUE) satellite, four of whose gyros had been known to fail several years before *Hubble* even reached the launchpad.

Fortunately for IUE, this much smaller craft can limp along with two gyros and a highly accurate sun sensor, which pinch-hits as a third gyro. (Such sensors track the Sun, to keep both the solar arrays facing it and the telescope aperture away from it.) Yet, to our further bewilderment, the much more expensive *Hubble* has on board a less sophisticated sun sensor, making it apparently less capable than IUE is of orienting itself accurately in space, should *Hubble* lose, as we expected, more of its gyroscopes.

To be sure, toward the end of the commissioning period, a third gyro experienced a sharp glitch (symptomatic of another radiation hit), its spin motor drawing excessive current, yet another hardware defect. (It, too, died after commissioning ended, leaving *Hubble* with no spares as it began its operational era.) So which is it? Gyros built some two decades ago for another spacecraft and another time when weekly shuttle flights were claimed to be able to service *Hubble* often and at will, or gyros that were not properly hardened by engineers seemingly oblivious to the radiation fields through which *Hubble* passes? Best I can tell, it's both: gyros that are old *and* insufficiently hardened.

What's more, we now know that the failed gyros had been previously tested for some 70,000 hours—or about eight years running—before *Hubble*'s deployment. Once more, our multi-billion-dollar spacecraft was apparently the recipient of used goods. Said a rightly indignant colleague, "Either they don't test these systems at all, or they test them to death. I just don't understand." Said another, "It's like driving a brand new, high-priced automobile out of the fanciest showroom in town, only to discover that it already has more than a hundred thousand miles on it." The mind boggles.

By contrast, the military-intelligence community has much experience with *Hubble*-class vehicles in low-Earth orbit, including passage through the South Atlantic Anomaly—at least enough knowledge to "rad hard" much of their crafts' vital electronics with additional metal. Such shielding is done partly to guard against what is called the electromagnetic pulse (or EMP), which is a damaging by-product of nuclear attack, but also because it makes sense to do so for a craft destined to travel repeatedly through an intense radiation zone. (Weight at launch is not a concern here; *Hubble* weighs considerably less than the largest *Keyhole* spacecraft.)

Much of this became apparent at an intelligence gathering in the spring of 1991, when while briefing on an upcoming experiment of the Strategic Defense Initiative Organization (SDIO), a defense engineer tried to squirm out of a tight argument by saying something to the effect, "The *Hubble* guys do it this way." At which point the room erupted in criticism, with some surveillance spacecraft designers yelling, "Don't tell us what the *Hubble* guys do. They don't even know enough to harden their electronics!" (Actually, the SDIO types were partly boasting, since on the same day that *Hubble's* second gyroscope failed, a small military satellite was temporarily incapacitated when one of its gyroscope-affiliated reaction wheels suffered damage, almost certainly caused by energetic particles streaming from the Sun.)

Nor did the castigation stop there, for these critics went on to claim, to some astonishment, that half of one of *Hubble's* solar arrays occasionally sways badly enough that it might break off any day! True or not, *Hubble* clearly had been imaged up close again by "nonexistent" national technical means and that imaging was displaying a Space Telescope potentially near catastrophic failure, for if an array broke off the resulting imbalanced spacecraft might begin tumbling uncontrollably—yet another way for it to become unserviceable by shuttle astronauts and perhaps lost as a functioning observatory.

(True to form, NASA roundly and authoritatively denied any such spying on its bird, shortly after which the SDIO declassified parts of a doctored yet revealing videotape of *Hubble* on orbit. Sometimes I think that the intelligencers were trying to make NASA look not just bad but foolish. They were laughing at us again, and perhaps for good reason. Admittedly, however, I could detect no such excess swaying in the on-orbit tapes that I saw at the time.)

The dark-side operatives continued their diatribe against fair *Hubble*, noting that years ago the covert community had abandoned such large and flexible solar arrays in favor of smaller and more advanced, rigid arrays that still provide plenty of energy to power *Hubble*-class spacecraft. Thus, I now had confirmed the hints and clues passed my way at two earlier closed-door briefings: The military, long aware of the adverse effects of the flimsy lightweight arrays, had redesigned them to be more robust and compact, above all less susceptible to jitter. And, once again, they were apparently unwilling or unable to tell the civilian-based *Hubble* project.

Combined with the growing sense that the *Hubble* telescope's sum and substance were breaking down at an alarming rate, several other incidents during the months following its first anniversary on orbit—and with commissioning still under way—contributed to the notion that I had had enough of this project.

On the science front, a heated dispute broke out among two competing teams regarding who was to get the credit concerning one of *Hubble's* most

Another *Hubble*-class spacecraft, representative of those better built, more powerful and versatile space telescopes orbited during the 1980s by the military-intelligence community. Notice the smaller, rigid solar arrays (made of gallium-
arsenide) toward the craft's aft, quite different from the flexible arrays (made of
silicon) on *Hubble*. The large cylindrical canisters at the sides are fuel tanks (also
unlike on *Hubble*) used to maneuver the spacecraft both to vary the craft's orbit
unpredictably, thus making it harder to track, and to take evasive action if necessary in the event of antisatellite attack.

interesting observations to date. One team of astronomers had used the faint-
object spectrograph to detect a handful of absorption features in the ultraviolet spectrum of a distant quasar—possibly the Lyman-alpha forest of lines
alluded to earlier, arising from numerous clouds along the line of sight. The
interpretation, thought to be revolutionary in mid-1991, suggested that the
newly found clouds were both nearby and primordial—a real puzzle given that
primordial clouds left over from the formative stages of the Universe are more
likely to be distant. This team elected to bypass recommended procedures for
releasing results to all media simultaneously (and thus fairly) and did so on
their own by granting selected interviews to friendly journalists—which was
their prerogative, however ill-advised.

Shortly thereafter, a second team of astronomers using the high-resolution spectrograph obtained much better and more convincing data toward the
same end, bolstering a genuine cosmological mystery. This team elected to
release its findings equitably to all journalists as part of a formal press briefing
during another of the institute's Science Writer Workshops. Even so, the
leader of the first team, incensed that the second team's findings were of better quality and likely to eclipse his findings, vigorously protested our granting
to the second team much visibility before the media. Standing over me in the
institute cafeteria with my deputy and an important European visitor, the disgruntled astronomer berated me for all to hear. I walked out when he became
so emotionally worked up as to begin involuntarily spitting on my food while
giving me hell. I found it the most disgusting moment of my career at the Science Institute. Offered a leading astronomer of the competing second team

who held the first team's tactics in disdain, "Even my grandchildren don't throw tantrums like that."

I had the feeling that professional jealousies were driving some people berserk, given the few morsels of breakthrough science sent Earthward by *Hubble*. (Jim Westphal, who had nothing to do with this episode, later e-mailed me some grandfatherly advice—namely, that it was a lose-lose proposition to try to play referee among such high-strung, competing teams.) Not long thereafter, the incident became moot, for most astronomers now think the absorbing clouds are probably not at all primordial in nature. Rather, they are more likely previously unrecognized gaseous halos of line-of-sight normal galaxies, in fact nearby and having little to do with the origin of the Universe. Much as during periods of inadequate science funding when competitors turn on each other, the science findings with *Hubble* were thin enough compared to prelaunch expectations to cause some scientists to display a truly ugly disposition.

On the engineering front, *Hubble*'s parade of hardware problems continued. Most notably, in July 1991, its high-resolution spectrograph suffered a major component failure, resulting in the loss of its unique, highest-resolution far-ultraviolet capability. The issue was traced to a defective solder joint in a simple power supply that provided electricity to one of the spectrograph's two detectors. More distressing, owing to the design of the instrument, reliable use of other near-ultraviolet modes on the spectrograph was threatened, since although each detector has its own power supply they both route their findings to the ground by way of a "data formatter" connected to the failed electronics. The whole affair reminded me of the old-fashioned, series wiring of bulbs on a Christmas tree—another example of questionable systems engineering.

The response from NASA was typical. During a regularly scheduled monthly project meeting at the institute, an engineering manager from Goddard professed himself unable to understand "what the big deal is," and went on to assure the assembled division chiefs that, "The spectrograph can easily be rectified on a future shuttle repair mission. The astronauts merely need to remove seventy-two screws, several of which are epoxied, take off the instrument bulkhead, and jump-start a dead power supply." Frankly, I doubt if this fellow could have successfully removed a dozen screws in the basement of his own home. A few days later, the astronaut office at the Johnson Space Center called me, wanting to know if this Goddard man was "for real." When I told them apparently so, their reaction is best not printed. As commissioning ended, the high-resolution spectrograph was still in a powered-down state, inoperable and yielding no science, although fortunately it was partly resurrected several months thereafter and resumed limited service.

NASA, too, was on the move, flush from the success of *Hubble*'s imaging campaign and its subsequent science advances, issuing once more statements that simply lacked the truth. In midsummer 1991, a senior *Hubble* manager at NASA Headquarters gave an interview to the Associated Press, claiming that Space Telescope was in fine shape—in fact, "making a discovery about every week," and that a "routine" servicing mission would allow *Hubble* to "fulfill

97 percent of its complete scientific promise." Every number given and virtually every technical argument attributed to the manager—including an order-
of-magnitude low-ball for the "added costs for [replacement optics] and other
fixes being held to $20 million"—was incorrect, all of it reminiscent of the
prelaunch hyperbole that had so inflated public expectations.

Days later, officials of both the Science Institute and the Lockheed Company caucused by phone, and together we confronted NASA Headquarters'
public-affairs officials in a subsequent telecon, arguing that it had been this
kind of misinformation and cheerleading that had gotten the project into
trouble previously. We pleaded with the space agency to come clean, with a
balanced appraisal of spacecraft woes and science successes. At about the same
time, Goddard's public-affairs people were also stepping out, releasing their
first educational product about Space Telescope—a computer software program culled from old press materials and produced by another of those belt-
way-bandit advertising firms—and again claimed, among many other dated
fallacies, that *Hubble* could see many times farther in the Universe than any
other telescope—a fact not true for a Space Telescope with perfect optics, and
most assuredly not true given *Hubble*'s considerably impaired state.

When I complained (privately, within the NASA system) about this and
other hype again beginning to emanate from the space agency, especially as
regards future repair missions, the Goddard public-affairs office responded by
filing a grievance with the *Hubble* project manager, accusing me and my institute educational group of "administrative harassment." Goddard's public-affairs
counterparts at headquarters joined in, claiming, in verbally abusive language
that violated even the most liberal telephone etiquette, that the institute was
"part of *Hubble*'s problem" by issuing media and educational announcements
that were full of "elitist science" and that were degrading to NASA. Although
the Science Institute directorate quickly came to my defense, and the grievance
was ultimately dropped as groundless, it was clear that the control-seeking
micromanagers were once again on the prowl regarding issues of credit, logo,
and self-preservation.

As we approached the end of *Hubble*'s long commissioning almost simultaneously with the quincentennial year of Christopher Columbus's maiden
voyage to the New World, by chance I came across a newly translated copy of
the old admiral's log. There I found, belying the myth representing his flagship, the *Santa Maria,* as a magnificent sailing vessel (and in contrast to its
smaller, faster, more sea-worthy companions, the *Nina* and the *Pinta*), that he
had actually described it as slow and bulky, distinctly clumsy to pilot, and not
very well suited to exploration: "Wednesday, 26 December 1492: . . . I had
been sailing all the time with the intention of making discoveries and not
remaining anywhere longer than a day unless there was no wind because the
Santa Maria was very cumbersome and not suited to the work of discovery.
The reason I took that ship in the first place was due to the people of Palos
[Spain], who did not fulfill to me what they had promised the King and
Queen. I should have been given a ship suitable for this journey, and the peo-
ple of Palos did not do that." A few months after making landfall in the

Bahamas, the *Santa Maria* ran aground, its seams opened, and was soon abandoned.

In the light of all the engineering and astropolitical troubles surrounding the *Hubble* mission, a young scientist at the institute inquired of our director late one afternoon in the fall of 1991: "When will *Hubble's* normal operational science era begin?"

Staring pensively out the window at the dark, damp autumn day, Riccardo Giacconi responded resignedly, "I'm afraid that this *is* the normal operational era." At which point, he sighed, "How are we ever going to attract talented, young people into space astronomy?" And then he waxed eloquent, in an uncharacteristically hushed voice, addressing nothing more than thin air: "We scientists are a curious species. Like children, we want to play with ideas, to feel protected in an atmosphere where there is indulgence for our craziness. Our business is to do astronomy, to attack nature, to answer the great cosmic questions and the grand intellectual challenges. If science can't be done because the system you're in prevents it, then you might as well go make shoes."

The commissioning of the Hubble Space Telescope officially and quietly ended on November 15, 1991, almost exactly a year after its scheduled conclusion, and nearly nineteen months after it began. Although *Hubble* was by no means fully calibrated, the space trials were declared done. There was no ceremony.

Miracle on Orbit

Therefore I declare (and my sincerity will make itself manifest) not only that I mean to submit myself freely and renounce any errors into which I may fall in this discourse through ignorance of matters pertaining to religion, but that I do not desire in these matters to engage in disputes with anyone, even on points that are disputable. My goal is this alone; that if, among errors that may abound in these considerations of a subject remote from my profession, there is anything that may be serviceable to the holy Church in making a decision concerning the Copernican system, it may be taken and utilized as seems best to the superiors. And if not, let my book be torn and burnt, as I neither intend nor pretend to gain from it any fruit that is not pious and Catholic. And though many of the things I shall reprove have been heard by my own ears, I shall freely grant to those who have spoken them that they never said them, if that is what they wish, and I shall confess myself to have been mistaken. Hence let whatever I reply be addressed not to them, but to whoever may have held such opinions.

Letter to the Duchess Christina
GALILEO GALILEI, Florence, 1615

Given both the unprecedented complexity of the vehicle on orbit and the idiosyncratic cast of human characters on the ground, it is a miracle that the Hubble Space Telescope works as well as it does. The many forces opposing a successful *Hubble* mission have been strong and varied: government and industrial engineers who seemingly have little regard for systems analysis and design; academic scientists who have even less regard for sharing the fruits of their work with teachers, students, and the taxpaying public; space-agency officials who have become boldly bureaucratic, roundly embracing mediocrity to the exclusion of program excellence; news media that are increasingly impatient, subjective, and sensationalizing; and a military-intelligence community that has much experience with complex space missions but looks on silently while civilians struggle to accomplish technically what has already been largely yet invisibly achieved.

Even so, I would judge *Hubble* to be the best telescope ever built by humankind—not the telescope of telescopes, or the "discovery machine" we had expected it to be but rather merely the latest advance in a long line of increasingly sophisticated instruments in the science of astronomy. Indeed, the really positive story about the *Hubble* mission is the amount of world-class science that has been extracted from such a cranky robot. The impaired *Hubble* can virtually match any of today's best ground-based telescopes in sensitivity, and can exceed any of them in resolving power. *Hubble* can also operate in the ultraviolet part of the spectrum, something that no ground-based telescope can do, no matter how large. But—and oh, what a big "but" this is—what *Hubble* might have achieved had its optics been better, its pointing stable, its onboard systems well built! Although it seems that the Great Observatory concept—orbiting huge, remote, complicated scientific devices—is prematurely finished for our generation, only time will tell if Space Telescope is the flagship of a preeminent line of grand science platforms in space, or merely a dinosaur in the sky.

What are the lessons to be learned, if any, from this, the most complex and expensive science project that humanity has yet undertaken? In what follows, I offer a half-dozen personal considerations, all stemming from having lived with this project morning, noon, and night for several years.

Regarding engineers: To my mind, no obvious fraud plagued the *Hubble* project, nor even maliciousness among the engineering teams that conceived, built, and now operate Space Telescope—just an abundance of technical deficiency and managerial mediocrity, amidst ample project disorganization, all decidedly inconsistent with the exorbitant price tag of this grand venture. Above all, *Hubble* quite simply suffers from unsound systems management, systems engineering, and systems testing—a flagrant lack of integrated thinking, design, and application. By this I mean that while small-scale parts of any large-scale enterprise need to be built and managed well, there is also need for a generalized oversight of the entire product, or "system."

Such a holistic approach ensures that a collection of many intricate parts works as a whole, and that no one failed part unduly interferes with the operation of other parts comprising the product under development or operation. Despite society's move toward greater specialization, in a world of increasingly complex technology one must also examine the larger picture. The "whole being greater than the sum of the parts" might mean, ironically, that a collection of well-made and functioning isolated parts can, *as a system,* not function well at all.

Of course, nearly all engineers claim that they regularly take a systems approach in practice, but the fact is that few of them really do. They are simply not trained to do so. In the case of the *Hubble* project, the very notion of systems engineering and systems management often seemed foreign. As best I could tell, no such integrated oversight was implemented within NASA or among its major industrial contractors—or, if it was, then *Hubble*'s track record during commissioning suggests that it was done poorly. NASA Headquarters, two NASA field centers, multiple major contractors, and myriad subcontractors, comprising a cast of some ten thousand individuals, toiled for more than a decade almost independently of one another to build the spacecraft and its ground systems. It was a little like team-teaching a course in a university department; seldom does any one instructor take responsibility for the course as an integrated whole, the result invariably being a lack of educational coherence and continuity.

Indeed, *Hubble* suffers most from an absence of technical coherence and continuity—twin problems inevitably amplified by the decade-or-longer lead time needed to build and deploy such a complicated device. Why should engineers and technicians who will probably be at work on parts of other projects (or even retired, and in some cases dead) by the time many of today's spacecraft reach orbit worry about puzzling test results or design inconsistencies? Why should anyone who is not an end-user become obsessed—and it is obsession that is necessary—about the viability of such a complex system?

The glaring hardware faults on *Hubble* derive from a case of engineering myopia, a clear and steady failure to heed the bigger picture. For example: telescope optics machined improperly and tested inadequately by overconfident engineers, with no meaningful technical or scientific input from outside the secretive contractor; solar arrays designed by engineers largely unaware that, although those arrays functioned flawlessly in the thermally controlled (hence artificial) ground environment, their metal parts would act differently when subjected to sudden and intense heating and cooling thirty times a day in the harsh (realistic) environment of space; an aperture door that swings rapidly open and slams shut, with no one apparently having given thought to the larger consequences of such sudden motion on a free-falling spacecraft; the incorporation into *Hubble* of used goods, such as decades-old gyroscopes and memory boards meant for antique space vehicles, all apparently without examining the longer-term implications of such actions; a ground-based computer system conceived and built by industrial contractors to acquire, calibrate, and display *Hubble* data with no initial input from the astronomers who would be using the system daily; onboard science instruments essentially designed and wired in series, so that if one part fails the whole instrument is jeopardized; unshielded electronics that work well on the ground yet become problematic in the charged-particle zones and magnetic fields surrounding the Earth. And so on.

Perhaps, in retrospect, the system that is Space Telescope is just too complex for humanity's current level of applied intelligence, too daring for human beings to have built at this time in the history of our technological culture. Perhaps the spacecraft and its myriad onboard parts and vast ground-support functions are too complicated for our relatively neophyte technological civilization. When the descendants of Noah tried to build in the ancient city of Babel a tower so high that it would reach the heavens, the Book of Genesis tells us that God punished them for their audacity. Perhaps a much less complex space telescope—a more efficient, evolutionary vehicle—would have been met with a less omnipotent rebuke.

*   *   *

Regarding scientists: Today's active scientist typically has little desire or ability to share science with the general public. Most have even less interest in teaching, or in working with educators to help ameliorate the appalling state of scientific and mathematical illiteracy in our world today. Despite all the rhetoric of recent years, very few working scientists are willing to make a meaningful commitment to aid our educational system, at a time when the United States, especially, desperately needs their help. It is not that most scientists resist disseminating the essence of science to nonscientists; most simply do not care, since, for them, talking about science publicly is irrelevant.

Rather, it is usually an elitist minority of scientists who are unalterably opposed to any interaction with the public or with nonscientific students, and these are the ones who tend to be vocal, powerful, and insistent. The academic scientist who stresses research to the virtual exclusion of teaching, who hoards results instead of sharing them, is often the one who percolates to the top of leading university departments. Scientists inclined to work with educators and the public to help the citizenry better understand the subject matter and methodology of science, or even to teach well in universities, are looked upon as second-class citizens in the science community.

The *Hubble* project's public image suffered badly early on, if less so today, partly because of the actions of a handful of scientists who bitterly opposed sharing early images from this unique space mission. In particular, a clique of aggressive astronomers and a few high-level NASA managers conspired to thwart early science data from entering the public and educational domains—the scientists because they feared they might not get credit for *Hubble*'s initial findings, and the managers because they sensed they might lose control of the project and that NASA would not get the credit. Only after the public perception of the *Hubble* project was irreparably damaged was the issue of public dissemination of early science results forced into the open and science findings released.

The widespread notion that the scientific method is unbiased and objective, that scientists are and always have been lacking in human emotion in the course of their work, is a farce. Today's science endeavor is as value-laden as most things in life. Even in the remote, wide-eyed, and largely impractical science of astronomy, elements of intense competition and concentrated ambition are ever-present, for some often all-consuming. Alas, there is no silver medal in science, no second-place accolade.

Clearly, the issue of proprietary data is the source of much difficulty within the conduct of modern science. Some would say that, aside from the plethora of engineering faults plaguing *Hubble*, such data rights are the single most corruptive influence on the Space Telescope mission. Proprietary concerns, protected targets, "legal ownership" of this or that star or galaxy are symptomatic of professional paranoia, and derive from an innate distrust among scientists—a fear that one scientist will steal another's data; all the while few scientists realize that the data are not theirs to own. *Hubble*'s science data are the property of the citizens of the United States and Europe, or should be.

This attitude of exclusivity on the part of leading scientists can only contribute to a widening of the celebrated culture gap, and possibly even a headlong slide toward a scienceless society. By that I mean one of these days, perhaps soon, the public will begin knocking on the front doors of science departments, informing those of us within that it neither understands what we are doing nor intends to support our work any longer. The upshot will likely be a steady, perhaps nearly irreversible, trend toward a society mostly lacking in scientific curiosity, the likes of which the world has not seen since the thousand-year doldrums of the Middle Ages.

In saying this, I am not suggesting that we are about to abandon technology and crawl back into the bush; indeed, all things technological, applied, and economically viable will be roundly embraced. But science itself—the study of nature for the sake of knowledge pure and simple—will become increasingly less valued in these days of rising fundamentalism and federal deficits. Already Americans, especially, are opting out of science in alarming numbers, the impression being that science is not worth bothering about.

What is needed is a greater sense of balance—an honest recognition by scientists that sharing is as valuable as discovering, teaching as honorable as research. In today's world of an active media and strong public participation, the capacity to explain and the garnering of public trust are vital parts of the science enterprise. Otherwise, as scientists, we shall continue to be the single greatest cause of a technically illiterate population.

Regarding NASA: The U.S. space agency is in desperate need of leadership—technically competent people who can set ambitious and productive goals for NASA and articulate those goals clearly and accurately before the American public. The taxpaying citizen has always welcomed the exploration of space, the adventure into the unknown. It may well be our manifest destiny to negotiate the next, and perhaps final, frontier. And NASA should be commended for attempting to move our species—intellectually and physically—in that direction. If NASA management were to explain effectively, based on technical truth and an absence of hype, a vision and rationale for the agency's many varied programs, it would not likely experience its annual budget crisis, its periodic congressional micromanagement, the ongoing lack of public confidence, and the deep and bitter frustrations within the agency itself. Lacking such ability—and a sense of trust that derives from it—the space agency instead has solidified into a bureaucratic morass that rewards mediocrity and self-preservation while damping creativity and innovation.

The controversial space station is a case in point. The public—and scientists, too—would largely support such a grand and daring construction project were it to have a sensible *raison d'etre,* yet no one in the government has set forth such a rationale to date. Surely, someday the world's nations will in fact construct a permanently manned outpost in Earth orbit. But quite aside from the question of whether NASA should build a viable space station now, a more immediate concern is whether it *can* build one now.

Given what I have chronicled in this book, as well as the pessimistic prognosis for NASA's future among so many of its own personnel, I do believe that the space agency could neither now, nor in the foreseeable future, successfully construct, maintain, and operate such a space station. NASA and its civilian contractors, in my view, simply lack the technical ability and managerial infrastructure to do so; they are caught up in a public-works program that produces more paper than hardware.

As for the *Hubble* project through commissioning, NASA could not have handled it worse. The agency oversold the mission to Congress and the American people; it underestimated and undertested the project in the critical early years when the spacecraft was designed and its optics fabricated; it aided and abetted the corporate secrecy that tended to exclude injection of fresh wisdom from experienced outsiders; it crawled into bed with the military intelligencers, but only partway and ever fearful of committing a Faustian bargain, all the while denying any such involvement; it managed the project poorly from multiple field centers and among multiple contractors in the years of delay leading up to launch; it hyped the project in the public domain before launch and performed ineptly in that arena after launch; it failed to embrace any kind of educational program designed to share *Hubble*'s adventure with schoolchildren and their teachers; and it squelched attempts to disseminate *Hubble*'s early science findings with the rest of the world until its own image and that of the telescope had been irretrievably damaged.

In sum and in my view, if the United States is to have a viable and vibrant space program, NASA needs to be thoroughly reformed or replaced. (At the very least, its educational programs that should propagate truth must be divorced from its public-affairs activities that spin cheerleading, for the former are clearly tainted by the latter.) Indeed, many knowledgeable individuals—including a surprising number of high-ranking NASA administrators—are privately of the opinion that the space agency may well need to collapse before its successor can be reborn. As of now, NASA's management structure is hidebound bureaucratized, its technical competence called into question, its leadership lacking, its communications apparatus floundering. To use a term of which NASA is so fond, it is the space agency itself that now needs a "new start," for the United States deserves a better space program in every respect.

Regarding the Space Telescope Science Institute: The Hubble Space Telescope is the first large observatory-class space facility to address the broad science needs of the international community of astronomers. To manage and fulfill those needs, the Science Institute was created as an organization of astronomers, run by astronomers, to serve astronomers. Its primary goals are to maximize the scientific utility of the telescope and to disseminate its science results widely. All else is secondary.

Unfortunately, the Science Institute was established too late to influence materially the fundamental design of Space Telescope—in fact, several years after its hardware had been frozen, its instruments nearly built, its crucial optical system already shaped. Even so, the institute concept has proved itself well, agressively advocating and often developing improvements in the ground system, thus maximizing the science return. And its personnel, having the greatest stake in the outcome, led the way during commissioning in the analysis of the telescope's faulty optics, in the application of techniques used to recover the intended resolution, in the initial imaging campaign so crucial for the telescope's early survival, in the planning to repair the telescope on orbit, and in the subsequent revival of the *Hubble* mission in the eyes of the scientific community as well as the general public.

In my view—a scientist's view—the institute concept is the way to conduct future large space-science missions. Only with consensus management among the missions' end-users can we expect intimate involvement in and minute oversight of every aspect of the copious and complex tasks needed to execute such a mission. Science institutes should accompany each major space-science project (regardless of duplication). Furthermore, and more fundamentally, these institutes should be the single prime contractor, ultimately responsible for all aspects of the particular mission.

Throughout its research and development phase, the *Hubble* project suffered most from poor technical leadership and faulty systems management, owing chiefly to multiple competing NASA sites interfacing with two major contractors, as well as insufficient scientific input. And, now in the operational phase, it continues to suffer from many of these same drawbacks, especially inadequate liaison between the astronomers and the administrators and engineers who manage the project at the highest levels. There was no prime contractor for the Hubble Space Telescope, no single organization with overarching responsibility for the end product, and there is no single entity with absolute authority for operating the spacecraft today. Complexity is not an argument against having a single-point contact in the design, development, and utilization of such "big science" apparatae. In fact, it is the complex space-science missions and not the relatively simple ones that most need a single science-based institution willing and able to guide and influence all aspects of the mission from birth to death.

Regarding the news media: With few exceptions, the media covering the *Hubble* story during the commissioning period largely abandoned objective reporting in favor of sensational journalism and gloomy editorializing. News reports were often long on exaggeration and short on accuracy. The result is a strong impression among the nontechnical citizenry, even today, that *Hubble* is an expensive albatross, symbolizing technical ineffectiveness in today's gadget-filled world.

Having said that, however, I wonder how the media might have reported the *Hubble* story if NASA had handled the information flow differently. The combination of early PR hype and later news vacuum was deadly. Surely educators were confused by some of NASA's oversold claims for this mission, but the real public damage was done when the space agency let the news media set the agenda during the troubled summer of 1990. Without credible information issued by NASA, the media may well have felt that it had few alternatives but to opt for attack journalism and go for the jugular.

Once *Hubble*'s ailments became known to the world, the science journalists were mostly taken off the job. Not until after the telescope began producing bona-fide science results were some of these knowledgeable reporters able to return to cover the mission. Perhaps if science writers were to become a bit more analytical in the course of their work—not accepting, for example, NASA's pronouncements at face value—then their editorial chiefs would permit them to continue their reporting on controversial science issues, even when news of those issues carries major societal implications.

The principal lesson is that little or none of the media distortion would have occurred had NASA been candid about the status of the *Hubble* mission and forthcoming in its relations with the press. A secondary lesson is that the media boardrooms should trust their professional science writers more, for this small community of journalists has earned a well-deserved reputation for informed, balanced, and accurate reporting.

Regarding the military-intelligence community: Although the Hubble Space Telescope has no direct engagement with the military-intelligence community, its origin and evolution clearly derive from a class of orbiting vehicles used to reconnoiter Earth's surface. NASA itself, whether it can or wishes to admit it publicly, has always had an association with this shadowy element of society. The astronaut corps is largely composed of military personnel; the administrator of NASA during *Hubble's* commissioning, and several of its highest ranking senior officials, were former military officers; and much of NASA's inventory of space hardware is shared with (and grants priority to) the cloak-and-dagger operatives, including, for example, the space shuttle and key relay satellites. In fact, the entire space program was born out of international military competition, during a Cold War in which human and electronic intelligence gathering was of paramount import.

An unholy alliance between parts of the civilian science-engineering establishment and the military-intelligence community has existed throughout the four centuries of modern science. Galileo himself helped make a living by tooling spyglasses for the Florentine city-state. And the alliance has flourished in the years following the Second World War. Indeed, I believe it is here to stay. So why fight what seems to be inevitable, perhaps even an integral part of human nature? As our civilization becomes increasingly technological, there will no doubt be heightened interest on the part of the military and intelligence cultures in basic research and applied science done by the civilian sector.

If this is the case and if the Cold War is now over, why must the veil of secrecy so heavily persist concerning technological activities ongoing within our nation's preeminent work force? Military-intelligence technology, to the extent that I have witnessed it, is clearly superior to civilian technology. And the technical people who ply the trade on "the dark side" seem uniformly more skilled, informed, experienced, and dedicated than their counterparts in the civilian world. Why? In my view it is because in the post-*Apollo* era, when NASA undertook few bold and exciting ventures in space science and technology, the best people were attracted to the institutions that were in fact doing so.

Among a whole host of defense-triggered and *Hubble*-pertinent hardware and software advanced in recent decades, the military-intelligence community has championed the development of charge-coupled devices and other precision infrared, optical, and ultraviolet sensors; pioneered efforts in adaptive optics to compensate for the murkiness of Earth's atmosphere; led the way in creating sophisticated computer codes to analyze image arrays considerably more robust than those now used by civilians; developed means to computer enhance remotely sensed images for greater clarity; conceived ever more adroit computer codes and impressively smart gear sufficient to point, aim, command, and control the shooting down of bullets with bullets; and successfully built and operated orbiting telescopes more cerebral, versatile, and powerful than *Hubble*.

Above all, the reconnoiterers know well the art of systems engineering; they highly prize the concept of integrated design, analysis, and testing; they practice holistic management, ever mindful of the complete product, the entire mission. Surely they have also witnessed techical failures and management shortcomings in recent years, such as when a huge mirror shattered on the floor of a military observatory in the Pacific or a *Hubble*-class spy satellite tumbled hopelessly in orbit. Yet, with the Cold War finished, shouldn't we begin to break down the classification barriers, allowing knowledge and experience to flow more easily between the two camps? The *Hubble* project, in particular, could have roundly benefitted from much greater technical input from those who years ago built its predecessor space telescopes.

Walking the perimeter of Walden, as I do now for mental refreshment if nothing else, I wonder what the great man Emerson might have thought of all this. Clearly, his compatriot, who knew these woods well while championing the "small and simple is beautiful" movement more than a century ago, Thoreau would have been appalled at the sheer size, scale, and audacity of the *Hubble* project. But what about Emerson who sought to put humankind's assault on nature into its broadest possible perspective? What might he identify as a single common denominator among all the lesser lessons learned? I judge that superlesson to be the need for better technical education at all levels.

If the *Hubble* project has highlighted any one issue in today's society, it is that our engineering curricula need to be integrated and broadened. In particular, engineers of all types must be made more aware of the importance of systems analysis and design, and increasingly so as our civilization embraces ever more complex and intricate devices in our technological world. Make no mistake, it is not managed teamwork or cooperative learning that I advocate but rather a greater awareness on the part of professional educators that the next generation of engineers, individually, must be able to ply the trade more broadly, more systemically.

Scientists, too, must become more genuinely involved in our educational enterprise. We especially need to realize that science education is as important a domain for research and application as is scientific discovery, and that without an improved means of delivering at least an appreciation of science and mathematics to our fellow citizens, we shall likely see a continued divergence of the two cultures—that is, between the minority that understands the essence of modern science and the majority that does not. The result may well be not only an erosion of our economic well-being (an erosion seemingly already under way), but also a lessening of our democratic way of life.

Afterword: The "Fix"

"Whether we like it or not, this [Hubble Space Telescope] program is going to become a national triumph or a national tragedy." So said Ed Weiler, one of the few senior *Hubble* managers still with the project by the time NASA mounted what *it* termed a straightforward servicing call—a statement for which he was heatedly rebuked and nearly sacked by NASA administrators. Having recently delivered to orbit a key weather satellite dead on arrival, watched helplessly as a billion-dollar Mars probe exploded while approaching the red planet, and had its soap-operatic space station put through the turmoil of its eighth annual redesign, crises-laden NASA had decided well before *Hubble*'s repair mission began that the perception (at least) must be that this mission was a success.

It was now the eve of launch again, late fall 1993, some three and a half years since *Hubble* had been deployed in the spring of 1990. Although the orbiting telescope had been roundly troubled, it had still managed to provide a wealth of interesting scientific insight simply unattainable with ground-based telescopes. On the other hand, the world's discouraged astronomical community had not kept the telescope in high demand—partly because data from *Hubble* had not been revolutionary and also because the (aberrated) data were so difficult to analyze.

Should a major repair be attempted, potentially bringing *Hubble* up to spec as originally designed? Or should the bird just be left alone, so that nothing new could go wrong in the process? Frankly, NASA had no choice. It had to mount an aggressive repair job for two reasons: The spacecraft's basic engineering systems had deteriorated so badly that the very health and safety of the craft were threatened (by 1993, it had no more spare gyros, its solar arrays were wobbling badly, and its observing versatility had worsened). But more important, in the political arena, the space agency had boxed itself into a corner: If it couldn't repair *Hubble*, then it surely couldn't build the space station—and without the station, however controversial, the current NASA has no reason for being.

In the end, the Hubble Space Telescope was partly repaired and much of its visibility improved, but, to my mind, not fixed—hence I put the word *fix* in quotes. The upshot is that NASA gets to live another day, and astronomers get to take another step forward—but not likely a leap—in the quest to decipher the secrets of Nature. What follows is a brief account of *Hubble*'s status as of early 1994, some two months after its celebrated repair mission.

During most of 1993, the astronaut crew of STS-61 had practiced for hundreds of hours in underwater tanks at the Marshall and Johnson Space Centers, thereby simulating through neutral buoyancy the weightless conditions in which they would have to move around equipment as massive as refrigerators and as unwieldy as pregnant flag poles. In all, seven members comprised the repair team, and each was needed if *Hubble*'s service call was to have a chance of success; four of them, in pairs of two, would conduct several extravehicular activities—EVAs, or space walks—of many hours apiece, the most extensive out-of-ship experience among American astronauts to date. Fortunately, they would have the extended life-support services of the newest entry in the space shuttle fleet, *Endeavour*.

    Although I had left the project several months before, I had occasion to brief some of the STS-61 crew at the Houston astronaut office in early 1993. Spacecraft commander Richard Covey was keenly aware that their noble efforts might cause more trouble for the crippled telescope: "Tell me all the ways we could conceivably make the bird worse," he said. Jeff Hoffman, an EVA specialist and old friend from Harvard graduate school days, was realistic in thinking that it was unlikely they would fulfill all of the needed repairs: "When I get out there, I'll have two issues uppermost in my mind—to avoid getting hurt personally, and to avoid damaging *Hubble* any more than it is already." All agreed that they would use common sense and care when spaceborne, regardless of the intricate plans orchestrated by ground-lubbering managers. "If we run into trouble, we'll caucus among ourselves and decide what to do next, regardless of the timelines," said Hoffman. I took this as a refreshing attitude and felt for the first time that perhaps they just might pull it off.

    The objectives of the repair mission were many-fold, the highest priority ones involving engineering: The astronauts would replace—or "change-out" in NASA lingo—all of *Hubble*'s faulty gyroscopes; presumably, this time, the gyros would be sufficiently hardened against radiation hits, and/or would not be used goods meant as spares for some other spacecraft. They would also try to strap on a coprocessor that would raise the puny (and partly failed) 48-Kbyte memory of its DF-224 computer to a more robust (but still small) 64 Kbytes—not exactly the full-scale lobotomy many of us had wanted, and not exactly a bargain at more than $4 million (for an 80386 chip), but an improvement nonetheless. Not last, but foremost among several other engineering-related issues to be addressed on-orbit, the EVA crew would be tasked with installing a new set of solar arrays. The original ones would be retracted into their cassettes and removed from *Hubble*'s hull—or jettisoned if necessary—after which new ones would be inserted and then deployed in much the same manner as in 1990.

    The solar arrays remained a point of contention throughout the years of preparation to service *Hubble*; they provide the life blood of the observatory and without their proper operation—electrically and mechanically—the telescope would be vulnerable or even die. Recall that these delicate winglike appendages had been built by ESA and thus came loaded with political overtones—on both sides of the ocean. Congress had kept up the pressure in the early 1990s to internationalize (and thus help pay for) our nation's big-science projects: failure to do so had been part of the reason the $8-billion Superconducting Super Collider was killed in 1993. Thus, it was tricky to use any other kind of solar arrays, even if they did exist—which, of course, they did, yet couldn't be used because they officially didn't. Alas, the "military panels" sat on the shelf at Lockheed while the astronauts installed a new set of ESA-made arrays having a cylindrical sleeve of aluminum insulation around its (still) stainless-steel bistems—an extravagant Band-Aid resembling the springy, accordian-like tube through which hot air vents from your clothes dryer.

Regarding an improvement in *Hubble*'s science output—indeed to attempt to restore its intended vision—two large devices would be implanted in the bird. One had been in the planning stages from the start—a replacement for the wide-field and planetary camera. The original camera, the size of a piano inside one of the radial instrument bays near the fine-guidance sensors astride *Hubble*'s main mirror, had a $68-million price tag, and its replacement clone was supposed to be a bargain at about $27 million, since most of the engineering design had been already accomplished. When the telescope's big mirror was discovered to be flawed, opticians quickly realized that one of the camera's small relay mirrors (a nickel-size sliver of glass near its pupil) could be polished with a steepened concave shape, thereby sharpening the telescope's vision by introducing an equal but opposite amount of aberration.

In the summer of 1990, NASA officials had estimated (in fact while testifying before Congress) that a mere few tens of thousands of dollars of additional money would be needed to make this revision in the small corrective mirrors, but in the intervening three years, those mirror modifications grew in cost by roughly a factor of 2,000, the rebuilt camera ultimately running the government more than $100 million. In 1991 alone, more than $30 million went to the manufacturer, a sum exceeding the entire annual operating budget of the 400-person-strong Science Institute. Yet with nothing more than paper designs to show for the spent money, the catcalls howled forth once more at the institute during another of those tense management meetings with the NASA hierarchy: "This is fraud," bellowed a senior astronomer who was, like me, about to resign from the project. It was the only time I ever heard that word used out loud during the *Hubble* mission. Even here, though, I judged it an overreaction and an untruth; I just considered this symptomatic of the shoddy management and industrial bilking we'd had all along.

Part of the reason for the increased cost was the added complexity of the new camera, now called WF/PC-2. Only because the aberrated beam of light reflected from *Hubble*'s big main mirror was sufficiently symmetrical—the "perfect mistake" noted earlier—could it be corrected, in principle, with a carefully crafted, 1-inch mirror (not a lens) inside the camera. This oddly prescribed optical element would be positioned *behind* the faulty mirror and would be expected, *after* the acquired light had been corrupted by the big mirror, to refocus into a star's core much of the light scattered across the unwanted halo—the way the telescope's main mirror should have worked in the first place. But, in practice, such an aberrated beam is known as a "dirty beam," meaning that the camera's corrective mirror would have to be aligned with hairline precision (relative to the main mirror) in order to clean up the scattered light before reaching the camera's CCD detectors. Thus, the camera opticians, their hubris now sufficiently damped as with all *Hubble* veterans, felt the need to incorporate tiny actuators behind the small corrective mirror, in this way allowing minute tip/tilt adjustments in its position and orientation to be made from the ground after the new camera was installed aboard *Hubble*.

As the cost spiralled upward to make WF/PC-2 an active-optics instrument (thereby catching up to where the intelligencers had been years before), managers were forced to descope the original camera. Now, they would use only four CCDs compared to the original eight, thus diminishing the new camera's efficiency and field of view considerably—in fact molding it into an odd chevron shape (known to some at the institute as the "bat wing viewing field"). This was one of many trade-offs meant to secure success of the repair mission: *Hubble*'s acuity would hopefully sharpen, but only within a narrower area. Whether you take this to mean less is more, or merely less costs more, depends on your viewpoint.

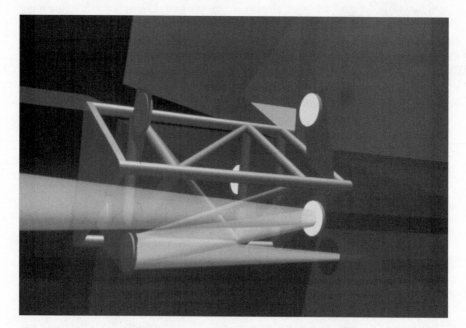

This simplified artist's conception shows the basic principles of the COSTAR package of corrective optics inserted into Space Telescope during the repair mission. *Hubble*'s main mirror is out of view at left, its instrumented aft bay at right. Aberrated light, traveling from the left and having already passed through the hole in the main mirror, is intercepted by a coin-size mirror (right) and reflected toward yet another small mirror (lower left) having a prescribed amount of opposite aberration built into it. This latter mirror then re-reflects the now-clean beam into the small aperture of a science instrument at bottom right. *[D. Berry]*

Its planetary camera mode of operation was especially hard hit, now having only one-fourth its original viewing field, never again to see clearly all of Jupiter, or Saturn, or many other celestial wonders that had so enthralled us early in the mission. For perspective, more than 3,000 exposures would have to be taken with the new planetary camera in order to build up a mosaic image of the Moon. Said a disgruntled astronomer whose planned observations were much compromised by this downscaling, "Even if it does work, we're getting half a camera for double the price. Looks like business as usual."

The second large science package to be installed inside *Hubble* was an ingenious device called by yet another acronym, COSTAR. During the 1980s, the project had built an empty, backup instrument box the size of a telephone booth, to be used aboard *Hubble* (with sufficient weights, and sand if necessary) in the event one or another of the axial science instruments failed the flight-readiness tests for the 1990 launch. This box, called STAR or Space Telescope Axial Replacement, would now be outfitted with a series of small mirrors designed to help some of the other science instruments regain their expected clarity. The prefix, CO, was added to denote "corrective optics," but its manufacturer, Ball Aerospace, preferred to think that it was meant to honor its home state of Colorado.

Briefly, COSTAR works like this. Four small metal arms (about the size of your forearm) are sandwiched into one corner of the otherwise empty telephone-booth-size housing. When installed aboard *Hubble* and upon command from the ground, a boom holding the arms would protrude into the heart of the telescope parallel to the incoming light beam already reflected from the faulty main mirror, the motorized arms would swing into place to intercept that beam of aberrated light, and several pairs of mirrors (again no larger than pocket coins, some attached to the arms) would compensate for the infamous error—much like for WF/PC-2, by effectively sweeping up the light in the fuzzy halo of a star and putting it back into the star's core image. Also, as with WF/PC-2, mirror alignments are critically important, hence the attachment of miniature ceramic pistons (the size of matchsticks) to the rear of some of those corrective mirrors. This insertion of a miniature robot within the much larger *Hubble* robot would be akin to a nearsighted human wearing contact lenses.

In this way, the intent was to correct the vision of several onboard (axial) instruments, all of which had been hampered by *Hubble*'s purblindness. But one of those instruments would have to be sacrificed to make room for COSTAR. The upshot of some astropolitical maneuvering not worth recounting—although also because it was scheduled to be used least frequently among all of *Hubble*'s original science instruments—was that the high-speed photometer would be permanently removed during the repair mission. COSTAR, in its place, should then be able to "clean the dirty beam" of light before it enters the faint-object camera, the faint-object spectrograph, and the high-resolution spectrograph. All this was promised for $32 million, and was eventually built for about $50 million—at a mere $18 million, one of the lowest cost overruns in the entire *Hubble* project.

Here, then, was another trade-off: *Hubble* would lose its specialized photometer, ironically the only perfectly working science instrument onboard the spacecraft. Accordingly, also lost were many important observations scheduled with this instrument, such as measuring the extremely rapid variations expected for light sucked inward by purported black holes, and the exquisite timing of the slow down of pulsars, which are rapidly rotating neutron stars beyond death.

For all its ingenuity, COSTAR did not prove to be a miraculous device; it forced yet other trade-offs, much as for WF/PC-2. Owing to the unavoidable location of COSTAR's corrective optics in *Hubble*'s aft bay, some aspects of the science instruments to which it fed light would be less robust than the originally intended instruments. For example, the two spectrographs' throughput would be knocked down by some 30 percent, and the faint-object camera's maximum viewing area (originally very small at 22 arc seconds across) would be about a third of its designed field—just for perspective again, now requiring more than 16,000 exposures to image the full Moon!

We astronomers began to realize that there was indeed no free lunch, much less a celestial feast at any price. Even if the intentionally misshapen small mirrors designed to compensate for aberration worked flawlessly, they themselves would introduce additional limitations. This, again, is an integral part of any *systems* approach: to increase the cameras' sensitivity by 2–3 magnitudes would unavoidably require the cameras' viewing field to shrink by a factor of 2–3. Compromises abounded, as in the repair of virtually any complex system today. Alas, there could be no real way to fully "fix" *Hubble* on orbit, no means to restore the design capabilities of the telescope (short of returning it to Earth)—and certainly no way to make *Hubble* "better than it was supposed to work in the first place," as was often claimed (both before and after the repair mission) by hype-seeking NASA officials.

Finally, on the science side, the astronauts would attempt to strap a power supply to the outside of the high-resolution spectrograph, thus restoring its ability to detect far-ultraviolet light with the finest possible spectral resolution—a key capability lost about a year after launch. (No, the astronauts were not asked to remove 72 screws in the instrument's bulkhead.)

In all, the astronauts' plate was so full that some of *Hubble*'s problems uncovered in its early years couldn't be addressed on this servicing trip. The fine-guidance sensors, for example, would neither be replaced with new ones nor improved by any other newly installed magic ocular. Hence, the sights seen by the three FGSs would remain fully aberrated, the telescope's pointing-control system somewhat compromised in its choice of guide stars.

In a spectacular nighttime fireball, the crew of STS-61 departed the Kennedy Space Center in Shuttle *Endeavour* from the same Pad 39B as had the original STS-31 *Hubble* astronauts years earlier. Within two days, on December 4, they rendezvoused with *Hubble* at about 356 statute miles altitude (it had fallen some 25 miles in 43 months), rather quickly grappled it with the Canadarm, and then attached it vertically to a specially built platform in the cargo bay of the shuttle. The orbiting observatory had been designed to be serviced, and this aspect of its construction worked well. The Marshall *Hubble* team—now seemingly a collective persona non grata regarding the repair mission—had years ago overseen development of a berthing dock to keep the telescope steady, a rotating pallet to enable access to different parts of its hull and appendages, and numerous railings (or "hand-holds") and foot restraints to permit the astronauts to crawl over the bird at will.

For five days, while the world watched enthralled, we all witnessed a milestone event in the history of space flight. The astronauts—and not just those outside the shuttle doing EVA tasks—put on a remarkable show high above the Earth. They managed to do what hardly anyone—and even most of the astronauts themselves—thought possible. They successfully accomplished all their main objectives with only a few hitches. An instrument door jammed at the latches, but a lasso-like strap was improvised to pull the latches into place and to bolt the door correctly. A screw got loose near the (closed) aperture door, only to be caught in mid-air (well, mid-space, really) by one of the astronauts perched at the end of the Canadarm. And they adroitly jettisoned the old port solar array when it wouldn't retract into its cassette—the result of a severe bend in the flapping array, a fact back-channel-reported to NASA by the intelligence community as early as mid-summer 1991 during one of those remarkable imaging sessions when *Hubble*'s anatomy was examined by an impressive black asset. ("Tell me again," said casually the NRO guy—the acronym for the National Reconnaissance Office having recently been declassified—"which part of your spacecraft did you want to see up close?")

All things considered, the STS-61 mission was a wonderful advertisement for manned space flight—exactly what most astronomers did not want to see.

An STS-61 astronaut, dangling at the end of the Canadarm at right, is about to perform some delicate surgery on the Hubble Space Telescope, seen here berthed in the cargo bay of space shuttle *Endeavour*. A second astronaut at bottom is unstowing some of the prosthetic gear to be installed on the spacecraft. *[NASA]*

We had before us a classic dilemma: Surely, and especially at the Science Institute, we had pushed hard for *Hubble* to be repaired to whatever extent possible, yet we were loath to see the role of astronaut space flight so prominently and well displayed. We all knew that a highly successful mission would likely be, in the long run, very bad news for space science. Much as the space shuttle program had cannibalized NASA's science projects in the 1970s and 1980s, government-sponsored space science would now once again take a back seat to crewed space flight, engineering demonstrations, and space spectaculars that excite the public to demand more space spectaculars. Even before we knew if any of the instruments installed by the STS-61 crew worked properly, the rallying cry within NASA had become, "Onward to the space station"—which to many of us scientists resembles an attempt to orbit a largely empty, hundred-billion-dollar garage in the sky.

The conduct of the *Hubble* mission post-repair in early 1994 was chronicled much as it was following deployment in 1990. Operationally, there were two categories of spacecraft status—those for public consumption and those closer to technical truth. For example, when asked about the solar arrays, NASA people (in press releases, press conferences, or upper-management summaries) would invariably respond that they are "no problem," are "working better than expected," or even are "fully fixed." By contrast, institute staff—especially astronomer-users of the telescope—would often answer that the arrays are "still causing some jitter," are "not working as well as hoped," or "sometimes seem to trigger loss of guide stars." That *Hubble* is still jittering after its repair mission is beyond doubt, although everyone agrees that array-induced spacecraft shudder has been damped to some extent. Only time—and spying eyes, perhaps—will tell if the new arrays work well enough to support the long-term objectives of the reinvigorated *Hubble* mission.

Regarding early science data acquired with the refurbished instruments, the project experienced the same sort of machinations surrounding the launch and deployment of *Hubble* years earlier. Unperturbed, I had begun two years before, in late 1991, another early-release observations program—EROs-2—to acquire some images and spectra as soon as possible after the repair job. These, once again, were meant partly as a contingency sample and partly as a means to apprise fellow astronomers of the extent of the "fix." Not least, and without hidden agenda, a notable by-product was the opportunity to share with the lay public and with educators alike the early fruits of the "new" Space Telescope.

Uptight committees debated the EROs merits well after I left the project, some engineers claimed months-long priority to test the vehicle on-orbit, and at least one astronomer objected to "my quasar targets" being prematurely imaged. The story was much the same as related earlier, few lessons having been learned from past bouts in the arena of pure astropolitics. In the end and amid much commotion and shouting as to how to handle the early public releases of *Hubble*'s new data, a small group of scientists and managers from the institute and Goddard prevailed, and a handful of early observations were scheduled on the telescope calendar mere days after the shuttle returned to

Earth. That the media were not alerted to these celestial observations was made obvious: "It is official NASA policy to insist on the immediate firing of anyone responsible for making an ERO[s] image available to a media representative. . . ," read one e-mail message broadcast to Science Institute staff. Only selected EROs-2 data would be released—the best ones. Perception was all-important.

NASA's foremost worry—that its mistake-prone image would be confirmed—surfaced soon after EROs-2 began. The first science instrument activated—the high-resolution spectrograph—balked, its newly installed repair kit apparently faulty. Worse, attempts to boot up the power supply strapped on by the astronauts to jump-start that part of the instrument which had crashed years earlier now caused the whole spectrograph to die. NASA, panicked for fear the first news of the "fix" would be that something else had broken, immediately abandoned all activity with the spectrograph, pulling all scheduled tests of it from *Hubble*'s timeline. It was "life or death" for the space agency. Positive news was absolutely necessary for its very survival.

The WF/PC-2 was tried next and, on its first image, fortunately showed definite improvement in *Hubble*'s eyesight. Much of the blurriness caused by aberration was gone, and fainter cosmic objects could clearly be seen. The corrective optics had worked! Not to flog a dead horse, it was the sensitivity—the previously impaired *Hubble*'s greatest loss—that was partly regained, even though the telescope's net resolution was not much improved. In astronomers' units, the new camera had won back not quite 2 magnitudes, enabling us to study well objects of twenty-sixth magnitude—and to detect, perhaps, light as faint as twenty-eighth magnitude, by exposing the camera for about an hour (or two useful orbits).

Technically, as of this writing (early 1994), the modified camera was sweeping up enough light from the dreaded halo to concentrate about 55 percent of a star's light into its pinpoint core. Literally, WF/PC-2 had still not met the pre-launch Level-1 spec of 70 percent encircled energy in a 0.1-arc-second radius, but this was quibbling. It was a welcome advance over the 15 percent we had to work with earlier, and additional tweaking of its corrective mirror might enable even better focusing. As for resolution, the repair was not very dramatic: WF/PC-2 was achieving about 0.1 arc second in its raw images—only a shade better than the earlier, computer-enhanced WF/PC-1 images (and, for subtle reasons, deconvolution techniques would help only marginally to sharpen the WF/PC-2 images). In all, WF/PC-2 had clearly improved *Hubble*'s eyesight, but with a diminished field of view and little more than half its captured light concentrated in the stellar cores, the new camera was surely not "working better than originally designed," as announced by spinning NASA officials.

Even so, some of WF/PC-2's early images were quite spectacular, clearly displaying the power of a spaceborne observatory to detect faint light—and, for the first time, to analyze that faint light quantitatively, in short to do not just observational astronomy but detailed astrophysics as well. For example, the 30 Doradus star-forming complex was revisited, including its central R136 star cluster, revealing even more embedded stars than seen with the "old"

Hubble, and enabling astronomers to track its faint nebulosity of gas and dust to much greater distances from the nebula's core. And a new image of the Orion Nebula that had earlier tantalized us all with unexpected structure, now showed with WF/PC-2 even more chaotic intricacy, ironically making this star-forming region even more difficult to fathom. Nature can indeed be devious: the closer we look, sometimes the more confused we get.

The resolution of these and other new images was not much improved over the previously acquired and deconvolved *Hubble* images, but the level of structure had much multiplied owing to the greater sensitivity of the new camera. All agreed that *Hubble*'s retrofitted capacity to detect and accurately measure truly faint gaseous features unrecognizable by any other means would now likely enable the telescope to probe a deeper level of cosmic exploration heretofore impossible.

Among the most stunning early targets for WF/PC-2 was a spiral galaxy, M100, a prominent member of the famed Virgo Cluster, an inordinately rich group of well over a thousand galaxies in the spring constellation Coma Berenices. Although detectable with a moderate-size amateur telescope on the ground, only *Hubble* could see its faint stars and exquisitely detailed spi-

This is the 30 Doradus nebula discussed earlier in Chapter 5, seen here at visible wavelengths through *Hubble*'s new wide-field and planetary camera 2. The nebula's central star cluster, R136, is boxed near the upper right of the chevron-shaped image, which is actually a mosaic of one small CCD chip (upper right) and three larger CCDs. For scale, this small box measures about 15 arc seconds across at a resolution of about 0.1 arc second. More stars can now be seen in R136 than with WF/PC-1, but the real power of this new, second-generation camera lies in its ability to detect the outlying faint nebulosity of the surrounding gas cloud. The camera's improved visibility will enable detailed studies of star-forming regions well beyond our Milky Way. *[NASA and J. Trauger]*

ral-arm gas clouds—in fact, and impressively so, down to a level of 30 light-years in a galaxy roughly 50 million light-years away. Indeed, this was the new image that caused professional astronomers at the winter meeting of the American Astronomical Society to sit up and cheer. For here, in M100, was a benchmark galaxy in which the Cepheid variable stars might now be individually visible—"standard candles" traceable all the way to the Virgo Cluster, upon which much of our knowledge of the size, age, and fate of the Universe depends. Only additional observations and much careful analysis would tell for sure if the expected twenty-seventh-magnitude stars could be found, but we now finally seemed to have within our grasp the chance to address some of *Hubble*'s grand objectives—the "big issues" for which it was originally built.

Other WF/PC-2 images were puzzling, such as the new one of Eta Carinae—puzzling because the nebula's strange ladderlike structure noted earlier in Chapter 7 was entirely missing. Opponents of the art of deconvolution made much of the disappearance, claiming that it had been an artifact of the controversial computer-enhancement process itself, but proponents pointed out that the peculiar ladder could be seen in *Hubble*'s raw data (without any

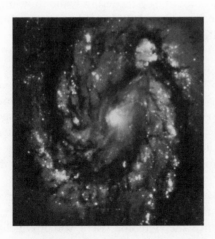

WFPC-2 : Wide Field Camera

WFPC-1 : deconvolved

The image at left shows *Hubble*'s new view of M100, a bright, face-on member of the Virgo cluster of galaxies, estimated to be some 50 million light-years distant. It was taken with the WF/PC-2 replacement camera and spans about 0.5 arc minute. The two frames at right are magnifications of a small part of one of the galaxy's outer spiral "arms," showing stars, gas, and dust in interstellar space. At right bottom is the WF/PC-1 view taken shortly before the repair mission—and computer-enhanced to reveal fine detail and some structure in those arms. At right top is the new image of the same field of view, taken with the WF/PC-2 camera. All three images have similar resolution (~0.1 arc second) and all display structure as small as 30 light-years across. However, as noted by the superimposed arrows at upper right, the newest image enables astronomers to see some faint stars and nebulosity (of about twenty-seventh magnitude) essentially invisible in the earlier data. These more appropriate comparisons of the enhanced before and after images (at right) more honestly illustrate the extent of *Hubble*'s incremental improvement after the repair mission. *(See also color insert.)* [NASA and J. Trauger] enhancement) acquired in 1990 with the old WF/PC-1. The mystery remains, although some think both that the ladderlike structure is real and that the new image is okay; each finding could be correct owing to slightly different filters used in the two observations. In any case, we were beginning to see some of the limitations of computer enhancement; there is no unique solution for deconvolution.

Of all the repairs, putting costs and trade-offs aside, to my mind the COSTAR device was the most impressive. Intercepting the aberrated light reflected from *Hubble*'s main mirror, COSTAR's many pairs of relay mirrors managed to refocus most of a star's previously fuzzy image into its point-like core; instead of having only 15 percent of the captured light in the appropriate core, the faint-object camera now had a whopping 80 percent, well exceeding the Level-1 spec. In astronomer's lingo, this would win us back about 2 magnitudes of sensitivity, enabling studies of some previously unviewable faint objects. "We nailed the prescription dead on," said Jim Crocker of the Science Institute. "Yeah, and we even got the sign right," deadpanned his colleague Chris Burrows, referring to the fact that multiple mirror reflections can sometimes play games on optical designers, indeed could have worsened *Hubble*'s eyesight.

The success of COSTAR is a lesson well-learned. The device's manufacture was led by engineer Crocker (who thought up its design while showering before a *Hubble* meeting in Europe in the fall of 1990) and by scientist Burrows (who had worked backward to derive the misshape of the main mirror). More important, these two men led a small team of mostly Science Institute personnel who oversaw every conceivable aspect of COSTAR's construction and pre-flight testing. For three years, these guys had become obsessed with COSTAR, foregoing vacations, working weekends, and often sacrificing family duties. This is the kind of attention to detail that Riccardo Giacconi had spoken of earlier as being absolutely necessary for the successful execution of space science—an obsession that can only come with the intimate involvement of the end-users, namely, the astronomers who would ultimately work the device in ways it was intended—to do science!

The tally, then, of the *whole system*, some six weeks after *Hubble*'s repair mission and on the day that NASA gathered a couple of hundred reporters at the Goddard Space Center to announce the outcome, went something like this. First, the positive features of the repair:

- new solar arrays seem more robust
- wide-field and planetary camera has reduced aberration
- COSTAR, with faint-object camera, has greater sensitivity
- health and safety gadgets apparently repaired—fuses, computer, gyros

And on the downside:

- telescope still jitters, sometimes enough to lose lock on guide stars
- all three fine-guidance sensors still aberrated, their guide stars blurred as before
- high-speed photometer removed to make room for COSTAR

- both cameras' field of view narrowed
- faint-object spectrograph throughput lowered
- repair kit for high-resolution spectrograph faulty, entire instrument apparently dead

NASA, in its early euphoria over *Hubble*'s "fix,"—indeed unduly eager to declare victory for *Hubble*'s improved vision—neglected to mention to the media any of the lingering problems or inherent trade-offs such as those on the second list above. Its press briefings, intimidatingly planned to suppress any statements other than in superlatives, also avoided comparing new images with old images that had been computer enhanced—the proper comparisons immediately sought by all professional astronomers. Photos that raced around the world showed only raw images taken both before and after the repair mission—dramatic, NASA-orchestrated comparisons that the media swallowed hook, line, and sinker.

The media, in fact, wanting to report success (for more failure was not newsworthy) were surprisingly docile, unquestioningly buying NASA's line that "*Hubble* is in perfect working order," and that "*Hubble* is now exceeding its original specifications." Admitted John Noble Wilford of the *New York Times*, recently addressing a group of astronomers in Washington, "For most science news to get printed these days, it has to involve big bang, big bucks, big screw-up, or big comeback—and with *Hubble* you've got them all."

None of this attempt on my part to present a more balanced appraisal of *Hubble*'s ups and downs should detract from the magnificent execution by the skilled astronauts of STS-61. Nor should any of the downside trade-offs subtract from the improved visibility of the orbiting telescope. Nonetheless, that balanced view says simply: Space Telescope is now no more "fixed" than it was originally "broken." *Hubble* has always been a world-class telescope, then and now, capable of sending Earthward a rich lode of fascinating scientific findings. It was largely the media, in responding to NASA's hyperbolic announcements, that caused the world to gain these skewed perceptions, then and now. And the media's herd mentality was again much in evidence, with some NASA managers openly amused at the ease of inducing them to stampede before the full verdict was in. Regrettably these days, the snappy sound bite does pass for informed judgment, yet we would all do well to remember that what we are dealing with here is, after all, rocket science.

To be sure, not everyone was pleased with the outcome. At the winter astronomy meetings where a few more post-repair *Hubble* images were released, however fascinating, one could hear and feel the early strains of internecine struggles among the glut of young astronomers (thanks largely to NASA's graduate-student subsidies in the 1980s), many of whom have trouble appreciating the astronomical cost and protected status of this celebrated megaproject. Even some NASA employees were put off by the overt fanfare and apparent lack of balance toward other worthy space missions. Wrote one NASA astronomer to me after the "fix" of *Hubble*, "I am discouraged with NASA (as is almost any NASA employee I know these days). We are getting . . . very poor leadership at NASA HQ. I am hiding in my office doing research until the future of NASA shakes out."

Has it all been worth it? Considering today's trend toward fiscal restraint and away from big science, can we justify the new-found hope for Space Telescope that the repair mission has won it? Given its huge, $4 billion price tag to date—$2.4 billion on the launch pad in 1990, $0.27 billion per year operating costs, and two dedicated shuttle flights at $0.45 billion apiece—*Hubble*'s science output has so far been admittedly slim. The telescope has yet to make a major discovery, and its "science-for-the-dollar ratio" is small.

Then again, only recently did I realize that perhaps science is no longer the essence of this mission. When asked by a reporter at a post-repair press conference, "Why should the average taxpayer care that we have clearer pictures of outer space?" Barbara Mikulski replied that the whole *Hubble* thing boils down to one word—at which point I held my breath, hoping that she would say something like "exploration," or "curiosity," or even closer to my heart, "education," for this powerful senator had previously and eloquently articulated better than any of us the objectives and rationale for this mighty venture. Instead, after holding up the first post-repair photos for all the world to see, she dashed off her newfound justification, saying simply, "jobs." The *Hubble* project, at least in the eyes of some of our nation's key appropriators, and much like for the space shuttle before and the space station to come, had apparently become a jobs program—a public works project for the aerospace industry. Maybe we really are heading toward a scienceless society.

Triumph or tragedy? I think neither. Much as claimed among the Epilogue conclusions (all of which I still hold dear), I judge *Hubble* to be both the best and the most expensive aid to our human visual sense, and thus a marvelous means to view the cosmos in dramatically new ways not otherwise perceived by us—a remarkable adventure, with unknown consequences, and one still very much unfolding.

If the Hubble Space Telescope represents a miracle on-orbit, it will likely require an even larger miracle to regularly repair and maintain it for years to come—not just to fix its operational necessities such as solar arrays, gyroscopes, computers, and the like, but to refurbish and upgrade its science instruments sufficiently to restore its designed capacity to address key astronomical issues. Only in this way will Space Telescope, the boldest experiment ever undertaken by astronomers, take its rightful place in our intellectual history by thrusting us into a whole new era—an era in which we shall likely map better the size and scale of the Universe, measure accurately the cosmic expansion rate, determine the age and fate of the Universe, explore an ancient epoch near the beginning of time, image up close the bizarre environs near suspected black holes, find planets around other stars in the Milky Way, and in the words of Galileo Galilei, resolve many other "wondrous things" in our richly endowed Universe.

For Further Reading

Allen, L. (ed.), *The Hubble Space Telescope Optical Systems Failure Report,* NASA, November 1990. A detailed analysis by a government-established board of inquiry to determine how and why Space Telescope's mirror was fabricated incorrectly.

Astrophysical Journal, vol. 369, no. 2, March 10, 1991, and vol. 377, no. 1, August 10, 1991. Two special issues of a scholarly journal highlighting *Hubble* science results and written by the principal scientists most involved in each of the early observations.

Bahcall, J., and L. Spitzer, "The Space Telescope," *Scientific American,* vol. 247, p. 40, July 1982. An account of the prelaunch science agenda for the orbiting observatory.

Brown, Robert (ed.), *Educational Initiative in Astronomy,* Space Telescope Science Institute, Baltimore, 1990. A call for a greater commitment on the part of the professional astronomical community toward improving science and mathematics literacy among citizens everywhere.

Capers, Robert S., and Eric Lipton, "Hubble Error: Time, Money, and Millionths of an Inch," *Hartford Courant,* March 31–April 3, 1991. A series of four articles capturing the tense working environment inside the Perkin-Elmer plant in the early 1980s when *Hubble*'s mirror was fabricated.

Chaisson, E. J., "Early Results from the Hubble Space Telescope," *Scientific American,* vol. 266, p. 44, June 1992. An account of the postlaunch science results from the orbiting observatory.

Chaisson, E. J., and R. Villard, "The Science Mission of the Hubble Space Telescope," *Vistas in Astronomy,* vol. 33, pp, 105–141, 1990. A technical account of the scientific goals and objectives of Space Telescope. (A popular version of the same article can be found in *Sky & Telescope,* p. 378, April 1990.)

Field, George, and Donald Goldsmith, *The Space Telescope: Eyes Above the Atmosphere,* Contemporary Books, Inc., Chicago, 1989. The best introductory book about Space Telescope for the general reader, but written prelaunch and thus pertaining to the intended telescope, not the realistic one on orbit.

Fuson, Robert H. (translator), *The Log of Christopher Columbus,* International Marine Publishing Co., Camden, Maine, 1987. An annotated translation of the personal diary of the admiral of the ocean sea.

Galilei, Galileo, *Sidereus Nuncius (Starry Messenger)*, translated by A. Van Melden (Chicago: University of Chicago Press, 1989). A first-person principle account of what Galileo saw the first time he turned his telescope toward the heavens; perhaps the greatest scientific popularization ever written.

Galilei, Galileo, *Letter to the Duchess Christina*, translated by S. Drake (Garden City, N.Y.: Doubleday, 1957). An abstruse account of Galileo's trials and tribulations while attempting to do science within a large and conservative bureaucracy; perhaps the greatest scientific polemic ever written.

Giacconi, R., "Space Telescope and Cosmology," in *Highlights of Modern Astrophysics*, S. Shapiro and S. Teukolsky (eds.), p. 331, Wiley, 1986. A prelaunch discourse of some of the grand cosmological objectives planned for Space Telescope.

Hall, D. N. B. (ed.), *The Space Telescope Observatory* (IAU Commission 44, 18th General Assembly, Patras, Greece), U.S. Government Printing Office, Washington, D.C., 1982. An early account of the vehicle design, its science instruments, and its goals and objectives.

Hubble, Edwin P., *The Realm of the Nebulae*, Yale University Press, New Haven, 1936. A wonderful summary of our knowledge of the distant reaches of the Universe, as understood in the first half of the twentieth century.

Longair, M. S., *Alice and the Space Telescope*, Johns Hopkins University Press, Baltimore, 1989. A readable, cute account of those key areas of modern astrophysics in which *Hubble* is expected to make major contributions.

Murray, Bruce, *Journey into Space: The First Thirty Years of Space Exploration*, W. W. Norton & Co., New York, 1989. An insider's account of the glory days of NASA's planetary exploration program and an indictment of its space shuttle program.

Smith, Robert W., *The Space Telescope: A Study of NASA, Science, Technology, and Politics*, Cambridge University Press, 1989. A scholarly history of the early years of the Space Telescope project, largely based on recollections of and interviews with key scientists, engineers, and officials who guided *Hubble*'s development in the 1970s and 1980s.

Index

Page numbers in *italics* refer to captions.

ABOUT THE AUTHOR

Eric J. Chaisson has published approximately 100 scientific articles, most of them in the professional journals. He has also written several books, including *The Life Era* and *Cosmic Dawn,* which won the Phi Beta Kappa Prize, the American Institute of Physics Award, and a National Book Award nomination for distinguished science writing. Trained initially in condensed-matter physics, Chaisson received his doctorate in astrophysics from Harvard University, where he spent a decade as a member of the Faculty of Arts and Sciences. While at the Harvard-Smithsonian Center for Astrophysics, he won fellowships from the National Academy of Sciences and the Sloan Foundation, as well as Harvard's Bok Prize for original contributions to science and Harvard's Smith Prize for literary merit. In 1992, after having spent several years as a senior scientist and director of educational programs at the Space Telescope Science Institute, Dr. Chaisson became the first director of the Wright Center for Innovative Science Education at Tufts University, where he is also Professor of Physics and Professor of Education. He is also affiliated with MIT and Harvard. He lives with his wife and three children in historic Concord.